CAMBRIDGE LIBRARY COLLECTION

Books of enduring scholarly value

Botany and Horticulture

Until the nineteenth century, the investigation of natural phenomena, plants and animals was considered either the preserve of elite scholars or a pastime for the leisured upper classes. As increasing academic rigour and systematisation was brought to the study of 'natural history', its subdisciplines were adopted into university curricula, and learned societies (such as the Royal Horticultural Society, founded in 1804) were established to support research in these areas. A related development was strong enthusiasm for exotic garden plants, which resulted in plant collecting expeditions to every corner of the globe, sometimes with tragic consequences. This series includes accounts of some of those expeditions, detailed reference works on the flora of different regions, and practical advice for amateur and professional gardeners.

A Selection of the Correspondence of Linnaeus, and Other Naturalists

After the death of the younger Carl Linnaeus in 1783, the entirety of the Linnean collections, including the letters received by the elder Linnaeus from naturalists all over Europe, was purchased by the English botanist James Edward Smith (1759–1828), later co-founder and first president of the Linnean Society of London. In 1821, Smith published this two-volume selection of the letters exchanged by Linnaeus *père et fils* and many of the leading figures in the study of natural history, revealing some of the close ties of shared knowledge and affection that bound the European scientific community at that time. Where necessary, Smith translates the letters into English, with the exception of those written in French, which are presented in the original. Volume 1 illuminates the epistolary relationships of Linnaeus senior with Peter Collinson, John Ellis and Alexander Garden, providing a very brief biography of each. Garden's letters to Ellis also feature prominently.

Cambridge University Press has long been a pioneer in the reissuing of out-of-print titles from its own backlist, producing digital reprints of books that are still sought after by scholars and students but could not be reprinted economically using traditional technology. The Cambridge Library Collection extends this activity to a wider range of books which are still of importance to researchers and professionals, either for the source material they contain, or as landmarks in the history of their academic discipline.

Drawing from the world-renowned collections in the Cambridge University Library and other partner libraries, and guided by the advice of experts in each subject area, Cambridge University Press is using state-of-the-art scanning machines in its own Printing House to capture the content of each book selected for inclusion. The files are processed to give a consistently clear, crisp image, and the books finished to the high quality standard for which the Press is recognised around the world. The latest print-on-demand technology ensures that the books will remain available indefinitely, and that orders for single or multiple copies can quickly be supplied.

The Cambridge Library Collection brings back to life books of enduring scholarly value (including out-of-copyright works originally issued by other publishers) across a wide range of disciplines in the humanities and social sciences and in science and technology.

A Selection of the Correspondence of Linnaeus, and Other Naturalists

From the Original Manuscripts

VOLUME 1

EDITED BY
JAMES EDWARD SMITH

CAMBRIDGE
UNIVERSITY PRESS

University Printing House, Cambridge, CB2 8BS, United Kingdom

Published in the United States of America by Cambridge University Press, New York

Cambridge University Press is part of the University of Cambridge.
It furthers the University's mission by disseminating knowledge in the pursuit of education, learning and research at the highest international levels of excellence.

www.cambridge.org
Information on this title: www.cambridge.org/9781108069700

This edition first published 1821
This digitally printed version 2014

ISBN 978-1-108-06970-0 Paperback

Selected books of related interest, also reissued in the
CAMBRIDGE LIBRARY COLLECTION

Babington, Charles Cardale: *Memorials, Journal and Botanical Correspondence of Charles Cardale Babington* (1897) [ISBN 9781108055604]

Banks, Joseph, edited by J.D. Hooker: *Journal of the Right Hon. Sir Joseph Banks* (1896) [ISBN 9781108029162]

Blaikie, Thomas: *Diary of a Scotch Gardener at the French Court at the End of the Eighteenth Century* (1931) [ISBN 9781108055611]

Bunbury, Charles James Fox: *Memorials of Sir C.J.F. Bunbury* (9 vols., 1890–3) [ISBN 9781108041218]

Darwin, Charles: *Journal of Researches into the Natural History and Geology of the Countries Visited during the Voyage of H.M.S. Beagle* (second edition, 1845) [ISBN 9781108038065]

Douglas, David: *Journal Kept by David Douglas during his Travels in North America 1823–1827* (1914) [ISBN 9781108033770]

Farrer, Reginald: *The Garden of Asia* (1904) [ISBN 9781108037211]

Fries, Theodor Magnus, edited and translated by Benjamin Daydon Jackson: *Linnaeus* (1923) [ISBN 9781108037235]

Haggard, H. Rider: *A Gardener's Year* (1905) [ISBN 9781108044455]

Hooker, Joseph Dalton: *Himalayan Journals* (2 vols., 1854) [ISBN 9781108029377]

Hooker, Joseph Dalton: A *Sketch of the Life and Labours of Sir William Jackson Hooker* (1903) [ISBN 9781108019323]

Hooker, Joseph Dalton, edited by Leonard Huxley: *Life and Letters of Sir Joseph Dalton Hooker* (2 vols., 1918) [ISBN 9781108031028]

Jekyll, Gertrude: *Wood and Garden* (1899) [ISBN 9781108037198]

Jenyns, Leonard: *Memoir of the Rev. John Stevens Henslow* (1862) [ISBN 9781108035200]

Johnson, George William: *A History of English Gardening, Chronological, Biographical, Literary, and Critica*l (1829) [ISBN 9781108037136]

Kingdon Ward, Francis: *The Land of the Blue Poppy* (1913) [ISBN 9781108004893]

Loudon, Jane: *Instructions in Gardening for Ladies* (1840) [ISBN 9781108055659]

Löwenberg, Julius: *Life of Alexander von Humboldt* (2 vols., 1873) [ISBN 9781108041775]

Mackenzie, George Steuart: *Travels in the Island of Iceland, during the Summer of the Year MDCCCX* (1811) [ISBN 9781108030212]

Mulso, John: *Letters to Gilbert White of Selborne* (1907) [ISBN 9781108038416]

North, Marianne: *Recollections of a Happy Life* (2 vols., 1892) [ISBN 9781108041300]

Oliver, F.W.: *Makers of British Botany* (1913) [ISBN 9781108016025]

Paris, John Ayrton: *A Biographical Sketch of the Late William George Maton M.D.* (1838) [ISBN 9781108038157]

Pulteney, Richard: *A General View of the Writings of Linnaeus* (second edition, 1805) [ISBN 9781108037303]

Pulteney, Richard: *Historical and Biographical Sketches of the Progress of Botany in England* (2 vols., 1790) [ISBN 9781108037341]

Raven, Charles E.: *English Naturalists from Neckam to Ray* (1947) [ISBN 9781108016346]

Raven, Charles E.: *John Ray, Naturalist* (1942) [ISBN 9781108004664]

Robinson, William: *The English Flower Garden* (1883) [ISBN 9781108037129]

Smith, Edward: *The Life of Sir Joseph Banks* (1911) [ISBN 9781108031127]

Smith, James Edward: *Memoir and Correspondence of the Late Sir James Edward Smith, M.D.* (2 vols., 1832) [ISBN 9781108037099]

Walters, S.M.: *The Shaping of Cambridge Botany* (1981) [ISBN 9781108002301]

White, Gilbert: *The Natural History and Antiquities of Selborne, in the County of Southampton* (1789) [ISBN 9781108138369]

Wilson, Ernest Henry: *A Naturalist in Western China with Vasculum, Camera and Gun* (2 vols., 1913) [ISBN 9781108030472]

A

SELECTION

OF THE

CORRESPONDENCE

OF

LINNÆUS,

AND

OTHER NATURALISTS,

FROM THE

Original Manuscripts.

By Sir JAMES EDWARD SMITH,

M.D. F.R.S. &c.&c.

PRESIDENT OF THE LINNÆAN SOCIETY.

IN TWO VOLUMES.

VOL. I.

London:

PRINTED FOR LONGMAN, HURST, REES, ORME, AND BROWN,
PATERNOSTER ROW.

1821.

London: Printed by John Nichols and Son, 25, Parliament Street.

TO

THE LINNÆAN SOCIETY

OF LONDON,

THESE VOLUMES,

THE REPOSITORY OF MUCH INFORMATION,

NOT ELSEWHERE TO BE FOUND,

AND OF MANY INTERESTING MEMORIALS OF PERSONS

WHO HAVE CULTIVATED,

AND EMINENTLY PROMOTED, THE STUDY OF NATURE,

ARE RESPECTFULLY INSCRIBED,

BY THE INSTITUTOR AND PRESIDENT

OF THIS DISTINGUISHED SOCIETY,

JAMES EDWARD SMITH.

Norwich, March 25,
1821.

PREFACE.

THE epistolary correspondence of eminent characters is generally an object of curiosity. This may arise from an opinion that the writers are there likely to appear with less disguise, and under a more easy and familiar aspect, than in studied compositions intended for the publick eye. Hence the letters of professed authors have always been perused with avidity; though possibly not invariably written without a latent expectation of their being one day seen and admired beyond the limits of confidential privacy.

The letters here presented to the publick were probably not written with any such expectation; or, if they were, it was an expectation of their being consulted as registers of plain facts and scientific remarks. The effusions of the heart, which they not unfrequently contain, were certainly not poured out in ostentation, or to display brilliant sentiments, rather than warm affections. These effusions will principally be found in the correspondence of genuine disinterested lovers of Nature: while the letters of academical dignitaries, of such at least as were little inspired by any of this pure and elevated taste, too often display that irritability which is characteristic of rivals, whether in fame, interest, or any other personal object.

The ample stores, from whence the following collection has been selected, are, in the first place, the epistolary correspondence of the great Linnæus and his son, which came into the hands of the editor, by purchase of every thing that belonged to those eminent men relating to Natural History or Medicine, in the year 1784. As Linnæus was fixed, for the greater part of his life, in the remote University of Upsal, all the particular communications which he received, on the objects of his studies, were by the letters of his friends, amongst whom we find almost every man of scientific rank in Europe, and every traveller of eminence, for half a century. It appears that Linnæus preserved all the letters he received. We have only to regret that he kept copies of but few of those he wrote. This deficiency is, indeed, partly supplied by the publication of all his letters to Haller, and of a few here and there to other persons; as well as by transcripts handed about in literary circles, and preserved by curious collectors. But the following collection is especially enriched by means of the correspondence of Mr. Ellis, the celebrated writer on Corals and Corallines, which was given to the editor by the worthy daughter of that excellent man. In this are a number of peculiarly interesting letters of the great Swedish Naturalist, which render the correspondence between him and Ellis, as nearly as possible, complete. Such is likewise the case with the epistolary intercourse of Mr. Ellis with Dr. Alexander Garden, and other friends. The collection of letters written to Dr. Richardson of North Bierley, in Yorkshire, the personal friend and learned botanical correspondent of Sherard, Dillenius, Petiver, and almost all the Botanists of their time, has been most obligingly communicated by Miss Currer, the great-grand-daughter and heiress of that learned man. Part of this, especially the letters of Sir Hans Sloane, and those of the illustrious Sherard, of whom the world had previously known little or nothing as a writer, have already appeared in Mr. Nichols's "Illustrations of the Literary History of the Eighteenth Century;" to which the present volumes are to be considered as supplementary.

Finally, the manuscript correspondence of the late Mr. Emanuel Mendes Da Costa, communicated to the editor by Mr. Nichols, has supplied some valuable materials, especially several unpublished letters of Linnæus, to himself and others.

Of all these collections, a great proportion still remains behind, much of it not less valuable or entertaining than what is here given. It may hereafter see the light, if the publick curiosity should be excited by the present specimen.

In the selection now offered to the English reader, the editor has given a preference to the letters of British Naturalists, and to subjects connected with England. In the next place he has chosen whatever might throw any new light on the history or character of Linnæus, or of his son. The originals are mostly written in English, and have received necessary corrections only, with some slight abridgments. The translations are distinguished by mention of their original language. The very few French letters it has not been thought requisite to translate. The editor has supplied such notes as appeared necessary, with the established Linnæan names of the various subjects of natural history; a work of some difficulty in the letters of Haller, who usually speaks of plants by vague phrases, from memory only, and often very imperfectly. These Linnæan names are either given in the notes, or placed, like some other remarks, between brackets.

Biographical Memoirs of some of the chief contributors to this collection are prefixed to their letters. These are principally those of Collinson, Ellis, and Garden, in the first volume; and of Solander, Dillenius, and Mutis, in the second.

Since the eighth page of this volume was printed, the editor has met with an additional proof of the specific identity of the peach and the nectarine, in the following account, communicated by the late Dr. Richard Pulteney to A. B. Lambert, Esq.

"Dr. Hancock purchased a tree of a gardener near Blandford, about the year 1750, which he understood to be a nectarine tree. Not finding it to bear for a year or two, he changed its situation twice. The year after its last removal, it began to blossom, and, to his astonishment, produced nothing but peaches. He supposed the gardener had deceived him, in sending a peach-tree instead of a nectarine. The next year he perceived nectarines on the tree, which surprised him not a little; but his surprise was considerably increased when he perceived peaches also growing on the same tree. Many persons came to examine and *detect* it, as they said, supposing that buds of these two kinds of fruit had been grafted on the same stock. To their great wonder, however, they saw this could not be the case, because from one large bough bearing peaches, small branches were derived which produced nothing but nectarines. Thus the tree continued twenty years, bearing at the same time, and on the same boughs, both peaches and nectarines. The fame of it extended so far, that many people came from the distance of 80 or 100 miles to see this curiosity, and, as they thought, to discover the trick. But they always went away satisfied that it was the most extraordinary *lusus naturæ* they had ever seen. The curiosity of the publick, and the consequent numerous applications to see this remarkable tree, became at last so troublesome, that the Doctor was induced to give it to his friend Lord Harborough. It died in consequence of the removal."

CONTENTS OF VOL. I.

LIST OF AUTOGRAPHS.

VOL. I.

VOL. II.

Biographical Memoir

OF

PETER COLLINSON, F.R.S. & F.S.A.

AND HIS

LETTERS TO LINNÆUS.

PETER COLLINSON, F.R.S. and F.S.A. one of the earliest and most constant correspondents of Linnæus, was highly distinguished in the circle of Naturalists and Antiquaries in London for nearly half a century. He belonged to the Society of the Quakers; and his upright, benevolent, active character did honour to his religious persuasion. His family is said to have come from Westmoreland.

He was born Jan. 28, 1693-4, in a house opposite to Church-alley, St. Clement's-lane, Lombard-street, according to a manuscript memorandum of his own, communicated by A. B. Lambert, Esq. V. P. L. S.; but he resided, for many years, at the Red Lion in Gracechurch-street, as a wholesale woollen-draper, where he acquired an ample fortune.

He married, in 1724, Mary the daughter of Michael Russell, Esq. of Mill-hill, Hendon. This lady died in 1753, leaving him two children — a son named Michael; and a daughter, Mary, married to the late John Cator, Esq. of Beckenham, Kent. They are said to have inherited much of the taste and amiable character of their father.

Mr. Collinson appears to have occupied, in the earlier part of his life, a country-house and garden at Peckham in Surrey (where his brother had also a garden); from whence he removed, in April 1749, to Ridgeway-house at Mill-hill, and he was two years in transplanting his collection. The English gardens are indebted to him for the introduction of many new and curious species, which he acquired by means of an extensive correspondence, particularly from North America. Among these was the *Collinsonia canadensis*, so called by Linnæus, who has given a beautiful engraving of this plant in his *Hortus Cliffortianus.* It was first imported in 1735.

The following Letters of Mr. Collinson evince his ardent and genuine love of nature, especially of the vegetable tribes; nor do they less display a character of true piety, cheerfulness, and benevolence, well suited to so virtuous and soothing a pursuit. He enjoyed, throughout a long life, the communications of most cultivators of science in general; for he interested himself about every new or useful discovery, and was one of the first who attended to the (then recent) wonders of electricity; on which subject the great Franklin was obliged to him for the earliest European intelligence.

Nor was his personal friendship less valued by people of distinguished character and abilities in various ranks; among which the names of Derham, Sloane, Ellis, and Fothergill stand pre-eminent; as well as those of the accomplished Robert Lord Petre, who died in 1742, and the famous Earl of Bute.

Mr. Collinson became acquainted with Linnæus when the latter visited London in 1736. He died August 11, 1768, after a short illness, in the 75th year of his age, in the full possession of all his faculties, and of all his enthusiasm for the beauties of nature, attended by far more important consolations and supports. All these are so well expressed in his last letter to Linnæus, that we shall not here anticipate the pleasure of our readers by any extract.

The Philosophical Transactions and the *Archæologia* are enriched with several of Mr. Collinson's papers. Dr. Fothergill published an account of his life. He has left, in the hands of his descendants, many interesting anecdotes relating to the introduction or cultivation of particular plants; which have been communicated by his grandson, the present Mr. Cator, to Mr. Lambert, and are now before us. The following especially deserves to be made public, as the result of so munificent an undertaking is worthy of inquiry. "In March and April 1761, the Duke of Richmond planted a thousand cedars of Lebanon, on the hills above his house at Goodwood;

plants five years old, that I procured for him at 18 shillings each. P. Collinson." — The garden at Mill-hill, so assiduously cultivated by this gentleman and his son, and for many years abounding with rarities and beauties, fell afterwards into the most barbarous and tasteless hands. After a transient restoration by an eminent Botanist, it is now, as far as we can learn, almost entirely stripped of its chief curiosities.

Correspondence.

MR. PETER COLLINSON TO LINNÆUS.

DEAR FRIEND, London, May 13, 1739.

I could not omit so convenient an opportunity, by my worthy friend Dr. Filenius, to enquire after your welfare, and give you joy on your marriage. May much happiness attend you in that state!

I am glad of this conveyance, to express my gratitude for the particular regard shown me, in that curious elaborate work the *Hortus Cliffortianus*. Something, I think, was due to me from the Commonwealth of Botany, for the great number of plants and seeds I have annually procured from abroad; and you have been so good as to pay it, by giving me a species of eternity, botanically speaking; that is, a name as long as men and books endure. This lays me under great obligations, which I shall never forget.

I am concerned I can make no better acknowledgments than by the small token of Pennsylvanian ores which the bearer will deliver to you.

My best wishes attend you; and if I can any way serve you here, you may be assured of the readiness of your sincere and affectionate friend,

PETER COLLINSON.

P.S. As Mr. Logan has had two Latin tracts published in Holland, I doubt not but Dr. Gronovius has sent them to you. The one is on generation. When a convenient opportunity offers, pray let me hear from you.

DEAR FRIEND, London, April 10, 1740.

I hope you had mine with Mr. Miller's dimensions of the cedars of Lebanon at Chelsea *

I now come to make good my promise to send you some North American seeds, which I herewith send.

It will be a pleasure to me to hear from you. I wish you health and happiness.

From your sincere friend,

P. COLLINSON.

Pray my respects to Mr. Filenius. I send you some English fossils, and the bark of the Lace-tree from Jamaica.

DEAR FRIEND, London, April 3, 1741.

Mr. Biork being so obliging to acquaint me of a convenient opportunity of writing to you, I could not forbear indulging myself with that pleasure; in the first place to enquire after your health, and next to know if you received mine with a parcel of American seeds. I also wrote to you by Dr. Filenius †

* This letter does not appear.

† This gentleman was Professor of Divinity at Abo, and afterwards Bishop of Lindkœping.

(for whom I have an high esteem), which no doubt came to your hands. But, my dear friend, you have not thought fit to return me an answer; however, none is yet come to hand.

I know your active genius cannot be idle. Pray what are you doing? Some new work, I hope, is ready for the press, to entertain and inform the curious part of mankind.

You know we have frequent varieties by the commixture of the farina of different species in flowers, but nothing so rare to be met with in fruits. But I have this day sent me two apples, the one a russet, or brown-coat apple, and another a green apple. They both were original fruits of the green apple, whose boughs mixed with the russet, and acquired such distinguishing marks of their adulterous intimacy with each other's blossoms, that one part of the apple is russet and the other green, the colours not by degrees going into each other, but there is a remarkable line where one fruit is divided from the other, as in the mixed orange and lemon. The complexion of these two sorts of apples being so different makes the mixture the more remarkable. What is farther remarkable in it is, that though it is originally from the green apple, and grew on that tree, yet its neighbour the russetting has impregnated more than two thirds of it.

Lord Wilmington has another instance of this commixture or blending of fruits, for he has a tree that produces nectarines and peaches, without any art, but quite accidentally. The fruit does not mix together as in the apple abovesaid; but complete peaches and nectarines, both distinct, are on the same tree *.

My best wishes attend you. From your affectionate friend, P. COLLINSON.

London, Jan. 18, O. S. 1743-4.

MY DEAR FRIEND DR. LINNÆUS,

I almost despaired of a line from your hands, for I have not heard from you since the 3d of August, 1739; but at last I had the pleasure of yours of the 25th July last. I was much concerned that so large a collection of American seeds was lost; I hope we shall have better success for the future. I have now made up a parcel of South American seeds, and hope to add some Northern ones, to come by first ship. I was delighted to find the *Coreopsis altissima* (query if not a *Rudbeckia)* and the *Collinsonia* were acceptable to you. I hope John Bartram our collector will send more this year. For his great pains and industry pray find out a new genus, and name it *Bartramia*.

* Of this several instances have since appeared; but the Editor had once a present of a much more curious variety—a fruit precisely half nectarine half peach, the size, colour, surface, and flavour of each being perfectly distinct in the respective halves. This was witnessed by several persons. It grew in the garden of the late A. Aufrere, Esq. at Hoveton, Norfolk, on a tree which usually bore some complete nectarines as well as peaches; but in two different seasons, at some years distance from each other, the same tree produced about half a dozen of these combined fruits.

Your system, I can tell you, obtains much in America. Mr. Clayton and Dr. Colden at Albany on Hudson's river in New York, are complete Professors; as is Dr. Mitchell at Urbana on Rapahanock river in Virginia. It is he that has made many and great discoveries in the vegetable world. I writ to him to know the reason for his name *Elymus* for a species of wild oats, and many other new names. I hope in a year's time you will see his essays on botany, in Latin, printed. I have the first part finished; but he intends to add another, so the printing of the first is deferred.

The death of the worthiest of men, the Right Hon. Lord Petre, has been the greatest loss that botany or gardening ever felt in this island. He spared no pains nor expence to procure seeds and plants from all parts of the world, and then was as ambitious to preserve them. Such stoves the world never saw, nor may ever again. His greatest stove was 30 feet high, and in proportion long and broad. In it were beds of earth, in which these plants as under were planted, and flourished wonderfully.

The *Hernandia* was 10 feet high, 5 inches round the stem.

Guava — 13 feet high, 7 inches round, spreading 9 feet.

Female Papaw — 17 feet high, 2 feet 3 inches round the stem, and bears plenty of fruit every year.

Anotto — *(Bixa orelana)* 14 feet high, 11 inches round.

Plantain or *Musa*, 24 feet high, the leaves 12 feet long, and 3½ feet broad; 3 feet 2 inches round the stem, and has abundance of fruit.

A large Palm, 14 feet high, 4 feet round.

Cereus *(Cactus)*, 24 feet high, 1 foot 4 inches round.

Male Papaw, 20 feet high, 3 feet 9 inches round, with several branches 7½ feet long.

A *Rosa Sinensis* or *Ketmia (Hibiscus)*, 25 feet high, 1 foot four inches round.

One Sago Palm, *Toddapanna* of *Hort. Malab.* 8 feet high, and 2 feet round the stem, a fine plant; with a great number of very large plants, whose names would be too long to mention here.

The back of these stoves had trellises, against which were placed in beds of earth, all the sorts of Passion-flowers, Clematis's of all kinds that could be procured, and Creeping Cereus. All these mixed together, and running up to the top, covered the whole back and sides of the house, and produced a multitude of flowers, which had an effect beyond imagination; nothing could be more beautiful or more surprizing. There was also a Bamboo Cane 25 feet high.

Next to this magnificent stove were two others, two degrees lower, but these were higher and longer than most that are to be seen. He had also several besides. His Anana stove was 60 feet long, and 20 wide. The collections of trees, shrubs, and ever-greens, in his nurseries at his death, I had told over; and they amounted to 219,925, mostly exotic. As this young nobleman was the greatest man in our taste that this age produced, I thought it might not be unacceptable to give you some account of the greatness of his genius; but his skill in all liberal arts, particularly architecture, statuary, planning and designing, planting and embellishing his large park and gardens, exceeds my talent to set forth.

We have now a wonderful fine season, that makes our spring flowers come forth. I am sure you would be delighted to see my windows filled with six pots of flowers, which the gardener has sent me to town; viz. great plenty of Aconites, white and green Hellebore, double Hepatica, Crocus, Polyanthus, Periwinkle, Laurustinus, vernal red Cyclamen, single Anemonies, and Snowdrops. This is my delight to see flowers, which make a room look cheerful and pleasant, as well as sweet. None of these were brought forward by any art, but entirely owing to the temperature of the season, though some years I have known things forwarder than now.

I communicated your complaints to your English friends. They promise amendment, in particular Dr. Lawson, from whom I doubt not but you have heard, before this comes to hand.

You may remember my repository, in which I have a collection of all sorts of natural productions that I can procure from my distant friends. Nothing comes amiss that furthers the knowledge of Natural History. One sort of seed I will mention to send me, the *Pulsatilla flore albo et rubro**, if

* *Anemone pratensis. Linn. Sp. Pl.* 758.

you have it; and pray send me your *Systema Naturæ*, in 8vo or 4to, for the large sheets are not so convenient for common perusal.

I herewith send you a box of cranberries or *Oxycoccus**. Perhaps, if sown on your mossy bogs, the seeds may grow. They came from Pennsylvania. Ours in England are very small, not bigger than red garden Currants.

Now, my dear friend, I shall tire you no longer than to assure you I am, &c.

Pray make my compliments to Messrs. Filenius and Celsius.

My dear Friend, London, March 12, 1744-5.

I am greatly obliged to you for the favour of yours of the 9th of October last.

I now send you again seeds of *Collinsonia*, *Rudbeckia*, and others, which I hope you will have the pleasure to see germinate this year.

I am glad you have the correspondence of Dr. Colden and Mr. Bartram. They are both very indefatigable ingenious men. Your system is much admired in North America. Those two gentlemen are much obliged to you for the honour that you intend them.

Dr. Lawson continues in Germany all this year. Dr. Dillenius is well in health. Your letter was carefully sent to him.

Sir Hans Sloane holds out to a miracle, and has all his senses and memory entire. His face has

* *Vaccinium macrocarpon.* *Ait. Hort. Kew. ed.* I. *v.* 2. 13. *t.* 7.

none of the lineaments of a man not far from 90 years old.

Mr. Ehret has sent you a picture of what he calls an *Agaricus;* a very surprising production.

I have sent you a parcel of seeds to the care of Mr. Morsach, merchant in Dantzic; he is a friend of Dr. Breynius. I am, &c.

My dear Friend, London, Sept. 1, 1745. O. S.

I think in one of your letters you enquire after the etymology of some of Dr. Mitchell's *nova genera,* in particular the name *Elymus.* This he much wonders at, when you object to Burmann for not knowing or remembering the ancient names; vide *Crit. Bot.* p. 117. He says *Elymus* is the name of *Panicum,* in Dioscorides, *lib.* 2. *cap.* 120. For that reason he uses it for a species of wild grain, or corn, in Virginia. He further observes, you make *Diodia* to belong to your 14th class; but the Doctor says it belongs to the 4th.

A few days ago I went to see Sir Hans Sloane, and I found the good man very cheerful and well, for his great age; and Dr. Dillenius continues in good health. A few days ago he was in London.

I am glad to hear your books are so near completed. Your great pains and industry are deservedly admired by all the curious part of mankind.

I wish you would send me the figures that belong to your *Iter Oelandicum* and *Gothlandicum;* for though I understand not Swedish, yet the figures will improve my ideas, and give me some knowledge of what those countries produce. As I am passionately fond of Natural History, this will give me great pleasure.

Your *Flora Lapponica* delights me much, because you have intermixed variety of subjects with botany.

The American Ginseng has flowered and fruited with me this year, and several other curious plants.

My dear friend, I affectionately salute you, and am truly yours, P. COLLINSON.

MY DEAR FRIEND, London, April 1, 1746.

I am obliged to you for your favour of Dec. 31 ult. and am glad to find the seeds I sent last year were acceptable. I have added another parcel this year with those from Dr. Dillenius; but he complains that you send no seeds to him, which he says is very discouraging.

Mr. Ehret desires your acceptance of a coloured print of a new *Cereus* that flowered at Chelsea.

Our valuable friend Dr. Lawson is daily expected over from Flanders. He was taken prisoner by the French, but was afterwards exchanged.

Here are variety of American plants to be bought, but there is a great deal of trouble and expence to get them on board a ship; and then, unless some careful person was to look after them, they might not be good for much when they come to you.

Dr. Mitchell lives in Virginia, and has described several *nova genera* not yet printed.

I love all books of Natural History, and every production God has made. Pray what sorts of land, river, and sea shells are found in your country? is any thing peculiar observed in their natures? What sorts of fossils are found in Sweden? Your animals, the rein-deer excepted, are nearly the same as ours. Have you any particular species of fish that are found in no other parts of the world? any insects peculiar to Sweden or Lapland? Send me specimens of them, or any other natural production, as minerals, &c.

The *Claytonia* is now nearly in flower, and the *Echium* or blue mountain cowslip*, also a pretty *Ranunculus* with a yellow flower, and a *Fumaria bulbosa* of Plukenet†. These four species are the first flowers from Virginia.

Our dear friend Dr. Bœck was so good as to send me your *Systema Naturæ* in 8vo, which I am mightily pleased with. Pray is he returned to Sweden?

Now I must take my leave, wishing you health and happiness. P. COLLINSON.

DEAR DOCTOR, Aug. 5, 1746.

I hope you have the seeds I sent you in the Spring, under the care of our good Mr. Biork.

I now send you a specimen of a new and rare plant, not yet described that I know of, except in Plukenet. I forgot to gather the first-blown flowers,

* *Pulmonaria virginica. Linn. Sp. Pl.* 194.

† Nothing answerable to these two last occurs in recent writers upon North American plants, nor has Plukenet a *Fumaria bulbosa.*

which are four times as large as the specimens I now send you. The leaves are so much like the leaves of Cos lettuce, in shape and colour, that for two years, before the flowers appeared, I took it for a lettuce. You have now the leaves, flowers, and seed vessel, so I hope you will find out what class to give it. Perhaps you may, if a new genus, call it *Bartramia*; for John Bartram found it growing behind the first ridge of mountains in Virginia. It is a most elegant beautiful plant, the petals being of a violet colour. Sow the seed immediately, and you may hope to see it in two years. It is perennial, and very hardy *.

Sir Hans Sloane continues hearty and well.

The *Leonurus canadensis* † is a charming flower, and is very finely blown in my garden.

I am much yours,

P. Collinson.

March 10, 1747-8.

I am greatly obliged to my dear friend Dr. Linnæus, for the curious books and seeds he has sent me.

I am greatly delighted with the *Hortus Upsaliensis*, to see every part of it so regularly disposed, and so well contrived to preserve the plants of every climate.

I hope the seeds I now send you, by the kind assistance of my ingenious and worthy friend Mr. Burmester (who takes great pains to qualify himself

* This appears to have been the *Dodecatheon Meadia*.

† *Teucrium canadense?*

for the good of his country), will be acceptable to you. If the French had not taken two North American ships, you would have had a great many more.

I enclose you some tracts; and if you did but read English, I should send you more. Pray give one of Mr. Logan's to Count Gyllenborg. It may be acceptable to him, as he is the great patron of Natural History. When will your *Iter Oelandicum* and *Gothlandicum* be printed in Holland?

March 27, 1747-8.

I herewith send you some of the early ripe Indian corn. I expect better seed another year, but this may be sufficient to make the experiment. As it grows, draw the earth up round its lower joints, from whence it makes new roots. It loves a rich soil and warm situation.

You promised me some books, but none are yet come to my hands. Pray send me *Corallia Baltica*, *Iter Oelandicum* and *Gothlandicum*, *Animalia Suecica*, and any other book of Natural History.

I have your botanic works, and also your *Flora Lapponica*, which Sir Hans Sloane says is the best book of all your works.

I shall be obliged to you for specimens of the various species of *Cochleæ terrestres et fluviatiles*, and *Nautilus testa recta*. *Fn. Suec. ed.* I. *n.* 1330, also 1331, 2, 5, 6 and 1347.

My good friend, I must tell you freely, though my love is universal in Natural History, you have been in my museum and seen my little collection, and yet you have not sent me the least specimen of either fossil, animal, or vegetable. Seeds and specimens I have sent you from year to year, but not the least returns. It is a general complaint that Dr. Linnæus receives all, and returns nothing. This I tell you as a friend, and as such I hope you will receive it in great friendship. As I love and admire you, I must tell you honestly what the world says.

I am, &c. P. COLLINSON.

MY DEAR FRIEND, April 16, 1747.

As we have lost our worthy friend Dr. Dillenius before your letter came to hand, I then gave his letter to our ingenious, learned friend Dr. Mitchell, who was so good as to return the answer on the other side. It is to no purpose to send seeds if you do not contrive a better and safer way of conveyance.

No doubt but you have heard of dear Lawson's death, who is greatly regretted. Catesby's noble work is finished. I drank tea a few days ago with Sir Hans Sloane, and he continues to admiration in good spirits and hearty. We often talk of you. Mr. Edwards has lately published two very curious volumes in 4to. of rare and non-descript birds and animals, all coloured after the life, price £4. 4*s*. 0*d*.

Dr. Martin and Mr. Miller are both very well. His Dictionary has passed several editions.

I should be glad to see the Itineraries you mention. I doubt not but they will be translated into English, for we are very fond of all branches of Natural History; they sell the best of any books in

England. Dr. Sibthorp M. D. is chosen Professor at Oxford, in the room of Dr. Dillenius.

When you see Dr. Bæck, pray thank him from me, for the many curious books and seeds that he sent me; my best wishes attend you.

I am your sincere friend,

P. Collinson.

London, Oct. 26, 1747.

I had the pleasure of my dear friend Dr. Linnæus's kind letter, by the hands of your good *Theologus*. I shall gladly do him all the service in my power.

I am under great obligations both to your Royal Society and to you, for the honour of your nomination for a member, but I think myself nowise deserving that favour: yet I shall submit myself to your disposal, and shall always retain a grateful sense of your friendship and good intention towards me.

I have sent you Mr. Logan's dissertation on the operation of the farina on the maize; as it is in Latin, I recommend it to be read at a meeting of your learned Society.

The treatise on gravitation, by our friend Dr. Colden of New York, is a new system, which he desires may be thoroughly examined. I wish it had been wrote in Latin, to have been more universally read. But, as a great many of your learned men read English, I hope it will be acceptable to some of them. Be so good as to present one of each of these books to the Royal Society in my name, and

I will send you more by first opportunity. Our good friend Mr. Biork says, he gave them to a Captain who sailed about a week ago.

I thank you for the several curious tracts you have sent me.

I am greatly delighted with the *Hortus Upsaliensis.* I hope to send you some seeds; but pray contrive that they may come soon to your hands, and not lay a whole year undelivered. I shall with pleasure send you a pot, with a root of *Collinsonia,* by first ship in the spring.

I am pleased to see your *Nova Plantarum Genera.* I wish you had sent two books, one for John Bartram, and the other for Dr. Colden; I know they would be acceptable to them. I thank you for the *Flora Zeylanica.* The seeds are not yet delivered to me. Now, my dear friend, farewell; and be assured that

I am truly yours,

P. Collinson.

P.S. I thank you for the seeds; there are several curious plants which will deserve a place in my garden.

Dr. Dillenius was prosecuting, or working at, the *Pinax,* when he died; what will be its fate *now* I know not, for the present Professor I do not think of skill sufficient to undertake it.

The north American *Ursus* I have often eat of in England, and think it is the most agreeable taste of all flesh. My friend a merchant, had young bears brought over every year, and fatted them with *dumplings* and sugar. It is really fine eating, and the fat is whiter and finer than the fat of lambs.

Dr. Mitchell's system will be published under the direction of Dr. Trew at Nuremberg.

My garden is in great beauty, for we have had no frosts; a long, dry, warm summer and autumn, grapes very ripe. The vineyards turn to good profit, much wine being made this year in England. Sir Hans Sloane is hearty, Miller is well, and so adieu.

MY DEAR FRIEND, London, Oct. 3, 1748, O. S.

I am glad to hear of your welfare, and that you are so happily employed in useful discoveries.

But pray consider what will become of the clockmakers, if you can find out vegetable dials.

Next, I am afraid you will be spoiled for a gardener, you will grow so rich with the breeding of oriental pearls.

These are subjects new and unheard of before, and I wish you may succeed in them.

The books wanting to make up Mr. Catesby's work complete, I have sent in a box, and another box from Dr. Mitchell, which our worthy friend Mr. Biork is so good as to take care of. They are put on board the Assurance, Capt. Tornland, for Stockholm, and consigned to Mr. Grill, merchant. The ship will sail in a few days.

From the seeds you sent me, I have raised several curious plants, for which I am obliged to you.

Sir Hans Sloane continues very hearty and well for his great age (88½), and daily entertains himself with reviewing his prodigious collection.

I am glad to hear the *Napæa* is in flower with you. It is a fine garden plant.

Your experiments on the food for animals will be very useful *; but remember *this*, that hunger will bring creatures to eat any thing. A peasant near me in the country observed my trees, at this time of the year, to produce abundance of horse chesnuts. He tried his hogs, but they would not eat the nuts for a long time. But he was resolved they should eat them, or starve †. At last hunger brought them to, and they grew very fat with these nuts; whose intolerable bitter is agreeable to some animals. The deer in our parks feed on them with greediness. Does not this fine flowering tree, the horse chesnut, grow in your climate? Our hardest winters, even that of 1739-40, never hurt it.

We have a very fine history of insects, with the plants they feed on, all curiously painted in their natural colours ‡, four subjects for five shillings. I believe about an hundred subjects are already delivered, with all the newest and most curious observations relating to their natural history. This fine

* These are recorded in the *Pan Suecus*, *Amœn. Acad.* v. 2. 225.

† An excellent economical hint.

‡ This probably alludes to James Dutfield's Natural History of English Moths and Butterflies, 4to, published at London in 1748 and 49, of which six numbers only appeared, and which is little known. See *Dryand. Bibl. Banks.* v. 2. 253.

work may deserve the inspection of that great Naturalist Mr. De Geer.

We have had a fine Summer. Great plenty of all sorts of fruits and grain, and a very delightful Autumn. It is now as warm as Summer; no bearing of fires. My orange-trees are yet abroad. My vineyard grapes are very ripe. A considerable quantity of wine will be made this year in England.

We have not had one frosty morning this Autumn. Marvel of Peru, Double-flowered Nasturtium, and all other annuals, are not touched. My garden makes a fine show.

Now, my dear friend, farewell. My best wishes attend you. P. COLLINSON.

MY DEAR FRIEND, May 8, 1749.

I had the pleasure of your favours of the 20th September and 30th October.

I am glad to hear you are safe returned from your expedition into Scania. Your history of that country will be very acceptable to the curious. The *Meadia* is a charming plant. I am delighted that it is in your garden.

I am glad to hear by yours of Oct. 30th that Mr. Catesby's noble and elegant work is like to grace your library. Our ingenious friend Mr. Catesby died the 23d of December last, aged 70, much lamented. His widow, to encourage the sale of his work, abates half a guinea on every book, so eleven books, at a guinea and a half, come to £.17. 6*s*. 6*d*.

On Thursday the 8th of February, at about half past 12 at noon, we had a smart shock of an earthquake, so violent that many ran out of their houses, thinking them falling down. How the Winter has been in Sweden I do not know, but at London the like warmth and mildness were never remembered. Our Autumn was long, warm, and dry, with a few slight frosts before Christmas; but we have had since fine warm dry weather, and no frosts or snow. Our gardens were in great beauty in January and February; almonds, apricots, and peaches in blossom.

Feb. 23d, I went into the country. The elm hedges had small leaves. Standard plums, almonds, and *Cornus* in full blossom. Gooseberries shewing their fruit. In short it would be endless to tell you the wonders of this season.

March 5, the fig in my London garden had small leaves, when peas and beans under South walls were in blossom.

Our worthy friend Sir Hans Sloane is in good spirits and memory. Dr. Mitchell is well. My last letter from our ingenious friend Mr. Kalm mentions his return from Canada in safety to New York, which gave me great satisfaction, for I was afraid of some wild Indians doing him a mischief.

Now, my dear friend, farewell.

I am much yours, P. Collinson.

Our learned friend Dr. Camper has conferred on

us much honour by the publication of the *Amœnitates Academicæ* (vol. I. at Leyden, in 1749).

MY DEAR FRIEND, May 8, 1753.

It is a very long time since I had the pleasure of a letter; but my friends in Sweden inform me of your health, and the great improvements and discoveries you are making in natural knowledge. Our great friend and promoter of this noble science, Sir Hans Sloane, died Jan. 11, 1753, aged 93 years, being born in 1660. His great collection of books and curiosities is purchased by our Parliament for the use of the Public, for 20 thousand pounds; but Sir Hans told me they cost him more than 50 thousand pounds.

June 26, 1753.

In the first place I shall compliment you on being so unanimously and deservedly elected a member of our Royal Society; an honour I wish you long to enjoy, which you have long merited, and I have long sought.

Mr. Miller has published a new edition of his Dictionary, in one large volume folio, price £.2. 8*s*.

MY DEAR FRIEND DR. LINNÆUS, Sept. 20, 1753.

As this may be the last opportunity by your ships, I take advantage of it to send you the enclosed book, as a present to you from John Bartram in Pennsylvania. At the end you will see his *addenda* of American plants.

I submit to your examination some very curious observations by that great critic in Botany the Right Hon. the Earl of Bute, under his own hand-writing. He begins with the *Menispermum;* the plant in his garden has different characters from that in the *Hort. Elthamensis.*

In his next letter you will see his descriptions of the *Cytisus* and *Colutea,* which are submitted to your examination. You see his Lordship's reason for his new descriptions. He thinks all names hitherto given to these plants are not sufficiently expressive.

His Lordship's next paper is some hasty remarks made on an *Apocynum,* that runs 15 feet high in my garden, and bears tufts of purplish flowers; but the specimen will give you a better idea of it.

I was glad to see the first book of your *Species Plantarum,* a noble work, that will for ever do honour to your memory. I pray God give you health to complete it.

THE EARL OF BUTE TO MR. COLLINSON.

DEAR SIR, Aug. 4, 1753.

A thousand thanks for the plants.

The climbing purplish-flowering *Apocynum* is certainly a new genus *. The nectarium is so sin-

* This is the *Periploca græca* of Linnæus.

gular I have ventured to keep your coachman till I writ the following description:

Nectarium triplex: *externum* petaliforme, monophyllum, 5-fidum, coloratum, erectum; laciniæ acuminatæ, nunc integræ, frequentiùs incisæ; lingulæ 5 ejusdem consistentiæ ad basin singulæ laciniæ: *nect. intermedium* compositum ex 5 corpusculis crassis, carinatis, truncatis, genitalia ambientibus, coronatis apice squamulâ minimâ, petaliformi, nectario interno applicatâ: *nect. internum* fit corpusculum truncatum, orbiculatum, et quasi lamellatum.

I could almost call this last nectary a stigma, and the basis the pointal, notwithstanding the two germs; but this demands more time than I can at present spare. Yours most sincerely, BUTE.

TO THE SAME.

Canewood, Tuesday, Aug. 10, 1753.

MY WORTHY FRIEND,

I have examined your compound plant. 'Tis the *Prenanthes, flosculis fere quinis, foliis lanceolatis denticulatis; Hort. Cliff.* and is well described in *Pontedera's Dissertationes*, but much better in Haller. Morison's cut of it is tolerable.

Amongst many specific descriptions of plants that I am very unsatisfied with, and have, I think, altered for the better, I send you these of the *Cytisuses* and *Coluteas* that are in my garden. Some of them I do not find Linnæus, Haller, or Royen have described.

Cytisus Neapolitanus—petiolis foliis brevioribus, calycibus squamula duplici auctis, ramulis striatis erectis.

This is the plant I got for the ever-green or Neapolitan Cytisus. I cannot find it described in any author; at least so as to make me certain of it. You know it well, so I need not tell you it grows from five to seven feet high. Towards the top the stalks and leaves are covered with whitish hair. So are the pods. The *pedunculi* are *foliosi* *.

Cytisus petiolis foliis longioribus, calycibus inflatis coloratis, ramis teretibus incanis.

This Gordon calls the tall Siberian Cytisus. It grows with me about four feet high, with slender twigs bending downwards, and large flowers thinly placed, seldom above three in a tuft. The whole plant covered with white hair. The pod acinaciforme, pedunculi simplicissimi †.

Cytisus floribus capitatis, ramis teretibus villosis erectis.

The dwarf Siberian Cytisus. This grows four feet high, with erect branches, and smaller flowers than the last, that are only placed at the extremity of the branches, commonly 10 or 12, often more. The leaves tufted under them like an involucrum ‡.

* Could his Lordship here mean the *Medicago arborea* ?

† This seems to be *Cytisus hirsutus.*

‡ Probably *Cytisus capitatus* of Jacquin.

Hortus Upsaliensis has Siberian Cytisuses, but I cannot distinguish either of these three.

Before I leave this genus, I must observe that Linnæus's character of the cup should be altered thus — Perianthium, &c. ore bilabiato; superiore integro vel emarginato, inferiori inciso. I say this from observing several with the upper lip entire, sometimes obtuse, and the under lip is often, in the *Laburnum*, &c. bifid or trifid.

Colutea foliolis ovato-lanceolatis obtusis emarginatis, alis vix calyce longioribus, squamulis (ad basin calycis) minutissimis.

Æthiopian Colutea. The alæ are just the length of the claws of the carina, which last is longer than the vexillum *.

Colutea foliolis obversè cordatis, calycibus nudis, alis carinâ brevioribus.

African and Oriental *Colutea* †.

Colutea foliolis ellipticis apice obtusis, calycibus squamulâ duplici auctis, alis carinâ longioribus.

This is Dr. Pocock's *Colutea*, the flowers larger than the bladder senna, of a fine yellow. The vexillum has some deep streaks of red on the outside; and there is a small circular red line near the alæ. I never saw it above four feet high ‡.

Colutea foliolis obversè ovatis apice obtusis setâ terminatis, calycibus squamulâ duplici auctis, alis carinâ brevioribus.

* *Colutea frutescens.*

† *Colutea cruenta,* Ait. Hort. Kew.

‡ *Colutea Pocockii,* Ait. Hort. Kew.

Bladder Senna *.

You see, Sir, I was resolved my paper should be covered, though with trifles.

I ever am, dear Sir, most sincerely yours, BUTE.

Dr. Mitchell comes out on Friday. Could not you look in as you go by ?

Characters of the *Menispermum* of Hort. Elthamensis, sent by the Earl of Bute to Mr. Collinson.

MAS.

Cal. Perianthium 6-vel 5-phyllum, persistens; foliolis ovali-lanceolatis, concavis, albescentibus.

Cor. 6 vel 5 squamulæ minimæ concavæ, obversè cordatæ albæ.

Stam. filamenta 10 vel 12, erecta, calice breviora; antheræ compositæ ex 4 globulis luteis coalitis.

The most common number is 6 and 12.

The *Female*.

The cup and petals of the female resemble the male exactly, only the cup leaves are white. The chives are 6 in number; white flat filaments, and white flattish summits, that appear to me to have no farina, and to be barren. A single, double, or triple germ, though 2 is the most common; hardly any style; 3 emarginated stigmata. Besides which I have found, though rarely on the same spike, 1 or 2 flowers entirely male, with 6 chives, but still barren, and no rudiment of germs or pointal. This

* *Colutea arborescens.*

female plant has green stalks, paler leaves, generally 5-lobed like the male, but the lobes are more rounded, end obtusely, and yet have at the points a longer *seta* than the other. The stalks of the male are reddish; neither have any entire leaves, as Dillenius in *Hort. Elthamensis* paints them; and indeed I doubt if either will at all suit his or Linnæus's description.

MR. COLLINSON TO LINNÆUS.

London, April 20, 1754.

I hope, before this comes to your hands, you are fully satisfied that you have had the honour to be elected a Fellow of the Royal Society. I very industriously promoted your election, and engaged my friends to support it, because I knew that you merited that additional mark of the esteem of the English *Literati.*

I have had the pleasure of reading your *Species Plantarum*, a very useful and laborious work. But, my dear friend, we that admire you are much concerned that you should perplex the delightful science of Botany with changing names that have been well received, and adding new names quite unknown to us. Thus Botany, which was a pleasant study and attainable by most men, is now become, by alterations and new names, the study of a man's life, and none now but real professors can pretend to attain it. As I love you, I tell you our sentiments.

I am glad to hear the *Collinsonia* thrives so well.

We are greatly obliged to you for the account of the curious and learned works printing in Sweden. It is really very wonderful how it is possible for you to carry on so many great works.

I am glad to hear our worthy friend Professor Kalm's Voyage is printed. I hope some ingenious man will translate it into either Latin, English, or French. All books of voyages and travels, printed in English, sell the most of any books in England.

By the last ships I sent you a letter, and a specimen of what we thought a climbing *Apocynum* *, but when it came to flower we found it a new genus. Mr. Ehret discovered it, so it should be named *Ehretiana*.

I am making a catalogue of my garden, in which I have great variety of rare plants; but whether I shall find time to finish it I cannot say.

MR. COLLINSON TO LINNÆUS.

April 10, 1755.

I am greatly obliged to my worthy friend Doctor Linnæus for his letter of the 20th June last.

I did all in my power to oblige and serve your friend, by recommending him to our mathematicians. — I thank you for so many pretty dissertations; I am mightily pleased with them; in particular with the Herbarium Amboinense (for I have

* *Periploca græca.*

the six volumes): but it is a great defect not to publish an Index. Dr. Stickman has in part supplied that to very good purpose.—It is a curious performance of Dr. Barck, to shew how the Spring advances in the several provinces; and I must not forget to thank the ingenious Authors of the *Flora Anglica* and *Herbationes Upsalienses.*

You will see by Mr. Ellis's title-page to his curious Dissertation on Coralines, Corals, and *Polypi*, that we are not idle whilst your pupil is very busy making discoveries in Italy. It gives all Botanists a true concern to see the *Pinax* sink into oblivion, and *lost for ever.* It is only you, my dear friend, can restore it. You have begun by your *Species Plantarum;* but if you will be for ever making *new names*, and altering old and good ones *for such hard names that convey no idea of the plant*, it will be impossible to attain to a perfect knowledge in the science of Botany.

You desire to know our botanical people. The first in rank is the Right Hon. the Earl of Bute. He is a perfect master of your method; by his letter to me you will see his sentiments, and those of another learned Botanist, on your *Species Plantarum.* Then there is Mr. Watson, Mr. Ellis, Mr. Ehret, Mr. Miller, Dr. Willmer, Dr. Mitchel, Dr. Martyn. These all are well skilled in your plan; and there are others. But we have great numbers of Nobility and Gentry that know plants very well, but yet do not make botanic science their peculiar study.

Dr. Mitchel has left Botany for some time, and has wholly employed himself in making a map, or chart, of all North America, which is now published in eight large sheets for a guinea, and coloured for a guinea and a half. It is the most perfect of any before published, and is universally approved. He will get a good sum of money by it, which he deserves, for the immense labour and pains he has taken to perfect it.

I am glad Dr. Bæck is well: I hope to write to him soon. The next thing I have to do is to thank you for yours of the 23d November. I am glad the packet came safe. Your letter I delivered to Mr. Miller, and he promised me he would give you all the information in his power relating to *Rivina*, *Duglassia*, and *Ligustroides*. Miller will, no doubt, tell you what he is now publishing; if I don't forget, I will put in one of his proposals.

It is by you, my dear friend, that the learned and curious Naturalist is so amply gratified, from every part of the world, with new and rare discoveries. Your agents bring you tribute from every quarter; we are to thank them for their observations on the Nile, and at Brazil. I am greatly obliged to you for these informations. The new fish must be very acceptable to young Gronovius, whom I much admire for dedicating his youth to useful knowledge. Your last letter to Dr. Ascanius came safe, and was delivered to him. Now accept of my best wishes for your health and preservation. I am, yours,

P. Collinson.

FROM THE EARL OF BUTE TO P. COLLINSON, ESQ.

Dear Sir,

I return the *Species Plantarum*, and am extremely obliged to you. I have taken a very small quantity of most of the seeds, that you may have enough and to spare to others for trying them in different soils, it being sometimes one man's fortune to raise what another cannot. — Dr. Linnæus has immensely changed his names and genera in this book; so that till he publishes a new edition of his *Genera*, it will be of small use. I cannot forgive him the number of barbarous Swedish names, for the sake of which he flings away all those fabricated in this country; witness the *Meadia*, the *Azalea*; that is become a Calmuck, or *Kalmia*. I own I am surprized to see all Europe suffer these impertinences. In a few years more the Linnæan Botany will be a good Dictionary of Swedish proper names. There are also many bold coalitions of genuses, that I would keep asunder if possible, merely for fear of making them too long. But pray what connexion has *Padus* with *Prunus*, more than a dozen others of the *Calicanthemi?* He wanted before to join the Cherry; so that by degrees we shall have more *confusion with order* than we had formerly with *disorder*. You will observe that, conscious of this, he has called in to his assistance Haller's method of subdividing his *genera*.

But I have been already too long. Give me leave to assure you that I am, with the greatest regard, my dear friend, most sincerely yours, Bute.

MR. COLLINSON TO LINNÆUS.

A celebrated Botanist * desired me to lend him your *Species Plantarum*. He returned me the books with the following observation:

"I have very carefully examined Dr. Linnæus's *Species Plantarum*, and do find this to be the most careless of his performances; and through the whole work he seems so vain as to imagine he can prescribe to all the world. The strange confusion of

* How much is it to be regretted that the name of this "celebrated Botanist" is not preserved; for the opinion here recorded would, doubtless, have rendered him much more famous! Whatever might be the alleged vanity of Linnæus, his good sense and good temper, in passing by such criticisms in unruffled composure, are highly commendable. His silence might originate, without much presumption, in some degree of conscious dignity; but the utmost stretch of self-conceit could not anticipate the future importance and reputation of his immortal work, which it is now as superfluous to praise or defend, as to refute the above most futile strictures of Lord Bute, or those of the anonymous "very careful" examiner. With respect to the accuracy of his Lordship's assertions, it will be necessary for the reader only to turn to the history of the genera *Azalea* and *Kalmia*. The former was not "fabricated in this country," but established by Linnæus himself, *anno* 1737, in his *Flora Lapponica*, and still remains undisturbed. The history of the latter, with generic and specific characters, as well as descriptions, first appeared in the second dissertation, entitled *Nova Plantarum Genera*, published by Linnæus, Oct. 19, 1751, p. 18, n. 1079; the full generic character being almost precisely reprinted in *Gen. Pl. ed.* 5. 185, and the rest of the history in *Amœn. Acad. v.* 3. 13. Gronovius had, indeed, referred one of the species, very erroneously, to *Azalea*, and the other, no less inaccurately, to *Andromeda*. Lord Bute has first made *Kalmia* a *Calmuck*. We leave *Padus* and *Prunus* to answer for themselves.

Synonyms, shews his want of knowledge; and his applying them in many instances to various plants, is a proof of his want of attention."

P. COLLINSON.

July 29, 1755.

The specimen* here enclosed, was collected from my garden. Two great adepts in botanic science are divided in their judgments about its true name.

Dr. Schlosser declares it a *Patagonula;* but a different species from that figured in the *Hortus Elthamensis.*

Dr. Ascanius pronounces it a species of *Lithospermum.*

Now, my dear friend, it is left to your profound judgment, what it is, and what name it shall bear, and to what class it belongs.

This I hope you will mention, by the first opportunity, which will oblige your friends here, and in particular your sincere friend,

P. COLLINSON.

This plant, the *Patagonula*, flowered in my garden last year, and again this year.

I have had two plants of the *Saracenia* flower this year. I keep the plants in an artificial bog, filled with moss, in which the roots spread, and do not strike into the earth.

* *Ellisia Nyctelea.* Linn. Sp. Pl. 1662. The letter is accompanied with a good detailed drawing of the fructification, by Mr. Ellis himself.

The *Acacia flore rubro**, of Catesby's last figure in his appendix, flowered this year in June, for the first time. I had the tree sent me from South Carolina. It is a most beautiful sight. Mr. Ehret will print and publish it, with many other new and rare plants that have lately flowered in my garden and others.

My dear Friend, May 12, 1756.

Without further ceremony I must tell you I was highly pleased with your friendly epistles of Aug. 10, and Oct. 17 ult.

Dr. Martyn is in good health, and has lately published an abridgment of the Transactions, to the year 1753, in 2 vols. 4to.

Mr. Ehret is fully employed with teaching the noble ladies to paint flowers, and has no time to spare. He has only published the *Beveria*, being what at Paris is named *Butneria*†. It is a charming *suffrutex*, and grows in my garden in the open air, bearing flowers abundantly every year.

Dr. Hill is publishing a history of plants, of which I send you a specimen. As he proceeds through the genuses, he criticises your method, but not like the foul-mouthed Germans. He treats you, like an Englishman, with decency and good manners; and although we cannot agree in all points, for no system can be perfect, yet we honour

* *Robinia hispida. Linn. Mant.* 101.

† *Calycanthus floridus.*

and esteem you for spreading arts and science, and increasing knowledge.

I can tell my dear friend with great pleasure that I have the *Rubus arcticus* now in flower. It increases with me, but as you have a readier method of doing it, pray tell me.

Dr. Browne has published his History of Jamaica, the plants drawn by Mr. Ehret.

Dr. Russel, a very learned man, and master of all the Eastern languages spoken about Aleppo, has lately published the natural history of that city and country about it; the nondescript plants all drawn by Mr. Ehret, in large 4to, price bound 17*s.*

I hear you have published your *Somnus Plantarum.* I have for many years observed these sleeping plants, of which there is great variety.

I but lately heard from Mr. Colden. He is well; but, what is marvellous, his daughter is perhaps the first lady that has so perfectly studied your system. She deserves to be celebrated.

Your friend John Bartram is also very well. His son is an admirable painter of plants. He will soon be another Ehret, his performances are so elegant.

You must remember I am a merchant, a man of great business, with many affairs in my head, and on my hands. I can never pretend to publish a catalogue of my garden, unless I had one of your ingenious pupils to digest and methodize it for me. It only serves now for my own private use.

Now, my dear friend, live happy and contented, as

you have so much the love of mankind; and believe me to be, as I really am,

Your sincere friend, P. COLLINSON.

April 1, 1757.

To his dear friend Dr. Linnæus, Peter Collinson wishes health and felicity.

I am glad to hear that a new edition of your *Systema Naturæ* is to be published. So great a collection, with so much industry brought together, will be very acceptable to all lovers of Natural History. But take time, and be not in a hurry, for new and rare subjects are daily discovered.

Dr. Hill has published proposals to publish your botanic works in English.

I did not doubt but my friend Ellis's curious work would meet with your approbation, as it opens new scenes of wonders.

Sir Hans Sloane's noble collection is now reposited in the British Museum, in a magnificent building, where there are proper apartments for all the branches of natural knowledge, and will be soon opened for public view.

My garden is delightfully covered with flowers. The *Arum betæ folio* of Catesby began to flower Feb. 7, and is not yet out of flower.

In the 2d vol. of Edinburgh Essays, is published a Latin botanic dissertation, by Miss Colden; perhaps the only lady that makes profession of the Linnæan system; of which you may be proud.

Ridgeway House, Dec. 25, 1757, finished 27th.

TO MY HONOURED AND LEARNED FRIEND SIR C. LINNÆUS.

Some time since a most magnificent book* was sent me, without a letter or any information whom I am to thank for it.

But when I consider the royal museum it describes, and who it is presides therein, I no longer doubt, but reasonably conclude it must be the noble present of my dear friend Linnæus.

This curious performance doth both him and his country honour. The engravings are fine, and the descriptions shew the learning and great knowledge of the compiler of that pompous work. All that I have shewn it to, deservedly admire it.

The ingenious Mr. Edwards has now finished his 5th volume of new and rare birds and beasts, and it is ready to be delivered.

Mr. Ellis continues to add new illustrations to his corals and corallines, to prove they are the work of Polypes, &c.

The extraordinary heat of our summer has ripened all sorts of fruits to perfection. In two gardens I saw this year pomegranates against south walls, without any art, ripened beyond what can be imagined in so northern a climate. They look extremely beautiful, and are of the size of some brought from abroad.

Our autumn has been long dry and warm, and so continues, for a few slight frosts have not stripped

* *Linn. Museum Regis Adolphi Friderici.*

the garden of flowers at Christmas-day. We have four sorts of *Aster* and *Virga aurea* in flower, and plenty of Leucojums, double and single, Chrysanthemums, &c.

The winter scene is not closed before spring flowers begin; for there are plenty of Polyanthus-narcissus, Pansies, and sweet Violets, *Primula veris*, Polyanthus, Aconite, Hepatica, Anemonies both double and single, and Laurustinus. You would have been delighted and surprized to see the large nosegay that was all flowers gathered out of the open garden, without any art, Dec. 27, 1757. Now the old year is making its exit. May the new year prove healthful and happy, is the sincere wish of your affectionate friend, P. COLLINSON.

Mr. Clayton writes to me from Virginia, Sept. 7, 1757, "at last I have completed a new edition of the *Flora Virginica*." But when, or where, or how it is to be printed, he says not a word.

London, Jan. 25, 1758.

To my honoured friend Dr. Linnæus I wrote, about a month since, to thank him for his noble present of the King's *Musæum*.

I have since found that Dr. Browne has left his specimens with one Mr. Millan, a bookseller, and have looked them all over*. There is a very great

* These composed the herbarium of Dr. Patrick Browne, author of the Natural History of Jamaica, the plates of which

number, and as they lie one upon another, they are more than 18 inches high. But you must think, in such a quantity, there are many very imperfect, and many laid together very confusedly and broken. It seems to me to be an endless work to put them in order. But such is your skill in these things, you will soon surmount this difficulty.

They are in the hands of a man who will have his price, and he says he will take no less than 10 guineas. This is a great sum of money; and yet there is a great number of plants, of all genuses and species that those countries produce. But I am not able to say if those specimens are amongst them, that are published in his book.

Mr. Miller gives his service to you. He has a long time expected a letter. Fine mild weather. If you determine to have the specimens, my next care will be, how to send them safe to you. Pray order some person to call on me for them. They must be put into a large chest, to secure them from damage by sea.

work were drawn by Ehret, from these very specimens. After Linnæus obtained them, it appears that he examined the whole, in company with his pupil Solander, in whose hand-writing are the Linnæan names, now attached to each specimen; though every plant is marked *Br.* by the pen of Linnæus himself. The above letter serves to correct a mistake, into which the editor was led, by his usually very accurate friend Mr. Dryander, and which is recorded in his Introductory Discourse to the Linnæan Society, *Linn. Trans.* v. I. 43, that Dr. Solander himself bought these specimens after he came to England, and sent them to Linnæus.

Pray send me a specimen of *Buxbaumia aphylla*.

I am, with much respect, your sincere friend,

P. COLLINSON.

P. S. Jan. 31st. This day came Dr. . . . . . and drank tea with me, and delivered your obliging present of 13 dissertations on many articles in Natural History, which you, my dear friend, know is my delight. You have my sincere thanks. I shall be extremely pleased and entertained in the perusal of them. *Vale.*

London, April 30, 1758.

As soon as I received my dear friend's commands, the next consideration was to try if I could get an abatement in the price of the specimens of plants. It is needless to tell the method I took to do it; but I could not get them lower than 8 guineas, the account as beneath. The balance of £2. 8*s*. I am ready to pay to whom you shall please to order me.

A very good opportunity happening, I could not refuse it, but have sent them by Dr. J. H. Jaenisch, who goes directly for Stockholm, and sails with convoy. He has promised great care of the box, and to deliver it himself into your hands.

Dr. Jaenisch is a very ingenious young man, and has been here some time to improve himself in the art of physick. He has been on board the ship two weeks, but the wind has been contrary ever since. The ship will sail the first fair wind.

Last week my friend Mr. Ellis wrote you a letter, recommending a curious botanic dissertation, by Miss Jane Colden. As this accomplished lady is the only one of the fair sex that I have heard of, who is scientifically skilful in the Linnæan system, you, no doubt, will distinguish her merits, and recommend her example to the ladies of every country.

Mr. Edwards sends his compliments, and thanks you for your obliging letter.

I am, &c. P. COLLINSON.

Rubus foliis ternatis, uniflorus is now in flower; also *Acer platanoides*, Norway Maple *vulgo*, which flowered April 5th.

P. S. May 2d. I am now drinking coffee with Dr. Russell. He thanks you for your kind letter of the 20th of March, and will answer it as soon as possible. Mr. Ellis is with us, and gives his compliments. He is very industrious to procure you new materials for Natural History.

PETER COLLINSON TO THE HON. I. TH. KLEIN, SECRETARY TO THE CITY OF DANTZIG.

London, March 6, 1758.

Read at the Royal Society, March 9, 1758.

I know not which to admire most in my dear friend Klein, his learning or his judgment, in compiling so many ingenious and instructive books in natural history.

But I must beg leave to dissent from my learned

friend in an article he takes great pains to establish in his *Historiæ Avium Prodromus*, which is, "that Swallows are not birds of passage, but at the time of their disappearance they retire under water, and live therein all winter."

This is not to be comprehended, being so contrary to nature and reason; for as they cannot live in that state without some degree of breathing, this requires a circulation of the blood, however weak and languid. Now as respiration is absolutely necessary for circulation, how is it possible to be carried on, for so many months, under water, without risque of suffocation?

Besides, if so remarkable a change were intended, the great wisdom of the Almighty Creator would undoubtedly be seen, in some particular contrivance in the structure of the heart of this bird, to enable it to undergo so very remarkable a change of elements. Yet my learned friend Klein has not attempted to shew any thing of this nature, in order to confirm his system.

An easy experiment may throw some light on this doubtful affair. At the time of their disappearance, take a swallow, and confine it in a tub under water. If it remains there for a week or two without any remarkable inconvenience, then there may be some probability of its continuing so many months in that state.

The conclusions that are drawn from some of the tribe of insects subsisting under water, are far from being decisive to found an analogy upon; as insects differ from other animals in so many particulars, that very little or nothing can be concluded or inferred of the one, from what we observe in the other.

Towards the end of September the swallows assemble on reeds in the islands on our river Thames, and have, no doubt, so done for ages past; and yet I never heard or read of any fisherman, or other person, that has ever found, in the winter months, a swallow under water in a torpid living state: for, if such a very marvellous thing had ever happened, it would have been soon communicated to the nation. Besides, these reeds, and willows, are annually cut down for several uses, and yet not a swallow has been discovered in his aquatic abode. Considering the multitudes I have seen on these reeds and willows in the autumn, if they took their winter's residence under water, it is most reasonable to think, in a river so frequented, and in so long a course of years, some swallows would have been found in that situation. Another circumstance I must add. In great towns, remote from water, where rivers and reeds are not near, I have frequently observed that, a little time before the swallows depart, they, every morning early, gather together in large flocks, on the roofs of houses exposed to the morning sun. This they daily do for some time, to collect themselves before they take their flight.

To confirm my opinion that the migration of some species of swallows is certain, I think I have some proofs. I have often heard Sir Charles Wager, first Lord of the Admiralty, relate, that in one of his voyages home to England in the spring of the year, as he came into soundings in our Channel, a great flock of swallows came and settled on all his rigging. Every rope was covered. They hung on one another like a swarm of bees. The decks, and the carving about the ship, were filled with them. They seemed almost spent and famished, little more than feathers and bones; but, being recruited with a night's rest, they took their flight in the morning.

Captain Wright, a very honest man, whom I could depend on, told me the like happened to him in a voyage from Philadelphia to London.

But a yet stronger confirmation of swallows being birds of passage, is M. Adanson's observation in his History of Senegal lately published, p. 67.

"The 6th of October, at half past 6 in the evening, "being about 50 leagues from the coast (between "the islands of Goree and Senegal) four swallows "came to take up their night's lodging on the ship, "and alighted on the shrouds. I easily caught all "four, and knew them to be true European swallows. "This lucky accident confirmed me in the opinion I "had formed, that these birds pass the seas to get into "the countries of the torrid zone at the approach of "winter in Europe. Indeed I have since remarked "that they do not appear at Senegal but at that sea-"son, with quails, wagtails, kites, and some other "birds of passage, which come thither every year "when driven by cold from the temperate countries "of Europe. A fact no less worthy of remark is, "that the swallows do not build in Senegal as in

"Europe: they rest every night, either in pairs or "solitary, in the sand upon the sea shore, which "they prefer to an inland situation."

This observation, as it comes from a professed naturalist, and one who went into those countries on purpose to collect what was curious, certainly puts the question out of doubt, that swallows are birds of passage; and the hearsay stories of ignorant peaşants, and credulous people, are by no means to be put in competition with it.

I have for many years been very watchful in taking notice of the times when the swallows leave us, and have twice seen them, undoubtedly taking their flight, in two different years, on the 27th and 29th of September. Walking in my garden about noon on a bright clear day, and looking up to the sky, at a very great height I distinctly saw innumerable multitudes of swallows, soaring round and round, higher and higher, until my eyes were so pained with looking, I could no longer discern them.

But as my learned friend Klein seems to be so positive that the *Hirundo riparia* at the approach of winter retires into the holes in which it bred up its young and made its summer residence, and there passes the cold season in a dormant state, as snakes, lizards, and some other animals do, I have been the more solicitous to come at the truth. But as the sandy precipices, in which these birds build their nests, are mostly inaccessible, some years have passed before I could find a situation where the experiment could be made without difficulty or danger.

Such a sand-hill I found at Byfleet in Surrey. The clergyman or pastor being my friend, and well qualified to make the trial, at my request he was so obliging as to undertake it.

I shall give his letter to me in his own words.

"DEAR SIR, Byfleet, October 22, 1757.

"I took a square of about 12 feet over that part "of the cliff where the holes were thickest, which "in going down from the surface would take in 40 "holes. I set to work, and came to the holes, but "found no birds, nothing but old nests at the end "of the holes, which were from 18 to 30 inches "deep, from the entrance. We carefully searched "40 holes but found no birds, but at least 30 of "them had nests. The passage to them was very "near in a straight horizontal line; the nests were "sunk about an inch and half below the level of "the passage. The materials next the bottom were "straws, then a course of fine grasses, the whole "structure of no great elegance. The few eggs "left behind were of a clear unspotted white, the "size of a redbreast's."

This careful examination being made by a gentleman of veracity and ability, is very satisfactory and conclusive. It certainly and clearly proves that the *Hirundo riparia* does not take up its winter abode in its summer habitations. There is therefore sufficient reason to believe, from all the observations above recited, that swallows are birds of passage.

P. COLLINSON.

To LINNÆUS.

Aug. 22, 1758.

I wish the foregoing dissertation were translated into Swedish, and read before your Royal Society, of which I am a member. Possibly it may excite some person of probity to give *matters of fact of his own knowledge* relating to the migration of swallows, for there is no depending on the reports of ignorant country people.

It gives me great satisfaction to hear by my dear friend's letter of July 18th, that Dr. Browne's specimens came safe, and afforded you so much entertainment. I have sent you a box and a paper packet, being a specimen of Mr. Warner's Jessamine *, by Capt. Fisher, in the Stockholm, which I hope will be acceptable. I am glad of your success with the *Nitraria* or *Cassia Poetica.*

I am surprised to hear Dr. Browne is in Portugal. Pray send me Hasselquist's travels.

I am, &c. P. COLLINSON.

London, July 20, 1759.

I have the pleasure to salute my dear Linnæus by the hands of my valuable friend Dr. Biörkin. The loss of his company I much regret, but cannot forget the many agreeable conversations we have had in examining my cabinet. Botany was not his favourite, so I could not tempt him to make my

* *Gardenia florida.*

garden a visit, which is enriched with many rare plants and trees not to be seen elsewhere.

You may blame me for not answering your queries concerning the *Homines Troglodytes*: but, as our friend Ellis undertook to do it, I thought it needless to repeat it, as he is a perfect master on that subject.

Dr. Fothergill is much pleased with your letter. Your plan of procuring the Chinese specifics is a good one, if it could be carried into execution. Many difficulties attend it from England, for we are not permitted to go up into the country, and it requires a great length of time to be well versed in their language. A Jesuit well skilled in medicine is the only man that can attain this great discovery, for he can have access to the Emperor's court, and may very easily get acquainted with the Chinese physicians. Or if the Court of Russia would adopt this scheme, and permit two skilful Swedish physicians to go in their caravan to Pekin; and procure them a licence from the Chinese Court for their tarrying there some years, to learn the language, and make acquaintance with their apothecaries and physicians, then we might hope to attain the knowledge of their art of healing, and of their *Materia Medica*. I do not see any other way.

July 25, 1759.

I lament the long absence of poor Solander. What can be the meaning of this delay? Is no certain advice come of his fate? Is not the name of the ship known, or the port from whence she sailed?

I thank you for yours of May 31. We had the mildest winter ever known. Our spring was early and very agreeable, and our summer the finest and warmest since 1750. Great plenty of all sorts of grain and fruits. New wheat of this year's produce has been the 21st instant at market.

For all sorts of news I recommend you to Dr. Biörken, who is well acquainted with the curious people here.

I have two species of what is called *Rhabarbarum verum*; the one with broad leaves, the other with long narrow curled leaves. The one was sent me from the Jesuits at Pekin; the other was from my dear friend Dr. Amman. But I observed a very different plant of Rhubarb in the Chelsea garden, which Mr. Miller tells me was sent him, I think, from Holland, and has all the characters of the true Rhubarb. I here send you some good seed gathered July 18th, in the Chèlsea garden.

In one of your letters you mention sending me your new edition of the *Systema Naturæ*; but I have not yet received it, which is a great disappointment to me.

I am glad you received the *Papiliones*. Let me hear from you, which always gives sensible pleasure to your affectionate friend, P. COLLINSON.

My address is to Mr. Peter Collinson, Merchant, in Grace-church-street, London.

Pray, my dear friend, do not forget to tell me what *certain discoveries* have been made on the migration of Swallows. What is said by yourself, and other curious people, to my paper on that subject? I fully expected to have heard something of it before this time.

London, May 20th, 1762.

Some years are passed since I sent to my dear Linnæus my dissertation on the migration of Swallows, but I have received no satisfactory answer. As you have asserted (I conclude from the report of others) that they live under water all winter, I did expect, and all the world expected, that you would verify the fact from your own knowledge, or from some of your pupils on whose veracity you could depend. My dear friend, your honour, your experience, your knowledge, call upon you to prove this fact, which most of the greatest naturalists of this age absolutely deny, as a thing contrary to nature and reason. What has been related by former authors, men of good credit, is doubted, because it is suspected they were too credulous, and easily deceived; but this enlightened age will not be imposed on; our belief must be established on undeniable proofs, before we can receive a relation so contrary to the course of nature, for real truth. Your reputation is so high in the opinion of the learned and curious of this age, that what you assert is taken and allowed to be a real fact; for when

I have been reasoning on the improbability of swallows living under water, it has been replied, "Dr. Linnæus says so; and will you dispute his veracity?" As it may be difficult to come at the certainty of this fact, I shall propose and recommend these two experiments; if they do not positively determine it, they will at least go a great way towards it.

I am, with great respect and sincerity, your affectionate friend, P. COLLINSON.

The first experiment I propose is, to verify that Swallows live under water by dissecting these birds, and demonstrating that there is an internal apparatus provided, and plainly to be seen, to enable them to undergo so great a change.

My next experiment that I recommend to be tried is, as near as can be to the time of Swallows going away or migrating, let half a dozen or more be caught, which may be easily done in the night with a net, whilst they are at roost on the reeds and willows. Have a large wide tub ready filled, a foot deep, with mud or sand, then fill it with water within a foot of the brim. Let a broad board float on the top of the water. On this board put the Swallows, and then cover the tub with a net; so leave them in quiet. This should be repeated every day with Swallows until no more can be found; and if after one or two month's time they are taken out of the water alive, then the fact is proved. As you are very certain of the period of Swallows going and coming, some persons of probity should be set to watch their motions. As their numbers are great it is unlikely all could conceal their going down in the water. If they do disappear in this manner some in so many must be discovered; and soon after some fishermen should be employed to drag them up again. In the spring, at their return, some persons should be appointed to watch their coming out of the water, in their languid and wet state, and how long afterwards they lie drying before they be fit for flight.

London, Sept. 2, 1762.

My dear Linnæus cannot easily conceive the pleasure of this afternoon. There was our beloved Solander seated in my Musæum, surrounded with tables covered with an infinite variety of sea plants, the accumulation of many years. He was digesting and methodizing them into order, and for his pains he shall be rewarded with a collection of them, which no doubt you will see. Afterwards at supper we remembered my dear Linnæus, and my other Swedish friends, over a cheerful glass of wine. My son, who is a great lover of plants, as well as myself, is greatly disappointed that the *Betula nana* is not sent us, as it is our favourite plant, and has been long promised to us; and what I do not take friendly is, that every friend has been obliged with your *Systema Naturæ* but myself. Must your oldest friend be served last? That is not kind.

I thank you for your letter of congratulation on

the taking of Martinico, and with it the true cinnamon—but alas! I am fearful this rich aromatic tree will be given up again with the island, without securing the tree to ourselves, for soldiers mind nothing but slaughter and plunder.

Solander is very industrious in making all manner of observations to enrich himself and his country with knowledge in every branch of natural history.

We have had a delightful warm summer; all the fruits of the earth very good, and in great plenty; and what crowns all, the blessing of peace is like to be added. That you may long enjoy health of body and mind, is the sincere wish of your affectionate friend, P. COLLINSON.

I am glad the bulbous roots please you. More species will flower next year.

My great *Magnolia* is now finely in flower.

MY DEAR LINNÆUS, London, Nov. 16th, 1762.

I am informed there are proposals from the Academy of Sciences at Petersburg to engage our dear Solander for a botanic professor. For the love and esteem I have for the doctor, I cannot forbear expressing my concern, for many reasons. First, from the uncertain situation of the public affairs in that kingdom; for it is impossible learning can flourish in tumults and riots. Who knows, in a revolution, which may soon happen, how far the person of a stranger may be safe, and his pension secure?

Next, if we consider, the russian empire is inland, and those countries sufficiently explored, and their vegetable and mineral productions discovered by the indefatigable pains of Gmelin, Steller, and others; besides, that country has no commerce with the West or East Indies, to bring new and rare productions from thence. Pray what is there to exercise Solander's great talents, which he has been with so much pains and industry improving? Am I not witness to his daily labours? do I not know of my own knowledge his unwearied application to attain competent skill, in every branch of natural philosophy? Must all these fine accomplishments be lost and sunk into supineness for want of proper subjects to exercise his aspiring genius? Can you, my dear Linnæus, advise Solander to be confined for years, no longer a free agent, but buried and lost in obscurity and confinement? If any good and advantageous station should offer, and a person of his abilities will not want friends, he is here confined, and cannot accept of it until his time is out: how hard is this! Should it please God, my dear friend, to take you from us, who is there you could wish to be your successor, but Solander? who is there could fill your place with so much propriety, and so well accomplished? But if he accepts the professorship on the old conditions, his hands will be tied, and his feet fettered; it will prevent his accepting any advancement. No doubt but you, my dear friend, know persons less eminent, but every way qualified in botanic science to teach Russian bears.

I heartily wish we could here give Solander the encouragement he deserves. Some of his friends have proposed a scheme for that purpose; the success will be known before the end of this year. If after all I have said, you have determined Dr. Solander for Petersburg, pray do not give a positive answer to the Academy there, until you hear from your affectionate friend, P. COLLINSON.

I presume my letter of the 20th of May, relating to the Swallows, you have received. I expect the favour of a satisfactory answer; for of my first letter, three years ago, on that subject, you never did me the honour to take the least notice.

TO THE RT. HON. BARON LINNÆUS, F. R. S. &c.

London, Sept. 15, 1763.

After reading my dear Linnæus's letter of the 23d Nov. 1762, how can I any longer doubt that Swallows live under water all winter? But it would confirm that surprising phænomenon, if my dear friend could any way contrive to examine anatomically what wonderful apparatus there is, at that period of time, of their going under water, in the structure of the heart of the Swallow, to qualify it to undergo a change of elements so contrary to nature, as for animals that are bred and live on earth and air, to continue to live and breathe so many months under water. If this desirable discovery is impracticable to be performed by yourself, I wish you could recommend, nay, injoin some of your numerous disciples that may have dwellings not too remote from these lakes, to apply themselves diligently to catch these birds, and examine them as near the time they abscond themselves as possible. I really think it is an inquiry worthy the dignity of the great philosopher, Linnæus, and it should be his ambition to give satisfaction to so many great and learned men, that donbt (others absolutely deny) the possibility of Swallows living so long under water, by shewing and explaining the reason of it anatomically. For undoubtedly the all-wise and powerful Creator hath occasionally (for that end and purpose of living in water) provided or substituted some vessels, or organs, near the heart, to supply a sufficient means for respiration, and to prevent the birds from drowning. If this can be happily found out, it will establish the veracity of your assertion, and be more likely to be met with in birds that have lain some time in the lakes. It may facilitate this inquiry, if rewards were offered to the fishermen, when they took up any Swallows in the water or from under the ice, to bring them to you or your disciples. I have another material reason to recommend to you for procuring Swallows in the winter from the lakes at any reasonable expense, that the particular species may be certainly determined that lives under water; for if one species only takes residence in that element, the hypothesis of another species migrating to distant countries will not then be doubted. I am as certain that I twice saw them taking their flight thither, as you are certain you saw them taken from under the water, and revive and live. It is a remark that strikes me, when I consider that all the rivers and lakes of England, Wales, Scotland, and Ireland, are not without their fishermen; and yet neither tradition, nor any relation, nor any of the books I have read of these kingdoms, have the least intimation or record that Swallows have been found, and taken up from under the water, in the winter months. Can you, my dear friend, account for this, or shew me that your Swallows are of a different nature from ours? I have proposed more than once two experiments to be made, in hopes of elucidating the subject in dispute. Possibly some interesting inferences may be drawn from them; but I am not so happy to be regarded or favoured with any account of them, which I shall beg leave again to repeat. At the time the Swallows are nearest absconding, they resort in vast numbers to the reeds and bushes on the sides of rivers and lakes; so may be easily taken in the night with a net, to make the trial. Experiment the first, take five or six Swallows, and tie a weight to their legs, and sink them under water. If they survive after lying therein seven days, who will doubt their living in the lakes! but it may be objected, this is forcing them against nature. To obviate this, make experiment the second. Take a large, deep, wide tub; put a foot deep of sand at the bottom; then fill it with water to within a foot deep of the brim. Then place a thin broad board on the water; on this board put some swallows; then cover the tub with a net. If they immerge under water and live, this will establish my dear Linnæus's assertion. You are, my dear friend, the great and good man whom your sagacious King and Queen "delight to honour." Your long life spent in the most arduous studies, your unwearied application to improve mankind as well as your own country, very deservedly entitle you to the high honours so lately conferred on you. May you live long to enjoy them, with health of body and tranquillity of mind, is the ardent sincere wish of your affectionate friend, P. COLLINSON.

P. S. It is now five years since my good friend told me I should receive his *Systema Naturæ*: what a tedious while it is for a philosopher to wait! and yet I have hopes my longing eyes shall see it. Pray when you see my dear friends Dr. Bæck and Dr. Biörkin, my love and respects to them. As soon as the great tall Siberian Larkspur * has done flowering, I then cut it down close to the ground. It soon shoots up new branches, and is now again in flower. This I do every year, and thus it flowers twice in a year. My great *Magnolia* hath been finely in flower this year. What is remarkable of the species of the *Magnolia*, they do not flower all at once and the same time like other trees, but continue flowering for two or three months.

Almost every day rain since the middle of July,

* *Delphinium elatum*. This practice invariably succeeds, without at all weakening the plant. *Ed.*

the spring and summer very dry to that time. Very great plenty of grass and all sorts of corn, but the weather unkindly for the harvest.

London, Sept. 5, 1764.

It rejoices my heart to hear by your disciple Mr. Combles of your recovery. I congratulate your country and the learned world on so happy an event, the prolongation of a life so dear to all men. I have wrote you letters on the Swallows, and congratulated you on your new honours; but you forget your old friend that loves you. It seems an age since I received a letter from you, but your sickness may have prevented me that pleasure. Now you are reinstated, delay not to oblige me with a token of friendship. As I sympathize with you in your calamities, so I take great pleasure to hear your son is honoured with the professorship. May he live long to enjoy it, with the high reputation of his father! I doubt not but our ingenious friend Mr. Ellis has sent you his curious discoveries and observations on the Sea-pen, or else I should have sent it you. I have a short time allowed me for these few lines, which I hope may excuse *errata* in them. I am, with sincere wishes for your health and preservation, your affectionate friend, P. Collinson.

It is now two years past that you were so kind to promise me the new edition of your *Systema Naturæ*, but I have not yet had the satisfaction to receive it; I hope it is not lost in coming hither.

London, May 1, 1765.

What a comfort it is to me that I hope I can salute my dear Linnæus in health, having so happily escaped a threatening sickness! You have, my dear friend, infinitely obliged me with your most acceptable letter of December the first. I felicitate you on your recovery, and on your daughter's nuptials. May happiness attend them! I am thinking what a many and variety of occurrences have happened since I saw my dear Baron at my house in London. It is amazing in our short lives what a few years bring about.

I did not in the least imagine but you had received the account of the *Penna marina* from our friend Ellis, because I know he sent it to you as soon as it was published; how it miscarried I know not. I am glad I sent it to Dr. Bæck, that you might have the opportunity to see it; but had I known you had been without one, I should have sent it to you. We all deplore the loss of the excellent Mr. Forskahl; but it rejoices me that he lived to send you the *Opobalsamum*, on which botanists were so much divided. It is surprising with what address you can dispatch such great, curious, and critical works, as the King's and Queen's musæum, &c.; but the exalted pleasure you enjoy in surveying the riches of the creation, is an ample gratification, suited to your noble mind, replenished with such eminent learning and knowledge. Pray are there any hopes of seeing published the non-descript insects engraved? for descriptions without figures convey but imperfect ideas of the subjects. You are happy, my dear Linnæus, that you can sit at home and receive the annual tributary collections from all parts of the world; may you long live to enjoy the fruits of your labours!

Dr. Kuhn is with us, translating your travels, which I long to peruse.

Dr. Solander goes on very successfully at the Musæum, and has been lately much engaged in surveying the Duchess of Portland's Musæum, where there is a very great collection of shells and marine productions, gems and precious stones. I desire your acceptance of some seeds from Maryland, which I hope will afford you some speculation. Where there is too much of a sort for the Upsal garden, you may divide them amongst your friends. Now is the delightful season. Flora appears bedecked with great variety of beautiful attire, altogether charming. The long approaches of our spring make our gardens very entertaining, for vegetation never ceases in our temperate climate. Somebody told me that you doubted the flowering of the *Sarracenia* in England. I can assure you, my dear friend, for many years past, both species of *Sarracenia*, the same that are figured in Catesby's Natural History, flower annually in my garden. The flower-buds now appear about an inch high. This is certainly one of the wonderful flowers. Both species are finely painted by Mr. Ehret. They are bog plants, and I know their culture. I have now the *Ledum palustre* very finely in flower, and that is a rare bog plant. If you desire it, I will give you my method. The leaves of the two species of *Sarracenia* are as surprizing as the flowers: for they are open tubes contrived to collect the rains and dews to nourish the plants in dry weather. A very curious work began to be published in the year 1763, entitled British Zoology, the principal editor Thomas Pennant, Esq. a gentleman of a good estate and great learning. There are three parts already done, a fourth will be finished next winter. The inclosed blue-covered book will give you some idea of the work, and the author's abilities. Each book containing 25 plates, is two guineas. Now, my dear friend, I wish you all happiness, and am affectionately yours, P. COLLINSON.

I am impatient for the arrival of your ships, that I may happily see your *Systema Naturæ*, which I have waited so long to see. The Martins always come two or three weeks before the Swallows; is not this a demonstration that they go not both to the same place? If they did, they would come away all together.

Can there be got no seed of the *Cimicifuga*?

As Mr. Edwards has presented you with a coloured print of Dr. Fothergill's great Pheasant from Tartary, I procured some of the feathers to give some idea of its wonderful plumage.

In the first place, the Chinese made an excellent painting of this most beautiful rare bird, on a sheet of paper six feet long. But lest we should suspect there was more art and ingenuity than nature, they most carefully packed up all the feathers of this wonderful bird, to verify their performance to be genuine. It is the size of a large cock Turkey; the two feathers of its tail are full three feet long; it has all the characters of the Pheasant. Mr. Edwards, from the eyes in the wings, names it the *Argus.*

As it comes from Chinese Tartary, it would thrive well in our climate. But what is remarkable, in all the Chinese paintings, neither I nor my friends ever saw this charming bird, which would make so glorious a figure. The size of the Chinese Argus or Luen is, from the tip of the bill to the end of the tail in a straight line, seven feet and a half; from the top of the back to the legs, one foot three inches and a quarter. The leg and foot 10 inches and a quarter.

London, Sept. 17th, 1765.

The sight of my dear Linnæus's well-known characters revives my heart, and gives me a pleasure I cannot express. To you, to whom Nature pays tribute from all parts of the world, could I expect to offer any thing new? I am glad to find the China Argus proved so.

As you so justly admire the *Sarracenia,* as one of the wonders in the vegetable kingdom, that you may have a more perfect idea of the wonderful reservoirs, which retain the water to supply the plant in great droughts, I send you two leaves. I have filled them with moss within, to keep them to their size and shape, which may be easily taken out; and I have packed them in moss, so hope they will come fresh and plump to your hands, and give you a satisfaction that the best description cannot so well do. I have added the seed-vessels of this year's flowers.

I think the *Dipsacus* hath perfoliated leaves down the stalk, that hold water to replenish the plant; and the Viscums * of the West Indies, in their concave leaves, retain a great deal of water to refresh the plant, whose roots spread on the old rotten bark of trees, and do not incorporate with the tree as our Viscums do. You, my dear friend, surprize me, with telling me of your cool and wet summer; whereas our summer has been as much in the extreme the other way. For all May, June, and July, were excessively hot and dry; but six or seven rainy days in three months, so that all our grass fields look like the sun-burnt countries of Spain and Africa. Our Fahrenheit's Thermometer frequently 84 and 85 in the shade in the open air, but in my parlour frequently at 95. I do assure you I have had little pleasure of my life this summer, for I cannot bear heat. I have longed to be on Lapland mountains. The beginning of August we had some fine rains, but they did not recover our usual verdure. Since, to this present writing, hot and dry weather, not a drop of rain for fourteen days past. Our hay is very short, and oats and barley but a middling crop; but of wheat, which we most wanted, good Providence has favoured us with a plentiful crop, and a good harvest, which began two weeks sooner than

* *Tillandsiæ.*

in common years. Peaches, Nectarines, Figs, Grapes, Pears, &c. are early ripened, and are richly flavoured, and many exotic shrubs and plants flower finely this year. My garden is now a paradise of delight, with the variety of flowers and plenty of roses now in bloom, as if in May or June. But to obtain all this pleasure, great pains have been taken to keep the garden continually watered every evening. That you, my dear friend, may long enjoy good health and tranquillity of mind, is the sincere wish of your affectionate friend, P. COLLINSON.

P. S. Must I not have the pleasure of seeing your noble work the *Systema Naturæ before I die?* which you have given me the expectation of some years ago.

Pray my respects to your son. I also have a son who loves Botany and Natural History, as well as myself.

Many leaves grow round the centre bud of the *Sarracenia,* which make a pretty appearance with their mouths open to catch the rains and dews; but many poor insects lose their lives by being drowned in these cisterns of water.

The two leaves now sent, are the leaves of last year, so are a little decayed. The leaves of this year are not attained to maturity

Inclosed are specimens of the *Erica Cantabrica,* &c.* now in flower in my garden, which was raised from seed sent me last year from Spain. It is an elegant plant, and makes a pretty show.

* *Erica Dabeoci. Engl. Bot. t.* 35.

London, Sept. 25th, 1766.

By the Spring ships 1765, May 1st, and by the last in the Autumn, Sept. 17th, 1765, I wrote to my dear Linnæus, and sent him the leaves of the *Sarracenia*, which he so much admired, that he might see its structure; and two specimens of the *Erica Cantabrica*. A year is now passed, and not a line from my dear friend! this is treating me unkindly, considering our long friendship. You have promised more than once to send me your master-piece of nature, the last edition of the *Systema Naturæ*. *Are not my eyes to be blessed with a sight of that universal Pinax, before I die?* Year after year I have not failed to remind you of your promise. If you doubt it, I will send your letters. It gives me concern to urge the affair so far; but pray excuse it, for it is the last time I will ever mention it to you. Some time ago I saw what I think a surprizing curiosity. On a large peach-tree full of fruit, there was a twig about two inches long, on one side grew a peach, and on the other side a nectarine. They grew so close together that they touched each other; I stood long with admiration viewing this wonder. The nectarine had the shining smooth surface with a red complexion; the peach was rough and downy, as peaches are*

We have had two more remarkable instances of peach-trees naturally, and without art, producing nectarines; so I reasonably conclude, the peach is the mother of the nectarine. Where this *Lusus*

* See p. 8 of this volume.

Naturæ has happened, ingenious men have improved the accident, by budding, or with grafting, from the nectarine branch; and thus the race of nectarines began. The variety we have in our gardens has been produced by sowing the stones, and I will tell my dear Baron an instance in my garden. Some person eating a nectarine, threw the stone away. Next year it came up. I suffered it to grow, supposing it to be a peach; but as it grew up to the fifth year, to my great pleasure it shewed it to be a nectarine; and this year, at this present writing, has near three dozen of ripe fruit on it, as rich and high flavoured as those against the wall. Dr. Solander came down to Mill-hill to feast himself with nectarines, and he saw this fine tree full of fruit, which ripens a week or two later than those against the wall. This accident confirms what many doubted, that a nectarine-tree can be produced from the stone, without grafting or budding. As you, my dear friend, have the largest Botanic library, pray let an ingenious pupil search it carefully, and answer the following queries, for I want to be satisfied about the origin of nectarines: 1st. Was the Nectarine known to the ancients? 2nd. Who are the first authors that mention it? 3dly. Do they take any notice of its origin, and mention the country whence it came?

We have had a most uncommon rainy summer, which was no way propitious to the growth of the wheat; but it pleased Good Providence to send us the finest hot and dry harvest ever known, yet the warm constant rains drew up the wheat so much to stalk, that the ears are very light. I hope there will be sufficient to support the nation, now we have prudently stopped the exportation; for so great are the wants, and the demand for foreign markets was so great and so pressing, that it advanced the price so considerably as to occasion insurrections in many parts of the kingdom, to stop by force the corn from being exported; but now a proclamation is come out to prevent it, I hope all will be quiet again. Much wet has made great crops of grass; so that every where we have had second crops of hay almost as large as the first, and a glorious autumn to make it. The fields have a most delightful verdure, and the gardens are in the highest beauty, being covered with great variety of autumn flowers, having not had the least frost to Oct. 4. I have housed none of my succulent exotics; for the weather is so hot, dry, and fine, they are better abroad than in the house. I survey my garden with raptures, to see the infinite variety with which the great Creator has enriched the vegetable world.

A few days ago Professor Hope, of Edinburgh, came to see my little paradise. He was highly delighted to see so many new and curious plants he had not seen elsewhere. I loaded him home with specimens. Were it possible to see my dear Linnæus, what joys it would give me! but if I cannot see the father, I hope the son will be tempted to make England a visit, and to see his dear father's old friend. In the Princess of Wales's garden at

Kew the *Protea major* is in fine flower; and the *Andrachne* has been finely in flower last May, in Dr. Fothergill's garden. Mr. Ehret has painted it. As soon as it is engraved I will send you a print. My tree is much larger than Dr. Fothergill's, but has not yet flowered. It will give me great pleasure and comfort to hear you enjoy your health; that it may long continue is the ardent wish of your sincere and affectionate friend,

P. COLLINSON.

Pray my respects to your son.

Your last letter was Aug. 15, 1765.

P. S. Oct. 4, 1766. As the Seals or *Phocæ* have an opening in their hearts called the *Foramen ovale*, and a passage called *Ductus arteriosus*, which being both open, they can keep a long time under water. If a provision or contrivance, something of the same nature, could be discovered in the structure of the hearts of swallows, then I should no longer *doubt* of their continuing under water all winter. This I have by letter after letter proposed to my dear Baron; but he turns his deaf ear to all my proposals to elucidate this dark affair, who has so many ingenious men under him capable to do it, and clear up this grand point, for land-birds to be qualified to live more than half the year under water.

We have several new animals from America, which I presume Dr. Solander will give you an account of.

Ridgeway-house, on Mill-hill, ten miles
North of London, March 16, 1767.

I am here retired to a delightful little villa, to contemplate and admire, with my dear Linnæus, the unalterable laws of vegetation. How ravishing to see the swelling buds disclose the tender leaves! By the public news-papers we were told that with you in Sweden the Winter was very severe, the Sound being frozen over. I have no conception of the power of that cold which could fetter the rolling ocean in icy chains. The cold was what we call severe, but not so sharp as in the year 1740. It lasted about a month, to the 21st of January, and then the thaw began and continued. February the 1st and 2d were soft, warm, sunny days, as in April, and so continued, mild and warm, with southerly winds, all the month. This brought on the Spring flowers. Feb. 8th, the *Helleborus niger* made a fine show; the *Galanthus* and Winter Aconite by the 15th covered the garden with beauty, among some Crocuses and Violets, and *Primula veris*, &c. How delightful to see the order of Nature! oh, how obedient the vegetable tribes are to their great Lawgiver! He has given this race of flowers a constitution and fibres to resist the cold. They bloom in frost and snow, like the good men of Sweden. These flowers have some time made their exit; and now, March 7th, a tenderer tribe succeeds. Such, my dear friend, is the order of Nature. Now the garden is covered with more than 20 different species of Crocuses, produced from sowing seeds, and the *Iris Persica*, *Cyclamen vernale*, and Polyanthos. The 16th March, plenty of *Hyacinthus cæruleus* and *albus* in the open borders, with Anemonies; and now *my favourites* the great tribe of Narcissuses, shew all over the garden and fields; we have two species wild in the woods that now begin to flower. Next the *Tulipa precox* is near flowering; and so Flora decks the garden with endless variety, ever charming.

The progress of our Spring, to the middle of March, I persuade myself, will be acceptable to my dear Baron.—Now I come to thank him for his most acceptable letter of the 8th of October last. I am extremely obliged for your kind intentions to send me the work of works, your *Systema Naturæ*. I hope it will please God to bless my eyes with the sight of it. I feel the distress you must be under with the fire. I am glad, next to your own and family's safety, that you saved your papers and books. By this time I hope all is settled and in order; so pray now, at your leisure, employ some expert pupil to search into the origin of the Nectarine; who are the first authors that mention how and when it was first introduced into the European gardens. It is strange and marvelous that a peach should naturally produce or bear nectarines, a fruit so different, as well in its exterior coat as flavour, from a peach; and yet this nectarine will produce a *nectarine* from the stone, and *not a peach*. This remarkable instance is from a tree of a nectarine raised from a stone in my own garden, which last autumn had several dozen of fruit on it, finely ripened. For more particulars I refer to my last letter.—Pray tell me who Perses was, what countryman, and who is the author that relates his introducing peaches into the European gardens?

That bats as well as flies lie as dead all Winter is true; but they do not change elements, and go and live all that time under water. Swallows cannot do it without a provision and contrivance for that end, which it becomes your great abilities to find out; for it is not sufficient to assert but to demonstrate the internal apparatus God Almighty has wonderfully contrived for a flying animal, bred on the land and in the air, to go voluntarily under water, and live there for so many months. Besides, we are not informed which species lives under water, as there are four species. You, my dear friend, have raised my admiration, and that of all my curious acquaintance, for we never heard before that mushrooms were of an animal nature, and that their eggs are hatched in water. We must suspend gratifying our curiosity until this phænomenon is more particularly explained to us here. Dr. Solander is also a stranger to it. Very probably some account has been published in the Swedish tongue; if that is sent to Solander, then we shall be made acquainted with the discovery.

I herewith send you a print of the *Andrachne*, which flowered, for the first time I presume in Europe, in Dr. Fothergill's garden in May last year. It was raised from seed from Aleppo, sent to the Doctor by Dr. Russell in the year 1756*. You see

* This original plant of the *Arbutus Andrachne* was sold, at Dr. Fothergill's auction in August 1781, for £53. 11*s*. A tree

1. Beginning of a Letter of Linnaeus to Ellis

Viro nobilissimo

D. JO: ELLIS

S. pl. D.

Car. Linnaeus

Eqv.

multum me in aere tuo esse ingenue fateor ob gratiam singularem quia excepisti solandrum meum, qui in literis nequit satis Tuas praedicare laudes; dum eum excepisti omni paterno affectu; et non tantum eum ipse fovisti, sed apud amicos commendasti.

2. Conclusion of a Letter of P. Collinson.

I am my Dear Friend with my sincere Wishes, for your Health & preservation, your affectionate Friend

P. Collinson = now Entred into my - 73 - Year in perfect Health & Strength, in Body & Mind = =
God Almighty be praised & Adored for the Multitude of his Mercies

3. Conclusion of a Letter of Ld. Bute.

You see Sir I was resolv'd my paper should be Cover'd, tho with trifles.
I ever Am - Dear Sir

Most sincerely Yours &c.

Bute

Canewood, tuesday morn: aug: 10: 1753

Dr Mitchell comes but on fryday, Could not you look in as you go by

R. Martin's Lithog.y Rochester St. Westr.

The material originally positioned here is too large for reproduction in this reissue. A PDF can be downloaded from the web address given on page iv of this book, by clicking on 'Resources Available'.

its manner of flowering is very different from the *Arbutus*. I have a large tree raised from the same seed, that stands abroad in the garden, but never blossomed. It is now beginning to shed its bark, as Belon or Belonius well describes; which is a peculiar difference from the *Arbutus*, and nearly agrees with the *Platanus*.

I am, my dear friend, with my sincere wishes for your health and preservation, your affectionate friend, P. COLLINSON, now entered into my 73d year, in perfect health and strength of body and mind; God Almighty be praised and adored for the multitude of his mercies! March 16th, 1767.

Pray present a print of the *Andrachne* to my worthy friend Dr. Bæck, with my sincere respects.

Pray send me seeds of the *Alstrœmeria*, figured and described in the *Amœnitates Acad.*

This eloquent concluding letter of the amiable Collinson, like the fabled voice of a dying swan, displays a mind ripe for immortality; just fitted to take its flight; fully prepared for, though not apprehending, its approaching happy removal to scenes more fitted to its improved nature.

"*Mentem tantarum rerum capacem, corpori caduco superstitem crede.*"

of the same species, full twice as large, long the boast of Chelsea garden, was killed by the cold Winter of 1796. — Mr. Aiton records, on the authority of Mr. Knowlton, who was gardener to Dr. Sherrard at Eltham, that the *Arbutus Andrachne* was cultivated there in 1724.

I cannot but subjoin the following remarks, written July 7th, 1808, by a very eminent and learned friend, to whom the preceding letters were communicated :

"Many thanks for the perusal of Collinson's letters, which I return with this. I have edified much on the subject of the Springs, which appear at that time to have been much milder than at present. We have now, for many years, had hard Winters occasionally, and an almost constant succession of ungenial Springs. The seasons are, I conclude, subject to these variations. The series of mild Springs, which ended about the year 1785 or 1786, seems to have begun at least as early as 1749, and to have lasted 36 years. Our present series of cold Springs has yet lasted only 23 years. Of course we have 13 bad years to come before we can expect Violets and Narcissuses in January, and Grapes ripe in the beginning of September."

These 13 years are now (1820) approaching to a conclusion ; and we may at least indulge in the "pleasures of hope," that the apparent deterioration of our climate (so generally felt, that any hypothesis is admitted, without foundation or examination, to account for it) may soon come to an end.

Biographical Memoir

OF

JOHN ELLIS, ESQ. F. R. S.

AND HIS

CORRESPONDENCE WITH LINNÆUS.

JOHN ELLIS, F. R. S. illustrious for his discovery and complete demonstration of the animal nature of Corals and Corallines, was a native of Ireland. In the only biographical account, or at least the first, that ever appeared, of this eminent character, in Rees's *Cyclopædia*, written by the author of the present memoir, he is erroneously supposed to have been born in London; where, however, he died Oct. 15, 1776, aged about 66 years. He appears to have been engaged in merchandize in the early part of his life, and was subsequently employed in one or more public appointments; being, through Lord Chancellor Northington's favour, made agent for West Florida in 1764, and for Dominica in 1770. His numerous foreign correspondents furnished him, through life, with supplies of curious subjects in Natural History from all parts of the world, especially North America, the West Indies, and China; nor could such materials fall in the way of a man more likely to turn them to the advantage or instruction of mankind. Ever intent upon the arts, manufactures, and natural productions, of remote countries, he laboured, with the most indefatigable perseverance, to discover any thing likely to be serviceable at home. The preparation of Sal Ammoniac in Egypt, of the fine Varnish of Japan, and the cultivation and improvement of Coffee, were among his chief œconomical inquiries. In Botany, and the more abstruse parts of Zoology, he took a wider and a deeper range. Every genus of plants that he established, *Halesia, Gardenia, Gordonia,* and *Dionæa,* have been confirmed by the unanimous suffrage of all Botanists; nor did he neglect the anatomy and physiology of vegetables, in his time a new and little explored path of science. His correspondence was extensive both at home and abroad, as the following pages shew. He will there be found to defend with ability, and to its full extent, his own original doctrine, of the entirely animal nature of the numerous and intricate tribes of Corals and Corallines. Even Linnæus hesitated to follow him to the utmost in this opinion; and has fallen into half-measures and ambiguities, which disgrace that part of his immortal *Systema Naturæ* where these productions are described. His definition of the Order of *Lithophyta* indeed is founded on the principles of Ellis; "a fixed calcareous Coral, constructed by animals thereunto attached." But his *Zoophyta* are defined as "compound, blossoming animals; consisting of a vegetating stem *(stirps)* changing, by metamorphosis, into a flowering animal." Mr. Ellis has shewn that this tribe of animals differs in no essential respect from the true Corals, or *Lithophyta*, except in having more gluten and less lime in their constitution, so as to be more compact, homogeneous, and flexible; nor would he give up this deliberate and well-founded opinion even to Linnæus himself, whose knowledge and judgment he otherwise almost adored.

The whole Correspondence of Mr. Ellis, with his own copy-book of many of the letters he wrote, and various other papers, were given, by his only surviving daughter, to the writer of this, for the purpose of turning them to any public use, that the latter might judge proper. We thus become possessed of the Correspondence, on both sides, between Ellis and Linnæus; as well as of many other letters, which throw light on what Linnæus received from other persons. The lady to whom we are obliged was Martha the second wife of Alexander Watt, Esq. of Northaw, Herts. "She inherited her father's taste and character, more especially his piety and sensibility of mind, with a considerable likeness to his person. She died in childbed at Northaw in the Spring of 1795. Her will, written entirely in her own hand, and a letter to her husband, found after her decease, are worthy of the pen of a Richardson, and the character of a Clarissa." *Rees's Cyclop. as above.*

Correspondence.

JOHN ELLIS, ESQ. F. R. S. TO LINNÆUS.

No date

It seems to be his first letter, and was written in 1756 or 1757.

SIR,

I have sent you inclosed a specimen of a Carolina Shrub, called by Catesby, vol. I p. 64, *Frutex Padi foliis non serratis, floribus monopetalis albis campaniformibus, fructu crasso tetragono.* This I received last year from Dr. Alexander Garden, M. D. of Charlestown, South Carolina, who collected it on the hills, 200 miles to the north west of that city, at a place called Saluda, desiring me to get it engraved and give it a name, but as I am very sensible your judgment in these matters far exceeds either the english botanists or those of any other nation, I submit the description of it to you, begging the favour of you to honour it with the name of *Halesia*, from that worthy man, Dr. Stephen Hales, author of the Vegetable Staticks. Dr. Browne by my desire had made an *Halesia*, but Mr. Collinson tells me you had remarked that it was only the species of a genus already described. You will find this a Monadelphia monogynia, and quite different from anything you have already described. The Stamina in general are 12, but often 16. Though there

4. Beginning of a Letter of Ellis to Linnaeus.

Sr

I have sent you inclos'd a Specimen of a Carolina Shrub called by Catesby Vol. 1. p. 64. Frutex Padi foliis non serratis floribus monopetalis albis campaniformibus fructu crasso tetragono. This I rec'd last year from Doctor Alexander Garden M D of Charlestown South Carolina who collected it on the Hills, 200 miles to the North west of that City at a place calld Saluda, desiring me to get it engrav'd and give it a name but as I am very sensible your Judgment in these matters, far exceeds either the English Botanists or those of any other Nation. I submit the description of it to you, begging the favour of you to honour it with the Name of Halesia, from that worthy man Dr. Stephen Hales author of the Vegetable Staticks.

5. Beginning of a Letter of Dr. Garden to Ellis.

Sir

South Carolina, Charlestown
25th March 1755

Some few Days after my arrivall again in Carolina your very kind favour was delivered me by Capt Curling, Since that time what with setling a fresh & what with absence on some country jaunts my time has been so taken up that I have prevented tile now from having the Honour of writing you in return to so Obliging an Epistle.—

6. Passage in Latin from a Letter of Dr. Garden to Linnaeus.

Paucis abhinc autem annis, concilio præsertim & Amicitia Dni Ellis multa quidem adminicula comparavi & nuper vel etiam Hortum Tuum Cliffortianum, opus illud excultissimum nitidissimumque ex Dono Clarissimi mei Amici comiter accepi. Hodie iterum Suo favore, concilio & patrocinio auctus, nullus dubito quin majores multo progressus in parvo tempore faciam.

R. Martin's Lithogr. Rochester St Westmr.

should be four *nuclei* in the nut, yet rarely but one comes to perfection. The sulcated nut when ripe is as hard as a cherry stone. If the war does not prevent it, I expect some of the ripe fruit soon. If I get them I will send you some. I study your sexual method of botany, as finding it preferable to all other systems. I send you a proof print of this *Halesia* designed by Ehret; it is not yet finished by the engraver; I shall wait for your description to perfect it. You will observe *folia leviter serrata sunt*.

I return you my hearty thanks for your good opinion of my endeavours to investigate the nature of Corallines, which you have so kindly expressed in your letter to our worthy friend Mr. Peter Collinson. I wish I had begun a correspondence earlier with you. I should have received great improvement from your universal knowledge, and many things might have given you pleasure through my hands; but now it is begun I hope you will continue it. Excuse my writing in english, though I read latin familiarly, yet I cannot write it to please me.

I am endeavouring as much as in me lies to establish your system here, and from some hints I gave out among friends how agreeable it would be to the public to have it in english, some booksellers, who are greedier of money than science, have proposed it to Dr. Hill to translate it. But as that gentleman's performances are not in my esteem, and indeed I believe with few of the Royal Society, I have proposed it to the publisher of Mr. Miller's gardeners' dictionary, to employ some skilful person, to translate as much of your botanical system as would tempt young beginners to study this art; for if there was too much to be learnt, few would attempt it.

The bookseller will spare no expense in having the plates well engraved, and I believe will have the best assistance among the few botanical gentlemen that are here. Miller's dictionary is the chief book that is read by gentlemen who study the art of gardening, and that author of late has found himself obliged to change most of his obsolete names and descriptions of plants, to your more intelligible and accurate ones. On this account it will be necessary that a compendious description of your method in familiar english should be published, to understand your terms.

I should be glad of your advice in this affair, as I promised the bookseller to write to you. I had thoughts of recommending the translation of your *Clavis Systematis sexualis* with the *Classium characteres* as in your *Systema Naturæ*, 6th. ed. published at Leipsick, with the proper plate shewing the 24 classes. To this I would add the two chapters of the *Plantæ* and *Fructificatio* with the proper plates to explain them as published in your *Philosoph. Botan.* from p. 37–85. And when the new edition of your *Genera Plantarum*, or the contracted characters in your *Systema Naturæ* come out, to publish them at the end.

Though our Mr. Miller is a good gardener, he is of opinion that he is a most excellent botanist, which all the world will not allow him. He has been a little severe upon the Abbé Mazeas, in a letter published in our Transactions last year, addressed to Dr. Stephen Hales, relating to the discovery of staining linen black with the juice of certain Rhus or Toxicodendrons. I have ventured to vindicate the Abbé Mazeas; how far I am right you will be a judge when I send you the account. I have in this differed from some synonyms you have adopted, but I hope to satisfy you I am right. I cannot agree that the *Oepata* of the H. Malab. is the same with the *Anacardium orientale*. You, I observe, have called it *Avicennia*. I have lately received some *fabæ* of this Oriental Anacardium, growing on a small fruit not unlike the *Acajou* of Tournefort; and I do believe it to be a species of Rumphius's *Cassuvium sylvestre, Vol.* I. *Tab.* 70, which you will please to turn to in his *Herbar. Amboin.* by Burmann. I have many other hints to give you, but am obliged to break off at present, and conclude myself, Sir, your most obedient humble servant, JOHN ELLIS.

Please to direct to Mr. John Ellis, Merchant, in London.

SIR, London May 31, 1757.

I received your agreeable letter *, for which I am much obliged to you. I thank you for adopting the *Halesia* among your genera. You desire my advice in the affair of the *Butneria* and *Beureria*. What you

* This does not appear.

say is very true. Call it which you will you will certainly give offence to one or another. Mr. Miller has called it *Basteria.* But if you will please to follow my advice, I would call it *Gardenia,* from our worthy friend Dr. Alexander Garden of S. Carolina, who will take it as a compliment from you, and may be a most useful correspondent to you, in sending you many new undescribed plants. I shall write to him by the pacquet next week, and let him know what you desire. He sent me this year great varieties of plants, but they have been all taken by the french; but Mons. Du Hamel has promised for the future to return Mr. Collinson and me whatever are taken.

I shall forward you by the first opportunity Dr. Russel's book, and Mr. Edwards has promised me some of his plates for you, that are not yet published.

I long to see your method properly established amongst us. Those people that used to laugh at it as chimerical, are glad to change their old names to your better chosen ones. I am sorry you changed the name *Meadia* of Catesby. It gave great offence to the lovers of botany, as you changed it to an obsolete one. I wish you would alter it. As I am well acquainted with the gentlemen who have the care of our British Musæum; I hope to persuade them to follow your method in the total disposition of it; when one may easily find what they want; but while different methods are followed, nothing but confusion must arise. I wait with impatience to see the collection of Fucuses there. I have a tolerable good one of all the English and

Irish ones, which as I have examined most of them microscopically, I have observed many things curious in them. I find the Confervas have blossoms and seeds. I have the drawing of two curious ones, that Ehret drew for me while we were on the sea coast. I find we are the least knowing in submarine botany. I find they are as ignorant in Holland as we are; for Dr. Job Baster of Zealand has very lately presented a paper to our Royal Society, endeavouring to prove that corallines are vegetables; and, in order to illustrate it, has given us drawings accurately painted, which demonstrate his subjects to be fucuses and confervas, instead of corallines. His paper is pointed at my book, but I believe my answer will convince him of the necessity of making himself master of that kind of botany, before he presents his other memoirs, which he has promised the Society. He describes a good many sea insects, which he found among the corallines, and several luminous insects, which he caught by filtrating the luminous sea-water through spongy paper.

Mr. Brander, a Merchant here, and my intimate friend, has promised to forward you what I intend, as he has already some things from Mr. Miller.

I took a great deal of trouble last summer in collecting and carefully examining all the Cratægus, Sorbus, and Mespilus found in our gardens, which are numerous indeed. I think the *Sorbus sativa* *, from its five seeds and coriaceous loculaments, to come near the *Pyrus*, besides the shape of the fruit. We have a

* *Pyrus domestica. Sm. Fl. Brit.* 532.

new curious american wild apple, different from the *Pyrus foliis serrato-angulosis.* This has *folia lanceolata serrato-angulosa,* and keeps its leaves on all the winter. It is as yet rare here. We have some young plants of the *Halesia* coming up from seed, which has been a whole year in the ground. It must bear our climate well, as the winters are often sharp in the North West parts of South Carolina, from whence it comes; it will be a valuable acquisition to gentlemen who delight in curious hardy flowering shrubs, which is the fashion here at present. The red-flowering *Robinia* is the ornament of our gardens. It blossoms when it is not a foot high, and thrives luxuriantly when inarched on the white blossomed *Robinia.*

I shall be extremely glad to do you any services here. I speak with sincerity, and beg you would be so free as to command, Sir, your most obedient humble servant, JOHN ELLIS.

LINNÆUS TO ELLIS. [Latin.]

SIR, Upsal, Feb. 8. 1758.

I learn by letters from London, that a Troglodyte, or *Homo nocturnus,* figured in Bontius, p. 84, and certainly very different from the *Satyrus* of Tulpius, is arrived in your capital. In order to learn the truth of this, as no subject is more interesting to me, I have not been able to think of any way so promising as to request your assistance. I therefore most respectfully beg of you to examine this animal with attention, and to compare it with the account of the above-mentioned author. The points on which I chiefly want information are the following: 1. Is the body white, walking erect, and about half the human size? 2. Is the hair of the head white, though curled and rigid, like a moor? 3. Are the eyes orbicular, with a golden iris and pupil? 4. Do the eye-lids lie over each other *(incumbentes)*, with a membrana nictitans? 5. Is the sight lateral, and is it only nocturnal? 6. Is there any whistling voice? 7. Is there any space between the canine teeth and the others, either before or behind? 8. What is peculiar in the organs of generation, whether male or female? I wish the excellent Mr. Edwards would make a drawing of this individual, as there is no more remarkable animal, except man, in the world. I earnestly intreat you to observe its manners with all possible attention.

If I had an opportunity I should be glad to send you the first volume of my *Systema Naturæ*, just printed, and I wish you would inform me how it may be conveyed. I am confident you would find many things interesting to you in this volume. It consists of 52 sheets, on animals only, treated in the most compendious manner.

This letter is inclosed to my excellent friend Collinson, as I am ignorant of your proper address. Be so good as to inform me of your precise place of abode.

I have just received Adanson's voyage to Senegal, a handsome quarto volume in french, published at

Paris, particularly rich in shells. I have also received, since the first volume of my *Systema* was finished, a specimen of the *Cymbium* of Adanson, which proves a true species of *Patella.*

I regret not having yet seen an english work, entitled, the Natural History of Aleppo. Pray tell me what it contains on the subject.

The distinguished Edwards is reported to be engaged in a continuation of his work on rare birds, and other animals. I am anxious to know what his fifth part, which Mr. Collinson has announced to me, will contain.

I have received the posthumous works of my pupil Lœfling, with very excellent descriptions of plants, and characters of new genera, made at Cumana in America.

Ellis to Linnæus.

Sir, London, April 25, 1758.

Our friend Mr. Collinson delivered me your obliging letter very lately, wherein you desire me to enquire whether there is such a creature here as a Troglodyte or Nightman, such as is figured by Bontius, which you think very different from the Satyr of Tulpius. I have enquired very narrowly after this animal, and cannot find that there is any such here. We had one of these kind of animals here about 20 years ago, which was called a Chimpanzee; this at that time I saw alive, but as it was habited like a young girl, for it was a female, I did not examine so particularly as to answer your questions. All I can remember is, that from its whole behaviour and actions, it appeared to resemble the human species more than any that I had ever seen exhibited here. I think it had but very little hair on it, and appeared nearer to Tulpius's than Bontius's figure, which last I think is but ill designed, and that of Tulpius placed in an aukward posture. For I examined in our friend Edwards's possession Dr. Tyson's anatomy of the Ouran-Outang, and find the Doctor had drawn both Tulpius' and Bontius' animal in the same plate with his own animal. This Ouran-Outang is very hairy, the hair thinly set, and of a reddish brown colour. The same kind of animal is still preserved in the College of Physicians, which Mr. Edwards has drawn, and is among his last plates.

I have got a plate of the Chimpanzee, which was published in the year 1738, and which I shall send you by the first ship that goes to Stockholm. The colour of this animal was of a pallid dusky kind, like the Mulattoes.

Mr. Edwards still goes on to draw what is rare among Birds or other animals that offer. When any new ones come out he has promised to send them to you. I am sorry that his plates and Dr. Russel's book were so long detained before you received them.

We have received lately a new animal of the *Concha anatifera* kind, but intirely fleshy and without any shell. There are two small tubes at the top, which open into the cavity where the animal lies, which, from a mouth or opening in front, sends forth cirrhi like the common *Concha anatifera.* This head or cavity, which contains the animal, is supported by a fleshy stem, a little spongy, not so hollow as in the forementioned. There were seven of these animals on a whale *Balanus*, which was taken off a whale that was thrown on the coast of Norway last year. They are about four or five inches long, of a yellow ferruginous colour, and will be described in our Philosophical Transactions next year. Upon the Keratophytons which I have received from Norway, there is another species of *Concha anatifera*, different from that of D'Argenvilles, Pl. 30 H. I. in having many shells to compose the whole. I have reckoned (in general) that they are composed of 14 shells; but it is very possible you are no stranger to them.

I have lately received a letter from Dr. Garden, of Charles Town, South Carolina. He has not received any letters from you, but is very desirous of a correspondence with you. If you send a letter enclosed to me, I will forward it to him. He is well worth your friendship, for he is very ingenious, and a sensible observer of Nature.

He has sent me some new Genera which he has lately discovered, but I wait for the specimens of the plants to examine them by. He writes in his character of the *Halesia* that it has four kernels inclosed in the stone, and ranks it among the *Monadelphia*, and says, *Filamenta infernè coalita in cy-*

lindrum; which agrees with my observations on it. He says it grows to the size of a Mulberry tree, and is found in plenty on the banks of the River Santè in South Carolina. It flowers early in the Spring.

Some of the seeds that he sent me are now growing, so that we hope to naturalize this plant in England.

If you want a correspondent here that is a curious gardener, I shall recommend you to Mr. James Gordon, gardener, in Mile End, London. This man was bred under Lord Petre and Doctor Sherard, and knows systematically all the plants he cultivates. He has more knowledge in vegetation than all the gardeners and writers on Gardening in England put together, but is too modest to publish any thing. If you send him any thing rare, he will make you a proper return. We have got a rare double Jessamine * from the Cape, that is not described; this man has raised it from cuttings, when all the other gardeners have failed in the attempt. I have lately got him a curious collection of seeds from the East Indies, many of which are growing, which are quite new to us. He has got the Ginkgo or *Arbor nucifera folio Adiantino* of Kæmpf. p. 811; this thrives well, and when he has increased it he will dispose of it.

Mr. Clayton of Virginia has lately sent Mr. Collinson his Flora Virginica greatly enlarged and improved. Mr. Collinson has put it into my hands to look it over. It is intended to be published im-

* *Gardenia florida.*

mediately, as soon as the plates can be got ready, which Mr. Ehret has undertaken, and to dissect each new Genus, to shew your system the better. I am now writing to Mr. Clayton to dispatch as soon as he can all the specimens of the rarer plants to New York, to come by the Packet, that we may not delay the work. Mr. Colden of New York has sent Dr. Fothergill a new plant described by his daughter; I shall send you the characters as near as I can translate them. It is called *Fibraurea,* Gold thread, and Mr. Collinson received from I. Bartram of Philadelphia another, which he calls Yellow root, which I have likewise put into your method.

Fibraurea, or Gold thread, from Miss Colden of New York.

Cal. Nullus, nisi foliolum lineare bracteæ formâ vocas.

Corol. Petala sex, ovato-lanceolata, paululum concava.

Nectaria tria vel sex, subrotunda, compressa, pedicellata, inter stamina et petala inserta.

Stam. Filamenta numerosa, dimidium corollæ longitudine, receptaculo floris inserta. Antheræ didymæ, subrotundæ.

Pistil. Germina quinque, ovata, utrinque acuminata, pedicellata. Styli totidem, incurvati. Stigmata fere bifida, permanentia.

Per. Capsulæ quinque, ovatæ, compressæ, pedicellis insidentes, uniloculares, bivalves, umbellæ formâ dispositæ.

Sem. Octo vel decem, parva, ovata, Valvularum

suturæ superiori longitudinaliter et alternatim adhærentia.

This small creeping plant *(Helleborus trifolius Linn.)* grows on bogs; the roots are used in a decoction by the country people for sore mouths and sore throats. The roots and leaves are very bitter. The flower is white, tinged with red on the outside.

This young lady merits your esteem, and does honour to your System. She has drawn and described 400 plants in your method only: she uses English terms. Her father has a plant called after him *Coldenia*, suppose you should call this *Coldenella* or any other name that might distinguish her among your Genera.

John Bartram's Plant is called Yellow Root, and as near as I can describe it from his character is as follows—this you may call as you please *.

Calyx. Nullus.

Corol. Petala tria, ovata, decidua.

Stamina. Filamenta numerosa, compressa. Antheræ obtusæ et compressæ.

Pistillum. Germina ovata, numerosa, totidem Styli, perbreves, conico ordine dispositi. Stigmata simplicia, lata et compressa.

Pericarp. Bacca composita acinis oblongis in capitulum collectis, singuli uniloculares.

Sem. Solitaria, oblonga.

This Plant is herbaceous, grows in rich soil near

* *Hydrastis. Linn.*

rivers, from six inches to a foot high, with a Vine or Maple leaf.

I have now filled up my paper, and therefore am obliged to conclude sooner than I intended. I return you my hearty thanks for your intended present; if directed for me to the care of Gustavus Brander, Esq. I shall receive it. Direct to me for the future, Merchant, in London. I am, Dear Sir,

Your obedient servant, JOHN ELLIS.

P. S. Mr. John Nourse, at the sign of the Lamb in the Strand, an eminent Bookseller, who sells your works here, would be glad of 50 of your first Tome of the Systema Naturæ; it is much wanted; he is a rich man and a man of honour.

Are P. Kalm's botanic works published by themselves? where are they to be got?

SIR, London, July 21, 1758.

I received your obliging letter of the 26th of May last, but was then so unhappy as to lose my wife, so that I was obliged to defer answering it till now. I lately received also your favour of the 23d of June.

As to your first letter, Mr. Collinson tells me he has received your money and paid for Dr. Browne's collection of Plants, which he says he forwarded to you by Dr. Janisch, but cannot tell you the name of the Ship, but believes it was a Danish Vessel; he is much troubled that you have not yet received those Specimens.

The book of Medical Observations is a present to you from that Society; I have desired Doctor Fothergill to thank them in your name.

I have seen Mr. Bierken, whom I find to be a very polite and ingenious young gentleman. I called at the place where the young woman that you think is a Troglodyte is kept. She is in all respects like a negro, but white like us; her hair is frizzled as the negro's is, but very white; she is near-sighted, and holds the book very close to her eyes when she reads; she is constantly moving her eyes to and fro, sideways, but this imperfection I have seen in other people; she cannot read in the dark, nor does the sunshine prevent it. She speaks English very well. She is near five feet high, and about fourteen years old; so that if she lives, she will be the common size of women of her own country.

I shall desire Mr. Edwards to send his last volume to the Academy of Upsal.

Doctor Russel sends you a few seeds he lately received from Madras; and as soon as he receives any fresh ones from Aleppo, he will send you some.

The animal like the *Concha anatifera* does not grow out of the two ear-like tubes, nor from the sides; there are seven distinct animals of this kind sitting on a large whale *Balanus*. I have sent you a print of it as done by Edwards, but I have got a better drawing which will soon be engraved.

The Plant of Kœmpfer's Amœnit. at No. 813, has not yet flowered; when it does I shall get a specimen for you.

You have plainly shewed me, that the *Fibraurea* of Miss Colden is already described. I shall let her know what civil things you say of her—her Christian Name is Jane.

I must wait for a Specimen of the Yellow Root from John Bartram before I can give any further description of it.

In answer to your Letter of the 23d of June: I have bought one of the same sort of Microscopes that I have used in my researches at the sea side, and have delivered it to Mr. Lindegren a Swedish Merchant here, whom Mr. Collinson recommended me to; he says the Captain of the Ship's name is Fescher. Mr. Collinson says he will pay for it. The Ship is to sail next week.

I am sorry to disagree with Mr. Miller, but he has an opinion that he knows better; and in a Public Society, I think, no man should advance any thing for certain truth without he is sure of it; but as it would be too long to give you an account of the whole of this controversy, I must refer you to our Philosophical Transactions, which I suppose you have in your Academy.

As soon as this China Varnish tree produces any blossoms, I will send you a specimen. I must own it is only conjecture in me to call it a *Rhus*. I called it so, because he had supposed it to be one, and from the habit and manner of its growth.

Mr. Collinson, Ehret, and I were the other day at Mr. Warner's, a very curious gentleman, at Woodford near this City, to see his rare plant like a Jasmine, with a large double white flower, very odoriferous, which he received about four years ago from the Cape of Good Hope. The flowers are so large, that a specimen, which he gave me to dissect, was four inches across from the extremities of the limb.

This Mr. Miller has described to be a Jasmine, in his Dictionary now publishing, and in his figures of Plants. As it is a double flower, the stamina, in many, grow into the sections of the petal; but this year it grows more vigorously, and the parts of fructification have appeared more distinct, especially in the specimen he gave me, where, upon opening the tube, I discovered six stamina adhering by the filaments to the inside of it, supporting as many antheræ; three of the antheræ were united together, but easily to be separated. These stamina corresponded to the same number of sections of the limb. The style has three stigmata, and is united to the germen, which is placed under the receptacle, and contains above 40 very small seeds, which, magnified, looked like the acini of a *Rubus*. The calyx has six alæ or angles, at equal distances, which end in so many erect foliaceous pointed denticles. Sometimes the limb of the flower is divided into five segments only, and then the stamina are but five, and the alæ of the calyx are but five, and the stigmata but two. There were three rows of petals or sections in my specimen, six in each row. Mr.

Warner by my desire dried a specimen for you last year, which Mr. Collinson sends you now, but I hope to get you a specimen where the parts of fructification are more distinct.

If you find this plant to be no Jasmine, but an undescribed genus, you will oblige me in calling it *Warneria* after its worthy possessor.

I have seen the first volume of your Systema Naturæ. I wish there had been figures to it. I suppose the books that are coming for the bookseller have the plates, which will greatly illustrate them.

I am obliged to you for methodizing the corallines mentioned in my book.

The Polype from Greenland, with the bony stem, differs much from the *Accarbarium rubrum* of Rumphius, as you will find when you receive the figure and description which I have sent you of the latter. If there is any service that I can do for you here, you may be assured I shall do it with pleasure.

I am Sir, your most obedient humble servant,

John Ellis.

Sir, London, August 1, 1758.

I wrote to you a few days ago, and sent you a drawing of the dissection of a curious Plant, called by Mr. Miller a Jasmine. It is in the possession of Mr. Warner, who has sent you a dried specimen by Captain Fescher, through the hands of P. Collinson.

Since, I have received another specimen, in order to be more accurate in the Seed Vessel, as the first was not come to so much perfection in that part. The calyx of this had but five alæ and laciniæ, and the limb in the exterior row was cut into five segments.

I called in Mr. Bierken to be present at the dissection, to assist me, and to testify to you what we observed. Upon opening the tube we found no stamina; but the style, instead of being one as described in my last letter, we found to be three distinct ones, each having a fleshy stigma and sitting on the germen, which we found divided into three loculaments, so that the base of each style was inserted into the top of each loculament. When we separated the germen from its place, the valves adhered to the hard spongy substance that inclosed them; these valves we examined in the microscope, and found to be of a strong fibrous or ligneous substance.

The reason why in this specimen the styles should appear separate, and joined in the former, is, because there were no stamina in this last, and therefore the styles had room to grow distinctly in the tube, being not so much compressed.

If then this plant appears to you after examination to be a new genus, I should be obliged to you to send me the characters of it, as you intend to describe them in your *Nova Genera*.

Mr. Warner begs of me to write to you not to

call it *Warneria*, and therefore I shall desire the favour of you to call it *Augusta*; a name, I hope, you will think highly suitable to the magnificent appearance of so elegant a plant, and in doing this you will much oblige Sir,

Your most obedient servant, JOHN ELLIS.

LINNÆUS TO ELLIS. [Latin.]

Upsal, September 29, 1758.

I yesterday received your packet so long and anxiously expected; and for its numerous contents, as well as for your various instances of friendship, and your abundant information, illustrative of the science to which I am devoted, I beg you to accept my most sincere thanks.

Your most excellent microscope, perfect in its kind, shows objects to the greatest advantage, and I shall take advantage of the first bright day, to examine all my *Sertulariæ* over again, under your guidance.

The Pipy-knobbed Red Coral, or *Isis ochracea*, Syst. Nat. (ed. 10.) 799. n. 3, is drawn by you most admirably. You have shewn me its *animalcula*, enclosed in their eight valves, which have enabled me easily to understand other corals. You, in these minute and almost invisible beings, have acquired a more lasting name, than any heroes and kings by their cruel murders and bloody battles. I congratulate you on this, your own stupendous victory, over the barbarous ignorance which hitherto has held the philosophic world in subjection.

Your *Alcyonium*, Phil. Trans. 49. t. 14, is a very singular and curious animal. I doubtless have it among my corals, which I shall enquire into the first opportunity.

The *Warneria* *, through your interest, has reached me, and I have examined it with admiration. It cannot possibly be a species of *Jasminum* or *Nyctanthes*, if the fruit be a capsule with many seeds. I wonder that the pistil should be in three divisions, as Miller represents a single cloven one. May I be allowed to insert the character of this genus, on your authority, in my *Systema*, as I am unable to make it out myself? I have, in the letter to Mr. Warner, which accompanies this †, requested him to inform me whether the young branches are milky or not, when broken. Be so good as to procure from him a speedy answer.

The figure of the Chimpanzee excites my high admiration. There can be no doubt of this animal being the *Simia Satyrus*, and the same which Tulpius has described. I doubt whether it be the *Homo sylvestris* of Bontius, nor can I persuade myself that Mr. Edwards's figure represents the same species.

The *Conchæ anatiferæ* found upon Mr. Edwards's

* Afterwards called *Gardenia florida*.

† This letter does not appear.

Fleshy Barnacle are unintelligible to me, as I certainly never met with any thing of the kind.

The female *Coriaria* has indeed rudiments of stamens, like several other dioecious plants; but they never become prominent, nor bear pollen, like those of the male; whose filaments are long, and the anthers double.

Your dissertation upon the *Rhus* proves its author a most acute botanist, which I rejoice to see.

Favour me with your determination, as soon as possible, whether I am to publish the character of the *Warneria* as you describe it; that is, with a three-cleft style, and a germen of three cells with many seeds.

Pray give my compliments to Dr. Bierken, thanking him for his letter describing the human monster. Please to inform him that Mrs. Ihre died yesterday.

ELLIS TO LINNÆUS.

Dear Sir, London, Oct. 24, 1758.

I have the pleasure of your favour of the 29th Sept. and am very glad you received the letters I sent you. My letter of the beginning of August gave you an account of the dissection of Mr. Warner's Jasmine by Dr. Bierken and me, wherein we found three distinct styles, each style sitting on a loculament, so that we plainly could, without glasses, perceive three loculaments full of seeds, of a roundish form, about 30 or 40 in each. The valves which adhered to the spongy substance of the pericarpium appeared to be hard and fibrous; and when they were separated from the seeds, we could plainly discover the partitions or *parietes* that formed the several loculaments. This account I sent to Mr. Warner at the same time that I sent you my last letter, but he has desired me to write to you to beg you would not call it *Warneria*, and which I believe I did in my last letter. I believe he is convinced that it differs from the Jasmine; but he has such an esteem for Mr. Miller, that he would not appear to differ from him in so capital a plant by adopting another name. Mr. Ehret is now engraving an elegant plate of it, and I suppose will give a dissection of it, and intends to call it by a name of his own, for we agreed when we examined only the stamina and styles at Mr. Warner's together, that it was by no means a Jasmine.

The lobes of the Corolla are obliquely bent, which I have never seen a Jasmine. There is but one flower on each small branch. It is not in the least milky. I carefully examined both the leaves, branches, and flowers, and could not perceive the least lactescent appearance when I cut them. In those flowers, where the stamina appear, there are as many of them (generally) as the lobes of the corolla. When the flowers come out first in summer, if the weather is warm, they have six divisions or lobes; but towards the latter end, the blossoms are smaller and the lobes but five. So that I believe it is Hexandria Trigynia, though most of the flowers have but five lobes.

I must therefore desire you would call this plant *Augusta,* which I think as well deserves that title for its elegance in every respect, as the *Methonica* does to be called *Gloriosa.* This will not offend our friend Warner's modesty, nor his particular delicacy to Mr. Miller. The description and characters you may collect from what I have wrote to you.

Your compliment to me on the print of the Pipy red coral is more than I deserve. I should not have shewn this specimen of coral to the Royal Society, but that I thought it would clear up the manner of the growth of these kind of bodies, beyond any other that I had met with.

I have inclosed you a proof print of the naked *Lepas* of the *Concha anatifera* kind, which is much better done than Edwards's. I have had all the species of that kind as well as the *Balani* added. I will endeavour to get you a specimen of this extraordinary animal. Something of this kind has been observed before, vide Act. Angl. N. 308. p. 2314, in a letter from Sir Robert Sibbald to Sir Hans Sloane; but the description and figure are so bad, that it is impossible to make any thing of them.

I thank you most heartily for your first part of the Systema; I am much pleased with it. I lately received it from a gentleman that is going to reside among the Swedes in Philadelphia.

I have lately wrote to Carolina to Doctor Garden,

to send me the Cochineal insect, together with the Opuntia that it adheres to; for I have been spoken to by several of your friends for that purpose. I wish we may be able to get it from Jamaica, for I am told they do not cultivate that plant now for this purpose. It chiefly is in the hands of the Spaniard. I can assure you there have been many letters wrote in order to procure it for you if possible.

I am at this time endeavouring to find out a method to bring exotic seeds from China, and other distant parts of the world, in a vegetative state.

I should be obliged to you for your opinion in these matters. You, no doubt, have given your pupils proper instructions on this head, when they went abroad.

Likewise the best method to preserve the plants alive, in so long voyages and so many different climates.

I am now smearing over the Acorns of the *Quercus*, that bears the Cork, with a thick solution of gum arabick, which soon dries; others I cover with wax; others I inclose in clay and gum arabick; each acorn is covered or smeared singly. I shall inclose others in clay and tow, or flax, worked up together and then dried. Others I cover with a mummy made of pitch, rosin and beeswax, in equal quantities. They are afterwards to be put into jars, some in sand, some in paper, and some in boxes, and then covered up close, and kept cool on board the ship. This is the method I propose, to bring seeds from China; and am now trying the experiment only for a short voyage to Charlestown, South

Carolina, to Dr. Garden; but am in hopes the gum arabick will preserve seeds from the most distant places, provided that when the seeds are enveloped in it, it is soon dried, and after placed in jars in the coolest part of the ship.

I hear your pupil Mr. Solander intends to come to England. Pray desire him to study English immediately, and in a month after he comes here; he will speak it fluently. I should be very glad to do him any services that lay in my power, as I find you have a great esteem for him.

I am, dear Sir, most truly your affectionate friend,

JOHN ELLIS.

LINNÆUS TO ELLIS. (Latin.)

Upsal, Dec. 8, 1758.

I received with the greatest pleasure your very friendly letter of the 24th of Oct. abounding, as usual, with valuable information on the wonders of Nature, which you are favoured, more than any other person of the present day, with the means of explaining, and thus of laying open what has hitherto been a secret to the learned world.

Your excellent plate of *Lepades* is hung up in my room, that I may daily contemplate it; for though I am continually learning something, it does but whet my appetite to know more.

Fig. 1, is truly a wonderful animal, of which I never before met with a delineation from the life.

2, seems distinct, in having 13 valves; whereas the common *anatifera* has but 5.

3, is the same as d'Argenville's t. 30. f. H. I, but I cannot make out, from your figure, how many valves compose the shell. I never saw this species.

4, is *Lepas Mitella.*

5 and 6, common *L. anatifera;* perhaps comprising two varieties, not distinct species.

7, *Lepas testudinaria.*

8, *L. Tintinnabulum.*

9, perhaps the same as the preceding.

10, may also be the same.

11, I do not understand.

12, *L. testudinaria.*

13, the same.

14, is its obliquity caused by peculiarity of situation? and is it distinct from f. 8?

15, perhaps a mere variety of f. 20.

16, unknown to me. Has it many valves?

17, possibly a variety of f. 20.

18, I have never seen. Has this many valves?

19, I never saw this, nor do I understand the figure.

20, *Lepas Balanus.*

I am happy to find you are intent upon a method of obtaining fresh seeds from China, on which subject, as a very important one, I have often had it in contemplation to write to you. No one has better opportunities than yourself for making this experiment, nor will any thing do you more honour, if successful. But I would especially recommend your attention to the bringing over a living plant of the Tea, from that country. This shrub can readily be purchased at Canton, for a very small sum; nor is it easily killed, though it often perishes from the heat of the sun in the voyage towards Europe. Osbeck brought a living tea-tree as far as the Cape of Good Hope, where it fell overboard in a storm, otherwise it would have survived. I am very sure this plant would bear the open air in England, as it thrives at Pekin, where the cold is more intense than in Sweden. Pray do not forget this request of mine.

Fresh seeds may with great facility be conveyed in the following manner from any distant country. Fill a glass vessel with seeds, so deposited in dry sand as not to touch each other, that they may freely perspire through the sand, tying a bladder, or piece of paper, over the mouth of the vessel. This glass must be placed in one of larger dimensions, the intermediate space, of about two inches all round, being quite filled with three parts nitre, one of common sea salt, and two of sal ammoniac, all powdered and mixed together, but not dried. This mixture will produce a constant cold, so as to prevent any injury to the seeds from external heat; as has been proved by experience.

With regard to your *Warneria*, there seems little doubt of its being the plant delineated in *Hort. Malab.* v. 2. t. 54, *Burm. Zeyl.* t. 59, *Rumph. Amb.* v. 4. t. 39, though these authors exhibit a wild specimen, in which the leaves are always narrower. They all assert it to be somewhat milky in its native soil. From this circumstance, as well as the whole habit, the obliquely abrupt segments of the corolla, the calyx, the minute stipulas at the insertion of the opposite leaves, the whitish flowers, and other particulars, I judge their plant to be closely allied to *Nerium*, belonging to the family of *Contorti*. Hence the flower, if single, must have a double capsule, though a simple style. I dare not therefore take my character of your shrub from a double flower, except on your exclusive authority. I had rather not meddle with this plant at all, till it is better known. It has no relationship to the Jasmines.

I received from America, some years ago, several cochineal insects on the Indian Fig *(Cactus)*; but the gardener, in my absence, mistook the young ones for some noxious insects, and cleared all away except two, which died without progeny.

Cannot you procure me Hill's letter on the sleep of Plants, which I have never yet seen?

Farewell my excellent patron and benefactor; and study that Botany may always be turned to some beneficial purpose.

LINNÆUS TO ELLIS. (Latin.)

No date. The London post-mark is Feb. 23, without any note of the year. This letter must have been written in Jan. 1759.

I have been for some time anxious to conform to your wishes, in establishing the new genus of plants called *Warneria*, founded on the beautiful double-flowered shrub, brought from the Cape of Good Hope; but the double flower being almost sure to mislead us, in the construction of the character, caused me to hesitate. Nevertheless, as you are so earnest on the subject, I could not withhold my consent; and I therefore agreed to publish a character of this genus, on your authority alone, feeling myself not qualified to decide on its certainty.

Meanwhile I had always present to my recollection a dried specimen, with a single flower, preserved in some part of my herbarium, which, as far as my memory would serve, I believed to be the same species. Many a time, in the course of the last half year, have I hunted for this specimen, but in vain, my thoughts having been so much taken up with the daily business of the University, my publick and private lectures, and numerous matters besides.

During the Christmas holidays, however, having found leisure to turn over all my dried plants, I luckily met with the specimen in question, which came from the East Indies, in very good condition. The flower is perfectly single, and by immersing it in hot water, I could clearly ascertain every part of its structure, so as to draw up, without any uncertainty, the following character of your genus *Warneria*.

Cal. Perianth of one leaf, with five angles, and five deep, sword-shaped, vertical, straight, nearly upright, permanent segments.

Cor. of one petal, funnel-shaped. Tube nearly cylindrical, longer than the calyx. Limb flat, in five deep obovate segments, the length of the tube, more straight at one edge than at the other.

Stam. Filaments none. Anthers five, linear, half as long as the limb, inserted into the throat of the corolla, but attached above their base, so that their lower part is concealed within the throat.

Pist. Germen below the receptacle. Style, thread-shaped, the length of the tube of the corolla, and terminating, beyond the throat, in a large, ovate, obtuse, emarginate stigma.

Peric. Berry of two cells.

Seeds. . . . . . .

Hence it is perfectly evident, that this shrub has no affinity to the Jasmine tribe, but belongs to the natural order of the *Contorti*, Phil. Bot. 31. n. 29, being akin to *Nerium, Plumieria, Cerbera, Cameraria,* &c.* The plant of Hort. Malab. v. 2. 106. t. 54. Rumph. Amboin. v. 4. 87. t. 39. Burm. Zeylan. 129. t. 59. Raii Hist. 1785, seems very near this, if not absolutely the same†. But a doubt arises from the small calyx, which all these authors attribute to their plant; whereas this part in ours, whether the flower be single or double, terminates in long sword-shaped segments.

Such being the case, our genus *Warneria* may be

* This genus *(Gardenia)* is now more properly referred to the *Rubiaceæ* of Jussieu.

† These synonyms belong to *Tabernæ montana coronaria.* Ait. Hort. Kew. ed. 2. v. 2. 72.

established as perfectly distinct, though its fruit is not yet well ascertained; and this genus may stand next to *Cerbera*.

A singular character is observable in the large ovate obtuse *stigma*, projecting, in the shape of a little egg, beyond the throat of the corolla. The *style* is absolutely single.

With regard to the name, I know not what course to take, as you seem, by your last letter, inclined to follow Miller, in calling another plant *Warneria*. What his plant is, I am ignorant. I have a bulb from him under this appellation, now just sprouting, and apparently akin to *Gladiolus*, *Ixia*, &c. But before it blossoms, I may have printed the whole of the 2d volume of my Systema, which treats of plants*. I therefore earnestly request your advice. Could you not send me a figure, or rather a dried specimen, of this new *Warneria* of Miller, in a letter, that I may be able to see what is necessary for its insertion in my work? I beg to have an answer as soon as your engagements will permit.

ELLIS TO LINNÆUS.

Sir, London, Jan. 23, 1759.

I am much obliged to you for your letter of the 8th. of Dec. last, full of most excellent remarks.

* Published in 1759. Perhaps this was Miller's *Watsonia*. Ic. t. 276.

I shall endeavour to clear up some difficulties that arise with you in my plate of the *Lepades*.

Fig. 3. is taken from d'Argenville, fig. H. and I. tab. 30. He describes it to have five valves, but I have never seen it. I have inserted it to shew it is not the same with my Fig. 2.

Fig. 7. and Fig. 7. a. is the *Pediculus Ceti* or *Diadema Turcorum*, and always of a subrotund figure, and a very different species from Fig. 12. which is your *testudinaria*.

Figs. 8. and 9. approach near to one another, but Fig. 9. is conical, and Fig. 8. subcylindrical.

Fig. 10. is a distinct species chiefly found in the Mediterranean sea, the summits of the valves are pointed and the opening wider than Fig. 8, differing much from it both in colour and habit, which is very difficult to describe, unless you saw both species together.

Fig. 11. is composed of six valves, as it is not easily distinguished without a lens glass, it could not be well expressed in the engraving. The valves are composed of minute quills or hollow tubes, their openings to be seen plainly at Fig. 11. a. This is evidently a distinct species. The latter I had from P. Collinson ; the groupe Fig. 11. from Dr. Fothergill.

Fig. 12. is much more compressed, even upon comparing it with different specimens of the same size with Fig. 13. this last being much higher, even twice as high in proportion as Fig. 12.

Fig. 14. I believe may owe its obliquity to accident, but this is their general form when they are found adhering to muscles, but their opercula differ much in having longer horns; and their shells in proportion to their size are more slender and thin than that at Fig. 8.

Fig. 15. These are very thick and more oval shaped than Fig. 20. with a much smaller opening. The valves of these are not pointed at top like that.

Fig. 16. I have not seen this, only I have described it from the figure of Mr. Borlace in his Natural History of Cornwall lately published. I have wrote to him about it in order to satisfy you.

Fig. 17. is the common small Lepas that grows on all our rocks and shells on the coasts of the British Isles. It is rarely ever found larger than here represented, and never so large and regular as Fig. 20. So that it must be a different species, especially as I have minutely examined and compared both with the best magnifying glasses. From the animal of Fig. 17. I have described the figure of the animal Triton, which I shall send you by the first opportunity in spring. It is drawn with great exactness. These animals, which we have in great abundance on our oysters at this season of the year, have the lower part of their shells filled with their *ova* or spawn, which possesses full half the shell. When these animals are separated from their shells, and put into the watch glass with salt put to the water, they will live many hours, and in that state I viewed and had them delineated from the microscrope.

These as well as most of the *Lepades sessiles*,

except those at Fig. 7. a.; and Figs. 12. and 13. have a base shell of a circular figure (which I have not seen remarked before), to which the six valves are united, as they are to one another.

Fig. 18. Is a true multivalve of six valves with the base. This was found adhering to a large species of muscle brought to my late friend Arthur Pond, F. R. S. from Greenland.

Fig. 19. Is a species of *Lepas* found adhering to *Keratophyta* near the Straits of Gibraltar. The base of this grasps and embraces the slender twigs of these Zoophytes, and as the succession of the young progeny arises up to encrease and extend the parent animals, they enclose these *Lepades* all but the openings, as you may observe in the three figures on the Keratophyte at Fig. 19. a.

The Fig. 19. is one of them laid bare from the animal of the Keratophyte.

Fig. 20. I found in the Musæum Britannicum collected by Sir Hans Sloane. It is extremely curious and perfect, and much more regularly disposed in its valves than any I have seen. It appears nearest to the specimen No. 8.

To conclude, I think you are extremely right in not adopting too many species, and I am afraid too many are only varieties, so shall leave this matter entirely to your judgment.

Since I wrote to you, I have had an opportunity of examining the cochineal insect, both male and female. The female I received alive from Carolina, from whence I expect a good many on the *Cactus*

Opuntia for you. I have desired Dr. Garden to be exact in the changes of both, and to be sure to send me some for you.

I made a few observations on these that I received. The females were alive, and the smaller they were, their legs and antennæ were longer in proportion. You have described them very well. Of the males I found two dead, one on the plant, one in spirits of wine. Their wings I believe must have been erect, when alive, but lay flat down when dead. The head is prominent with longer *Antennæ* than the females, and *moniliformes*. There is a small trunk like the beak of a bird from the head, but depressed; the body longer and slenderer, and a point arising from the tail or abdomen. I dissected a little bag in which was a male Cochineal fly. This explained the form of the beak, and confirmed me that the *Antennæ* of the sexes differ in form.

The females after impregnation grow torpid, and spin themselves a covering of a fine white web, where they lie till they are brushed off to be dried for use, or bring forth their young. The males that I found were very small. I suppose they die soon after they have answered the intentions of nature. The females when grown to the maturity of bringing forth their young, have six very small legs, and two antennæ, with a trunk, between the two forelegs, all so small that the microscope must assist us to discover that this oval mass is an animal. I had several of them preserved in spirits, and could easily discover that it spins its threads double like the silkworm. I cut open several of them, and found them full of eggs, impregnated and ready to send forth young animals. I put several of the cochineal insects from New Spain, used in dying scarlet, into warm water, and after they had lain twelve hours to soak, they swelled up to nearly the same form with the Carolina insects. When I cut these open, they afforded from 60 to 100 eggs or young ones. The richness of the colour seems to depend on the young insects in the eggs, as appeared to me in the microscope.

I am obliged to you for your method of preserving of seeds. I have sent some copies of it to the East Indies for trial to be made.

I must now freely give you my opinion. I tried my thermometer in the mixture of salts you were pleased to fix the proportion of, and I found it did not affect it, for as soon as the salts became of the same temperature with the air in the room where the experiment was tried, the thermometer did not move higher or lower; though covered by the salts. I imagine the use of the salts is to prevent a putrefactive fermentation in the seeds. For, unless during the dissolution of salts, I cannot find any cold air generated by any kind of salt.

I lately tried some experiments on the acorns of the common *Quercus*, and found that wax, rosin, or a mixture of wax, rosin, and pitch, were the best coverings to preserve seeds that perspire much; for gum arabick, senega, and clay, used separately on different acorns, did not answer the purpose, the

Cotyledons or kernels being quite dried up, and rendered unfit for vegetation. The others secured the Cotyledons perfectly sound, as if just fallen from the trees. I kept the acorns in a dry box in a room where a fire was kept every day, from the latter part of October to the present time. N. B. the perspiration of the acorns prevented the wax, rosin, or mixture to stick; so that the coverings came off easily. This method may preserve Tea Seeds, Mangoes, Spices of several sorts, sound, in long voyages, if kept cool in the ship's ballast, in your mixture of salts. Sir, your obedient servant,

JOHN ELLIS.

SIR, London, March 2, 1759.

I received your letter relating to the characters of the *Warneria,* taken from a dried specimen which you found among your oriental plants, and am glad to find it comes so near to the description which I formerly sent you, especially the first. I am well satisfied that when the limb of the petal is divided into five segments, there are five antheræ, and am convinced that the filaments adhere to the inside of the tube, because I separated them myself. But these I find you do not consider as filaments unless they are loose from the substance of the tube. As to the style in that flower that had six segments in its limb, it supported three distinct stigmata of an oval figure and fleshy substance. But Dr. Bierken easily separated the style into three distinct ones, by taking hold of each stigma, and drawing it asunder from the others. These three parts of the style were inserted into the upper part of a particular loculament full of seed, one part of the style terminating in the top of each loculament. I do not doubt but those that have only two stigmata, have also but two loculaments, as you have described them.

We both concluded that the narrowness of the tube had occasioned the three styles to grow together, but I am persuaded you have very properly distinguished this matter, and have made it very clear to me. As to the seeds, Mr. Warner has promised to let me dissect a flower next summer before him, to convince him of what Mr. Gustavus Brander, Dr. Bierken, and I have seen; for Mr. Miller will not admit that it has any seeds, nor that any seeds have been seen by any one. I have shewn your letter to both Mr. Warner and Mr. Miller, and Mr. Miller says he will write to you, and send you the specimen of a new Genus which he has called *Warneria*. The plant you mention is his *Watsonia*, which I find you remarked on the outside of your letter. Since then Mr. Warner desires this rare plant which I had called *Warneri*, may not bear that name. I hope you will have no objection to the calling it *Augusta*, by way of eminence among flowers, as you have called the Methonica *Gloriosa*, for I know you stick firmly to the rules you have laid down.

But if you have any material objection to this, be

so kind as to call it *Portlandia,* after that eminent patroness of botany and natural history the Dutchess of Portland, who is a great admirer of your excellent and learned works, by which you have opened the eyes and understandings of mankind to contemplate and properly arrange the works of nature.

I have now a complete collection of all the english Fucuses, which I wait to lay before your disciple Mr. Solander, that he may assist me in your method of classing them. I have many observations on their fructification to communicate to him, and shall send you some specimens where I have duplicates of the rarer kinds.

I am now making some further experiments to preserve seeds in long voyages, and have singled out the chesnut as one that is soon liable to rot, to see if it is possible to preserve them in a vegetative state for the next year. If this succeeds, it will give me great pleasure, and I hope be of great use to the better introducing all the rarer seeds from abroad. I was led into the following experiments by a friend of mine, who some time ago received some seeds of the *Magnolia altissima* of Catesby, preserved in an earthen vessel in tallow; these grew very freely, when scarce any seeds of the same kind, that were sent in sand or earth, came up. These seeds that were thus preserved in tallow were in it four months, and were sent from South Carolina.

Some of the chesnuts I have poured melted tallow on and over them, in a cylindrical earthen glazed vessel. The heat of this liquid and the following were no hotter than to make it fluid, which is not much hotter than the human blood. Others I have put into the same kind of vessels, and poured melted bees-wax, others melted tallow and bees-wax in equal quantities, others melted rosin and bees-wax. These vessels I have placed on a wooden shelf in a cellar, and inverted them. When the summer heats are over I shall examine in what condition some of them are.

I believe if these vessels were immersed in your saline mixture, it would be of service to prevent putrefaction, and perhaps to keep them cooler. I have given these methods to some gentlemen just gone to China, to preserve the Tea, Mangoes, &c.

I should be glad of your sentiments, and that you would study this great point. I have requested all my friends to send over the Tea-plants in the manner Kæmpfer describes. Adieu, dear Sir, I am most heartily your sincere friend, JOHN ELLIS.

LINNÆUS TO ELLIS. [Latin.]

Upsal, May 30, 1759.

Yesterday I received your truly valuable book, and a box from Dr. Fothergill. For the former accept my most grateful thanks. I have written to Dr. Fothergill, in a cover to Mr. Collinson, that I might not encroach upon your time or pocket.

No doubt my much-loved pupil Solander has, ere this, found a tranquil asylum in your friendship. I have recommended him to your protection, as I would my own son; and I now repeat my most earnest entreaties to the same effect. I wish he might be able to pass some time in London, and enrich himself from the treasures of your learning. I only apprehend that your country may be too expensive for him.

Please to forward the enclosed to Dr. Garden. The plant he has called by your name is a genuine species of *Swertia*; nor can any new genus be called *Ellisia*, that of Browne being perfectly distinct *. If a genus, once well established, be liable to alteration, there can be no certainty in botany. If Dr. Garden will send me a new genus, I shall be truly happy to name it after him, *Gardenia*. I could not but adopt the *Halesia* which you communicated, as being new and distinct; but the *Halesia* of Browne is otherwise, having previously been defined by me, under the name of *Guettarda*.

Be so kind as to communicate to Solander what is written on the other side. As I have now, through your favour, a duplicate of your excellent book, I shall present it to my only son, remaining not the less obliged to you, &c.

Upsal, Nov. 6, 1759.

It is long since I wrote to you, or received any of

* Linnæus afterwards thought otherwise, and reduced Browne's *Ellisia*, with great propriety, to *Duranta*.

your letters; nor shall I be quite free from anxiety till I hear you are in good health, and retain your regard for me. Perusing lately the second part of the 50th. vol. of the Phil. Trans. I met with your remarks on *Balani (Lepades)*, which are worthy of your abilities. In the beginning of April, my friend Solander took leave of me for England; but ever since that time he has remained at the extremity of Scania, towards Elsinore. Whether he is detained there by indisposition, as his letters indicate, or whether he is afraid to venture on shipboard on account of the war, I know not. Having written by him to you and other friends, I have been in daily expectation of hearing of his arrival, and at the same time of learning how you were. I have just had a letter from him, saying he hopes to be with you in about a week.

The fourth volume of my *Amœnitates Academicæ* is very nearly printed; and I hope in the spring, if it please God, to publish the third volume of my *Systema*, treating of fossils.

Mr. Forskall, an excellent pupil of mine, just appointed Professor at Copenhagen, is to be sent next year, at the expense of the King of Denmark, to the Cape of Good Hope, and Arabia Felix. If God preserves his life, we may expect from him many curious discoveries. In the history of insects he certainly excels, nor is he deficient in any other part of Natural History.

Dr. Biercken is returned from his visit to you, loaded with innumerable favours; but I have not yet had an opportunity of conversing with him, as he is detained by his patients at Stockholm.

Among the dissertations I am about to publish are, *Genera morborum, Aer habitabilis, Flora Jamaicensis, Sus Porcus, Anthropomorpha, & Generatio ambigena.* In the last of these I shall shew that the brain and spinal marrow only proceed from the mother, and the rest of the body from the father.

Nobody is writing any thing at present except your countrymen, who are every day discovering and publishing something new.

Make my best compliments to Mr. Edwards, to whom I am obliged for his valuable plates of birds, &c. Your country seems, as it were, the kernel of the whole globe.

Our Queen has commanded her collection of insects to be drawn in their natural colours; nor can any thing be more beautiful than what are already done. Her collection of insects, especially butterflies, is truly stupendous. Above 60 are finished; and I hope all the rest will be completed in the space of a year.

The first volume of Clerck's Plates of Insects is published. This contains above 100 of our *Phalænæ,* scarcely to be met with elsewhere, or at least not represented in any other work. The figures are coloured to the life.

Pray favour me with an answer as soon as possible, for I cannot exist without the pleasure of your letters.

I hope you will pardon me for troubling you with the enclosed for Mr. Guy, who is famous for the cure of cancers. He sent me a pair of Chinese gold fishes alive; but I know not how to address my letter of thanks, so that it may reach his hands. Please to seal it with your seal.

Upsal, April 29, 1760.

I have long regretted the want of your letters, and the silence of a friend I so highly value. Nevertheless I have just heard from Professor Ferner, with great pleasure, that you enjoy your health, and continue to advance our science by your labours.

The same friend informs me that you are displeased at my not having admitted your new genus, by the name of *Augusta*, into the second volume of my *Systema*. Allow me to state my reasons.

I am not without scruples respecting the genus itself, the fruit not being well ascertained, and the number of segments and stamens, whether five or six, uncertain. From its affinities indeed I should take this tree to be pentandrous, though not positively. But the name alone, were there no other reason, would have prevented my adopting this genus. A teacher may be ashamed to commit the very fault he condemns; as I have told you in another letter, to which I have received no answer*.

* That letter seems not to have reached Mr. Ellis.

I have laid down a rule in my *Critica* and *Philosophia*, that no adjective should be admitted as a generic name. On this ground I have expunged several names of other authors; but, that I might not carry innovation too far *, I admitted *Mirabilis* and *Gloriosa*, for which I have so often been blamed by my adversaries. Every one knows that the Harlem florists give this kind of names to their Hyacinths, Tulips, &c. such as *superba, augusta, incomparabilis, pulcherrima*. But if I were to adopt such a name, I should sin against my own laws. Neither do I presume to give a name of my own choosing to your genus. If *Watsonia*, or any other appellation be chosen, this genus may find a place, with several others, in an Appendix to the third volume of my *Systema*.

I lament that ill health has kept my dear Solander a whole year from the benefits of your society. He was detained at the further part of Scania. I hear he has just taken his passage for London, and I hope he will find you well. Pray remember me most kindly to Mr. Collinson, and the rest of my friends.

Will you, if in your power, favour me with a dried specimen of Mr. Miller's *Morea*, of which I have no idea?

What is become of Dr. Browne, who gave us the Natural History of Jamaica?

* What illiberal censures has this forbearance of Linnæus brought upon him!

I never passed a more severe winter than the last. The earth is still bound in frost in many places, and the woods are covered with snow.

Dr. Bierken has sent me a figure and short description of a marine animal, of the *Mollusca* kind, of which you shewed him a specimen. I have a correct delineation and description of the same from Italy.

Mr. Jacquin has many new genera, above 100, not noticed by others. He has learnt the art exercised by the Psylli and Marsi of preventing the bite of serpents, which has been so long a subject of inquiry.

If you see Professor Ferner, pray tell him I will very soon answer his obliging letter.

Farewell! may you long enjoy life, and preserve your affection for me, who have been, and will be, entirely yours as long as I live!

ELLIS TO LINNÆUS.

SIR, London, June 13, 1760.

I have sent you, by a friend of Professor Ferner's, a collection of seeds which I received from our mutual friend Dr. Garden; some of them are covered with tallow, and some with myrtle-wax; I believe the latter will preserve them better than the former. Since Mr. Ferner went to Scotland, I have received from Dr. Garden some specimens of dried fish, with their characters, in a letter to you; also the specimen and characters of a new plant. These I shall forward to you by the first opportunity.

In answer to your letter of the 29th of April, I desire you would please to call Mr. Warner's Jasmine *Gardenia*, which will satisfy me, and I believe will not be disagreeable to you.

What you say is right in regard to the keeping up to the rules you have laid down in your *Philosophia Botanica*, and therefore I submit.

I shall endeavour to get you a specimen of the *Morea* of Miller, and send it with the other specimens.

I shall write to Dr. Garden this day, that I have desired you to give the name of *Gardenia* to the Jasmine, which I am persuaded he will esteem as a favour; at the same time I shall send him Mr. Ehret's curious print of it, coloured by himself. If the dried plant which I shall send you from Dr. Garden proves to be a new genus, I hope you will have no objection to the calling it *Schlosseria*, from my worthy friend Dr. John Albert Schlosser of Amsterdam.

I am yet endeavouring at new methods to preserve seeds on long voyages. I am persuaded your method of immersing the bottle or jar (in which they are inclosed in sand) into a mixture of salts is a good one. I shall soon, I hope, have a trial of it completed, for I expect a friend from the East Indies, to whom I communicated it, and who has promised to try it, and several other methods which I have proposed.

I sent you in the last parcel some of the *Opuntiæ*, with the fruit on them, from Dr. Garden. There were on them some webs wove by the female cochineal insect; and I hope that you will find some of the insects themselves. Dr. Garden has promised to send me some of the male insect. I observed, among some that he sent me a year ago, the male insect had two long setæ proceeding from their tails.

I propose to continue my discoveries in the Corals, Corallines, &c. as soon as I have finished some experiments I am making on seeds that were preserved after different manners.

It is a great loss to the curious here that we are so long before we can receive your works; they are delayed full six months in Holland. I should be glad, when you write to your bookseller, Mr. Lawrence Salvius, that you will recommend him to correspond with the booksellers here directly. Mr. John Nourse, a bookseller of great reputation here, who sells most of yours and other foreign books, has wrote to him, and has received no answer. He has desired me to write to you, that we may purchase rather the Stockholm edition than the Dutch or German editions of your works. If you mention this when you write to Dr. Bierken, he will speak to Salvius, and perhaps write his answer to me.

I am much concerned that your friend Solander is detained by sickness. If he comes, I shall introduce him with pleasure to all my friends, who indeed have long expected him; for it now is above a year since you wrote so warmly in his favour.

A friend of mine has sent over a leaf of the true Cinnamon with some of the bark from Guadaloupe. I expect to receive a specimen of the blossoms. No judgment can be formed by the leaves, nor is the taste of the bark sufficient to distinguish the genus: but there is a probability that it is the true Cinnamon of the shops.

There were shewn at our Society for Arts and Sciences, &c. some very strong filaments like hemp, said to be found on the Muskito shore, near the gulph of Honduras. The plant is said to be a species of *Aloe*, or Pinguin of the Ananas kind. But I believe this is a mistake; for I think it must be a species of *Yucca*, this kind of plant being used for the making thread of various kinds in North America, by macerating the leaves a few days in water, and afterwards beating and combing it, treating it in the same manner as we do hemp. — It is said they have a species of this kind in the East Indies, which they dress so fine as to form filaments as glossy as silk. I shall send you a specimen of it with the rest of your things, I mean what was laid before the Society. Dr. Garden is of opinion that the arborescent *Yucca* of Carolina is a different genus from the common *Yucca*. I have sent with the seeds a fruit of the tree *Yucca*, which I hope you have received long before this. The Dutch, as Van Royen, call the Yuccas *Cordyline;* and some species are called in America Silk Grass, which points out the uses they have been before applied to.

Dr. Browne, of whom you enquire, I hear is at Santa Cruz, one of the Danish Islands in the West Indies. He informed me that they use the fibres of the *Agave*, or great American Aloe, instead of Hemp, to place between the sheathing of their ships; and that the bitter juice of the leaves, mixed with pitch, would prevent the *Teredo* eating into the planks. I have not heard it confirmed; but it is worth enquiring into. Perhaps upon enquiry we may find that all the species of *Agave* have filamentose leaves, and that this Silk Grass from Honduras may be from one of them. If you have met with any observations on this subject, be so good as to communicate them to me.

I am, Sir, with the utmost regard, your most obedient and affectionate humble servant,

JOHN ELLIS.

I cannot as yet get you a specimen of the *Morea;* they were out of flower before I received your letter.

I have just now heard that my friend Mr. Peter Woulfe the chemist is collecting fishes in Guadaloupe for you. I expect many curious specimens from him. Whatever is curious you shall know concerning them as soon as I receive them.

I hear a Swedish ship is soon expected here. By her I intend sending your things if she sails first.

LINNÆUS TO ELLIS. (Latin.)

Upsal, Aug. 11, 1760.

I acknowledge myself greatly indebted to you for the peculiar kindness with which you have received my friend Solander; who, in his letters, cannot sufficiently praise your hospitality, in not only welcoming him yourself with paternal affection, but in recommending him also to your friends.

I have at length received the seeds packed in wax, which you sent me early last Spring. But the person who brought them stayed a very long time at Amsterdam, and reached Upsal only three days ago. I hope, however, that many of the seeds may keep till next Spring.

I shall obey your orders as to the names of plants; but if I may without reserve lay open my mind to you, I could have wished that the supposed Jasmine might have been called *Warneria*, after the person who has first cultivated it in Europe; *Gardenia* being applied to some genus first discovered by Dr. Garden. I wish to guard against the ill-natured objections, often made against me, that I name plants after my friends, who have not publicly contributed to the advancement of science. If therefore I confer this honour on those who have discovered the respective plants, no objection can arise, nor can I be charged with infringing my own rules. Still, if my opinion displeases you, pray say so without reserve; for my attachment to you will not easily permit me to go contrary to your determination.

I have not yet seen either the new genera of Dr. Garden, nor the flower of *Morea*, which I earnestly beg of you to send me.

If you have made out any thing more concerning the fruit of the said double-flowered Jasmine, pray inform me, that I may know where to place this genus in my System, and not be under the necessity of putting it into the Appendix, among such as are imperfectly known.

Please to give the enclosed to Mr. Solander, as I am not yet informed how to direct to him in London.

It remains to be tried whether the seeds you have sent will retain their vital power through the whole summer. Your experiment, if successful, will certainly be very important for Botany, as seeds are so rarely brought uninjured from the Indies.

Present my compliments to all your scientific friends, my benefactors.

FROM THE SAME TO THE SAME. (Latin.)

[Without a date. The London post-mark is Nov. 4, on which day Mr. Ellis received this letter; and the year may be presumed, by the subject, to be 1760.]

I received yours of Sept. 12th, and return you my best thanks *.

I had given the name of *Gardenia* to an entirely new and very singular genus, the *Catti marus* of

* This letter does not appear.

Rumphius, *Amboin.* *v.* 3. 177. *t.* 113 *, in order so far to conform to your wishes. But as you still persist in your decision, that the Jasmine so often mentioned between us should be called *Gardenia*, I will comply, though I cannot but foresee that this measure will be exposed to much censure. I find it impossible to deny you any thing. All that I beg of you, my dear friend, is, that you would publish the genus and its character in some loose sheet, or some periodical work, or transactions; in which case I promise to adopt the name. I wish to learn from you what Dr. Garden has written in Botany, or what he has discovered, that I may make mention of it. Do not therefore indulge any more suspicions of my regard and devotion to you, who esteem you among the chief of my friends.

I am glad that you take so much pains about preserving seeds. If your attempts should succeed, the whole world will be obliged to you, for our gardens will, through your means, be enriched with abundance of Indian plants, whose seeds have hitherto been imported to no purpose. Thus we may, before long, see the Tea growing in Europe, as well as numerous plants besides, never yet seen among us.

I write in haste, not being willing to defer the above matters; any more than my most grateful acknowledgments for all your favours to my dear Solander. I only fear that his residence in your part of the world should be too expensive to allow of its being long continued.

* Now *Kleinhovia hospita* of Linnæus.

ELLIS TO LINNÆUS.

[No date.]

Mr Solander informs me he has acquainted you that I have now in my possession a curious recent *Encrinus*, lately fished out of the sea near Barbadoes in the West Indies. It seems to agree with the petrifaction described in your *Systema* under the title of *Hemintholithus Medusæ ramosissimus*, &c. &c.

This animal being entirely composed of crustaceous joints, united by fine fleshy ligaments, renders it so difficult to handle, that it is not so perfect as we could wish, many of the ramifications being broken off, particularly those large ones that compose the head.

I have desired him to describe it in your method, for your better understanding it; I propose to write a description of it, and all the class that we find fossil, in order to make a little dissertation of it for our Society.

Stirps radicata, crustacea, pedalis, adscendens, 5-angularis, glabra, ramosa, articulata: articulis densè approximatis, supernè ad latera plana sulco notatis, punctoque impresso distinctis.

Rami quini, simplices, verticillati, teretes, glabri, articulati, patuli, apicibus acutis reflexis. Spatium inter verticillos unciale, supernè angulare. Rami in medio caulis palmares, supernè et infernè paulò breviores.

Stirpem terminat flos, seu caput, eodem modo ac *Asterias caput Medusæ*, in ramis per dichotomiam divisum. Rami patuli, articulati, supernè emittentes ramulos simplices, teretes, adscendentes, inflexos.

Centrum tegitur membranâ crustaceâ concavâ, in cujus fundo apertura oblonga, minima, in sicco specimine.

Obs: Singulus caulis articulus substantia interna fibris stellatis constat, uti in fractura apparet.

As soon as it is engraved I will send you a print of it. Mr. Solander is going with a danish gentleman to visit Oxford, Bath, Bristol, and the mines in Cornwall. I have given him some letters to my friends, and do not doubt his being very well received by every body.

I must now inform you that my experiment in order to the preserving of the Tea seeds has so far succeeded, that by this time you will have received a capsule inclosed in wax, which when you come to open you will find in a vegetative state, as I have experienced by sundry which I cut open before the Royal Society. At the same time I cut open several acorns of our english Oak, that had been inclosed in wax since October 1759, which was the same time that the Tea seeds were inclosed in China. You are to remark that these Tea seeds were collected at Limpo, or Ningpo, near the latitude of 30 degrees north, in China, and I am well informed that the weather there from the middle of December to the middle of February is attended with severe frost and snow, but that the summer is excessive hot.

I remember in a former letter where you press my getting Tea plants from China, you remark they are so hardy as to bear the cold even of Sweden. I find that wax is preferable to tallow and wax; for unless the fat is taken from the kidneys, which we call suet, and melted directly with the wax in equal parts, it is apt to putrify in long voyages. There were many of the Tea seeds spoiled by using the ship's tallow, and mixing only a quarter wax; so that if wax is to be got, I should prefer it to all other substances. I have, by the desire of our Society for the Encouragement of Arts, Sciences, and Commerce, transmitted a capsule of Tea seeds inclosed in wax, to each of our governors of provinces from New England to Georgia; so that I hope to establish Tea in America, by that means, in time.

I have got a few seeds of the *Croton sebiferum*, which grows in the same place with the Tea, which I expect to raise here. Great care is to be taken when the capsules or seeds are put up that they are perfectly ripe and dry.

I suppose Mr. Solander has informed you that I have presented the characters of the *Gardenia* to the Royal Society here.

I expect some seeds soon from America; as soon as I receive them Mr. Solander shall have a share for you.

If you can procure me some seeds of the true Rhubarb, I shall be obliged to you, and any rare Siberian or Tartarian seeds, particularly those of trees or shrubs. I am, with the greatest respect, dear Sir, your much obliged humble servant,

JOHN ELLIS.

LINNÆUS TO ELLIS. [Latin.]

Upsal, April 3, 1761.

By your last letter I perceive, not without astonishment, that you have discovered one of the greatest curiosities perhaps in nature, the hitherto unobserved original of the *Entrochi*, or Columnar *Asteriæ*, on which pledge of lasting fame I congratulate you with all my heart. But I cannot conceive how a Medusean head can terminate the stem of a zoophyte plant. I wish you had given ever so rude a sketch of it in your letter, for my more perfect information.

Entrochi are either larger or smaller. The larger are surrounded at each joint, as well as in their cortical substance, with a row of tubercles; but their centre is perforated. I have never been able to understand the nature of those tubercles. But I received last autumn a very singular stalk of a petrified *Entrochus*, as you may see by the figure annexed *, of the natural size. The stem consists of compound joints, formed of the larger kind of *Entrochi*. The tubercles surrounding each joint have been numerous whorled branches, which gradually fall off, as in trees, and have disappeared. The upper branches still remain. These are narrow and forked, destitute of joints, or rather of elbows, and smooth. In this specimen, however, the stem is not angular, nor are the branches five together in a whorl. These branches are what have been termed

* This figure does not appear.

antediluvian corals, or coralloides; nor has any one ever suspected their belonging to the same species as the *Entrochi*. The hollow of their stem is obscurely pentagonal: but it has on the inside a lateral pore, at each of the tubercles which form whorls round the stem; which pores are likewise visible in common, or distinct, *Entrochi*. Moreover, the branches, present in the upper part of the stem, appear to have been hollow or tubular, in the living subject, though now filled with stony matter. Nevertheless, I cannot but refer this specimen to the genus *Isis*, which perhaps may be the case with yours, though distinct in species.

Have you never, my friend, met with traces of the animal which forms the *Lapides Judaici* of the shops? Whence are these stones, which are certainly petrified, and have as certainly been hollow? Has the nature of these escaped you, who have scrutinized the bottom of the sea, and the depths of the ocean itself?

I return you thanks for the Tea seeds; but they are not yet come up, nor perhaps ever will. I had last year a whole handful of these seeds from China, eight of which only were sound enough to sink in water; but they never vegetated; and I fear your seeds will not grow, not even those sown in America.

May I enclose seeds of the true Rhubarb in a letter to you? I know not whether your post would allow of this. I must however run the hazard, as the spring is fast advancing, and there is not time to await your answer. I therefore now send you these seeds, as well as those of *Robinia Caragana* and *frutex*.

Tell Mr. Solander I received his letter, enclosing seeds of the *Magnolia*, for which I am obliged to you. Has he received my letters of the 27th of January, and 2nd of March; as well as one of the 20th of March in a cover to Brander and Spalding?

Several of your seeds which were inclosed in wax, are come up, as the *Hamamelis*, one plant of *Stewartia*, and the *Halesia*, with which I am much pleased.

I am employed every day with my academical labours, and the new edition of *Fauna Suecica*, now in the press. I have nothing new, but what you send me. Tell me whether your Tea seeds have vegetated. Farewell.

ELLIS TO LINNÆUS.

SIR, London, June 2, 1761.

I have been favoured with your very kind letter of the third of April, inclosing some seeds of the true Rhubarb from China, together with the *Robinia Caragana*, both which grow very well. I should be much obliged to you for a pound of the seed of the true Rhubarb by the first Swedish ship, but beg that it may be of this year's growth. Please to send it to the care of my friends Brander and Spalding, and I will pay them or Mr. Solander the charge of it.

I now send you a proof sheet of the figure of the recent *Encrinus*, which I told you in a former letter

I received from Barbadoes. The figure which you proposed to send me of your fossil one, I believe you forgot to inclose, for it was not in the letter; so that I wait in hopes to see it. I am endeavouring to find out all the variety of fossils of that kind, in order to describe them, but fear I shall not be able to do it so well as I wish.

This is most certainly a testaceous animal, and in my opinion not allied to the *Isis*. Mr. Solander, who is very expert at these matters, will give you his description of it. It is very possible it adheres to rocks in great depths of the sea, and for ought we know, this may be but a branch or part of the entire animal. The branches which compose the summit or head of this animal are well described in Rosinus, Pag. 84. Tab. x. f. ii. No. 1.

The animal of the *Isis Hippuris*, as it comes to us divested of its outward covering, must remain unknown to us, till some curious person will take the trouble to examine it recent, or bring it home carefully in spirits of wine.

The *Isis ochracea* I have taken some pains to examine and describe in our Philosophical Transactions, part 1st of vol. 50. page 188. It appears to me to be totally different from the *Hippuris*.

The *Isis Encrinus* is not fixed to rocks, but floats freely in the sea. This appears from an examination of the base, which had the skin intire when I first received it. The bony stem of this is not in the least jointed.

Mr. Solander has promised to examine all my

described corallines over again, and also all the varieties of *Algæ (Fuci)*. We intend soon to go to the sea side for further discoveries. ***Lapis Judaicus*** is certainly the spine of an ***Echinus***. This appears plain from sundry specimens in the public as well as private Musæums, where the spines and ***Echinus*** are found in the same mass. I have never heard of any of this kind that has been found recent.

Your letter to Dr. Garden shall be carefully forwarded to him.

Governor Ellis of Georgia, who is now here, says, they have a Tortoise *(Testudo)* in the river Savannah in Georgia without a shell, except a small one on its breast; that the feet and neck of this animal are longer than the other kinds.

I will endeavour to get one preserved from thence from the present governor, as Mr. Ellis is appointed governor of Nova Scotia, one of the most northern colonies belonging to us.

Dr. Job Baster of Zurichsee in Zealand, has lately presented a memoir to the Royal Society here, to prove that the *Sertulariæ* are partly plants and partly animals; that the outward skins or cases of the *Sertulariæ* are of a vegetable nature, and the internal part animal. And quotes your description of a Zoophyte to prove that he is right. I so far agree that the external part vegetates, but, it is probable, it is after the manner of our hair or nails, or as the horns of animals. Yet I am convinced from experience, that it is of an animal nature as much as the shell of the Tortoise. He supposes they shoot out roots like vegetables, but I am going to the sea side with Mr. Solander, to shew him that what appears to be roots, is only the animal in its first state, (like the larvæ of insects). The young polype or *Hydra* drops from its vesicle or matrix, and begins immediately to extend its horns and body, and to fix itself to some Fucus, shell, or stone, by its hinder part. After it has secured itself there, the fore part becomes, or raises itself, erect, and proceeds growing on, and sending forth little heads in denticles, which are variously disposed according to the several particular species we meet with. If you examine these root-like bodies, you will find them cylindrical and wrinkled, like the dried guts of animals. From one and the same small gut-like root, or first beginning, of these polypes, I have seen several stems arise: but this is agreeable to the manner of the increase of the *Hydra* of Trembley. I hear that Dr. Baster is desired by his friend Mr. Philip Miller (who totally disbelieves the animal nature of Corallines or *Sertulariæ*) to proceed, in order to overturn this new doctrine. How far he will meet with success a little time will discover. I think I have half converted my opponent, who sets out in his memoir printed in our Transactions, vol. 50, part I. page 258, by telling the Society that *Corallina non magis à polypis fabrifieri, quam diversa fungorum genera ab illis fabricantur animalculis, quibus, æstivo tempore, quasi repleta inveniuntur.*

I made some observations some time ago on the *Alcyonium digitatum,* and find it a middle being between some of the Madrepore corals and some Sponges, coming near to the Sponges, which indeed, for want of good opportunities, have never been properly investigated.

The stupose part of this *Alcyonium*, divested of its gluten and magnified, is a perfect sponge. Nothing then appears but this reticular part, and this is the case with Sponges, which are always found, when recent and alive, filled with this glutinous substance: the regular holes, which we observe in dry Sponges, strongly indicate their being once filled with animals, I mean those parts of the animal which comprehends the *tentacula*, mouths, stomach, eggs, &c. But I shall defer this subject till I send you drawings of my observations, and hope to meet with some objects at the sea side, which may better elucidate this matter.

The plates of the *Gardenia* and *Halesia*, which were doing for my memoir to the Royal Society, are now almost finished; it will be published in our Transactions next month, when I will send you the memoir and plates.

Your calling our *Cicuta*, *Conium*, together with your differing from others as to its being the *Cicuta* of the ancients, has exercised the pens of many of our people here, particularly Dr. Watson, who is of the contrary opinion, and endeavours to support Dr. Storck of Vienna. For my part I cannot think that Storck's description agrees with our *Cicuta* in the poisonous qualities of the roots of theirs; nor can I think that two or three degrees of latitude can make so great a difference in its venomous quality; especially as it is a native of this country, and the most common weed we have. I should be glad of a letter from you on that subject to lay before our Royal Society, who greatly depend on your judgment and opinion in all physical and botanical affairs.

I am in fear that our Tea seeds will not grow, though we did not expect them to appear till the middle of this month. Those that were placed in our hot-house are perished; those in the natural ground (or open air) some gardeners have still hopes of. If they had been taken out of the wax at the Cape of Good Hope, and put into earth covered with land moss, I should think they might have had a better chance. But I planted acorns of the common english oak, which grew after having been preserved in bees-wax 14 months, which shews that bees-wax is the best substance to cover them with. These acorns were put into an earthen vessel, into which the melted wax had been first poured, and before it was hardened the acorns were put in; being first well dried on the floor of an airy room for a month: whereas the Tea seeds were only covered with half an inch round them of wax, which might be too little; however, I am not discouraged from proceeding in such useful experiments.

I had almost forgot to mention that Dr. Baster says, "that all the Corallines, properly so called, are *Confervæ;*" and adds "he is inclined to suspect, from very solid reasons, every species of Coralline enumerated by Linnæus; though he has not had an opportunity of examining them all. I shall be obliged to answer this ingenious doubting philosopher, by shewing the Royal Society that he does not know the difference between a Conferva and a Coralline, which is a task I by no means like. At present I shall refer you to his *Opuscula subseciva*, in two parts, which I suppose he has sent you; and am, dear Sir, your much obliged and most obedient servant, JOHN ELLIS.

P. S. As the print I enclose to you has not yet appeared in public, I must desire you would not give a design of it to your friends till ours appears. This copy which I send you is not yet finished; but as you seemed pressing for it, I send it to you just as it is.

LINNÆUS TO ELLIS. [Latin.]

Upsal, September 16, 1761.

I am in your debt for a letter of June 2d.—I have gathered the seeds of the Tartarian Rhubarb, and of *Robinia Caragana* in plenty, and shall send you both, by the first opportunity. The new Rhubarb from China has not flowered this year, having been sown late in the preceding season; but it may be expected to bear seed next year.

The Tea seeds communicated by you, and by two other friends, though preserved with all possible care, have not grown. Surely if your countrymen, who can accomplish any thing within the bounds of possibility, are not able to bring the Tea alive to

Europe, it will never be done. This shrub grows with the Lilac, the Horse Chesnut, and other things that bear our climate perfectly well. It might certainly be brought alive in a pot with a hole at the bottom. One of our ship captains actually brought a Tea-tree alive as far as the Cattegat, but in a single night the mice stripped off the bark entirely, and the tree perished. So adverse is fate on some important occasions!

Your *Lithophyton*, formed of columnar *Asteriæ*, has afforded me great pleasure. It is a wonderful discovery, concealed from our predecessors, though they sought after it with all possible care. You need not fear my communicating it prematurely to any one. I am not accustomed to deprive others of their discoveries.

You believe the *Lapides Judaici* to be the spines of an *Echinus*, beyond all possible doubt. Pray inform me to what species they belong. That they have been hollow, appears from their internal substance being sparry, and spar is generated in a fluid *.

I have read Storck's pamphlet on the *Cicuta*. His *Cicuta* is my *Conium maculatum*, as I learn by letters from Vienna. I have never before known the root of an annual plant brought into medical use, because all such roots are exhausted in producing the stem. The *Conium* is slightly fœtid or acrid; nor do quadrupeds feed on it, because its qualities

* These bodies, whatever their substance may be, are doubtless cast in the mould formed by the spines of the *Echinus*.

do not, for the most part, agree with their constitutions. Acrid and fœtid plants are what are usually termed poisonous. These are hostile to the vital principle of the brain and nervous system; and therefore, when taken into the body, there is an immediate effort to expel them by vomit or by stool, by perspiration, urine, &c. These acrid substances, taken in a smaller dose, are denominated expellents; and if applied outwardly, repellents; because our nature tries to avoid, or withdraw from them. Such, in our days, have afforded physicians their most valuable medicines. They overcome obstructions, and relax the rigid nervous fibres, almost even to a paralytic state; but I have treated this subject elsewhere.

To the tribe in question belong the Poppy with its Opium, *Belladonna, Phytolacca* gathered in autumn, Powder of Misseltoe, though the qualities are imperceptible in the powder; in a word, all heroic medicines. Perhaps such may be serviceable in gouty habits, and possibly in cancers, schirrus, and internal tumors, the causes of chronic disorders. Hence Colocynth, *Elaterium*, &c. are so powerful. Of the same nature is *Conium*, but of all such active medicines, this is the weakest. I wish physicians would try the virtues of *Cicuta (virosa), Aethusa, Actea*, &c. all undoubtedly more important. There could be no danger in such experiments, if made with caution, by extremely small doses in the beginning.

You inform me that Dr. Baster attacks you on

the subject of the animals of corals, declaring that corals are not fabricated by these little creatures, but that the latter inhabit the corals. I suspect that there is really nothing at all in this controversy, and that it consists in your not understanding each other. That shell fish, whether univalve or bivalve, fabricate their own shells, by laying over the inside a mucilaginous coat, which becomes calcareous, is so evident a fact, that no one, who has ever looked at an oyster-shell, can deny it.

As to my *Lithophyta*, the animals of the *Madreporæ* and *Milleporæ*, form their shells under them, in a similar manner, though they are compound (or aggregate) animals.

Zoophyta are constructed very differently, living by a mere vegetable life, and are increased every year under their bark, like trees, as appears from the annual rings in a section of the trunk of a *Gorgonia*. They are therefore vegetables, with flowers like small animals, which you have most beautifully delineated. All submarine plants are nourished by pores, not by roots, as we learn from *Fuci*. As Zoophytes are, many of them, covered with a stony coat, the Creator has been pleased that they should receive nourishment by their naked flowers. He has therefore furnished each with a pore, which we call a mouth. All living beings enjoy some motion. The Zoophytes mostly live in the perfectly undisturbed abyss of the ocean. They cannot therefore partake of that motion, which trees and herbs receive from the agitation of the air. Hence the Creator has granted them a nervous system, that they may spontaneously move at pleasure. Their lower part becomes hardened and dead, like the solid wood of a tree. The surface, under the bark, is every year furnished with a new living layer, as in the vegetable kingdom. Thus they grow and increase; and may even be truly called vegetables, as having flowers, producing capsules, &c. Yet as they are endowed with sensation, and voluntary motion, they must be called, as they are, animals; for animals differ from plants merely in having a sentient nervous system, with voluntary motion; nor are there any other limits between the two. Those therefore who esteem these animalcules to be distinct from their stalk, in my opinion, founded on observation, deceive and are deceived.

Pray let me know if my friend Solander is well. I have not had a letter from him these eight months. I am anxious to hear, as soon as possible, how he is, and whether he is still with you.

I wish you would procure me, if possible, some seeds of *Hamamelis* and *Spigelia.* Your *Halesia* grew from the seeds you sent me, and I have now two plants.

Whilst writing this, I have received from Stockholm your letter (see the following), accompanying that from Dr. Garden, with his observations on fishes. I also learn that there are some parcels for me arrived at Stockholm, which will soon be here. I will very soon reply to your last letter. Farewell.

ELLIS TO LINNÆUS.

DEAR SIR, London, August 13, 1761.

I have just now received two small boxes from Dr. Garden for you, containing a great variety of fish, snakes and other reptiles, and have delivered them into the hands of our worthy friend Mr. Solander, who has shipped them on board a Swedish vessel that will sail in a few days. We have been at the sea shore on the coast of Sussex, and have seen some new animals, and some that Mr. Solander never saw before: he will write you his opinion about the Corallines. We did not meet with Sponges enough to examine their nature thoroughly. What we saw, that had holes in them, shewed no appearance of insects to extend and contract, while under examination in sea water; but we observed the holes to open and shut; however I shall continue my observations in Autumn.

I must beg the favour of you to remember to send me a pound of the true Rhubarb seed by the first Swedish ship for this place. Mr. Solander tells me you can procure it without much trouble, or I should not have asked that favour.

If you continue your correspondence with Doctor Garden, I believe it will be agreeable to you both to write oftener to each other. I mention it to you, because he thought, from his not hearing from you so long, you had dropt the correspondence; but I have lately wrote the contrary, and begged of him to continue to collect every thing that is rare. I long much to have the pleasure of hearing from you, with your opinion of the *Encrinus*.

I am, dear Sir, with the greatest esteem, your most obedient humble servant, JOHN ELLIS.

ELLIS TO LINNÆUS*.

SIR, London, May 29, 1762.

The honours you have lately received from the King of Sweden could not add greater lustre to your name, than they have given pleasure to all your friends in this corner of the world. Long may you enjoy them, and long continue to enlighten the understandings of mankind, hitherto prejudiced and groping in the dark; till your excellent works brought Nature into a method to be understood; and cleared away those difficulties, which frightened mankind from attempting to investigate her.

I rejoice at hearing of a new edition of the Genera. Your Queen's Museum, I understand, will be an elegant work, worthy of her and of you.

You delight me in telling me of your success in getting a living and thriving plant of the Tea tree from China. Our friend Peter Collinson says, he has seen two plants, about 25 years ago, in England, which grew freely and blossomed; but they were destroyed through the ignorance of a gardener.

I shall communicate your intentions of disposing

* Received after the following two letters.

of it to some of our principal nobility, and do not doubt but they will be glad to purchase it; but desire you would be so free as to communicate your terms to me. Here it certainly will be taken care of, and from hence you may depend upon having some plants raised from it.

I am persuaded if the seeds of *Opobalsamum* had been full ripe, and inclosed in bees-wax, they would have vegetated. The seeds of Tea which I received from China were too soon covered with wax, before the superabundant moisture had exuded from them; which rotted the outward coat of the seed. The oak acorns and the chesnuts, which I put into wax in the beginning of February, were, a year after, planted, and are now growing; but those that I covered with wax, soon after they fell from the trees, were rotten at the end of a year. Pray try this experiment, and you will find it succeed.

I have heard lately that our friend Dr. Garden has not been well. He had some thoughts of coming to England. I believe he has fixed it for next year.

I wish you joy of your insects from the Cape; they must be rare and curious.

Pray send me a few seeds of my friend Cl. Alstrœmer's plant, that I may have something to look on agreeable like him. He leaves this country in a day or two.

When you write to me, direct to John Ellis, Esq. King's Agent for West Florida, London. It was a great misfortune to me that I did not receive your letter till last week.

My best wishes attend your Excellency.

I am your most assured friend, and humble servant,

JOHN ELLIS.

I shall in a short time send you a plate of a new *Pennatula*, with a short dissertation on it.

LINNÆUS TO ELLIS. (Latin.)

Upsal, July 16, 1762.

It is long since I received any of your letters, though very desirous of hearing from you. The seeds of *Magnolia*, *Liriodendron*, &c. sent enclosed in wax, have none of them vegetated.

I am greatly and eternally obliged by the innumerable favours which you shower down upon my much-loved Solander.

I hope to be able to send you, this autumn, seeds of an entirely new species of *Mesembryanthemum*, with fine yellow flowers and a hairy stem, from the Cape of Good Hope. The root seems to be annual*.

Please to forward the enclosed to Mr. Solander. I send it in your cover that it may the more certainly reach him.

This summer has afforded very few insects. I never collected so small a number.

Has nothing new been published lately by Mr. Edwards, who has enriched our science with so many rare birds and animals?

* *M. pomeridianum. Linn. fil. Dec. t.* 13.

What news have you, to whom Nature is every day revealing some secrets?

I am entirely occupied in publishing the 2d edition of my *Species Plantarum*, almost twice as copious as the first.

May God preserve you, for the benefit of natural science! Farewell.

FROM THE SAME TO THE SAME. (Latin.)

MY DEAR FRIEND, Upsal, Nov. 23, 1762.

It is long since I had a letter from you, which I much lament, for I always learn something from your acute and valuable observations.

I continue to be employed in the edition of my *Species Plantarum*, which will be by far more copious than the former.

You have doubtless observed that Jacquin in his work, has reduced Browne's *Ellisia*, to Plumier's genus *Duranta*. Having never seen the *Duranta* myself, I requested of Jacquin to send me one of its flowers in a letter, and am obliged to confess that *Duranta* and *Ellisia* not only form one genus, but scarcely differ in species, so that the latter cannot be kept distinct. This being the case, I began to look about for a new *Ellisia*, that you, who deserve so eminently of our science, may not be forgotten. Being particularly desirous to fix on some plant known in the gardens of Europe, I have thought of what is called in my *Species Plantarum* (ed. I.) page 160, *Ipomœa Nyctelea,* which Morison has badly delineated in his *Historia,* v. 3. 451. sect. 11. t. 28. f. 3. This grows in Mr. Collinson's garden, and probably in other English collections. I have seen, by the fructification, that it constitutes a perfectly distinct genus. Write to me as soon as possible, whether this meets your wishes, or whether you would prefer any thing else.

A report has lately arisen here, that my friend Solander labours under some difficulties in your country. I earnestly entreat you, by your regard for me, to give me some account of him by return of post. I wrote two months since to inform him of his appointment to the Professorship of Botany at Petersburgh; but, to my surprize, have had no answer. My letter went in a cover to Brander and Spalding.

What tribute have you lately received from the kingdom of Neptune?

Have you any news of Dr. Garden? Your *Gardenia* stands in its proper place in my new *Species Plantarum.*

ELLIS TO LINNÆUS.

SIR, London, Dec. 21, 1762.

I am ashamed that I have deferred answering your former letter, which I received in October last. I have now the favour of yours of the 23d of November, and have consulted with my good friend Mr. Solander about what he intends to do. I find that after maturely weighing what you so kindly intended for him at Petersburgh, and after consulting many friends here, he is determined not to accept of that professorship for many reasons, which he tells me he has wrote to you at large. I hope in time he will get something honourable to employ and distinguish himself. I am greatly obliged to you for the honour you intend me, in giving me a new plant. I believe I met with it some years ago in Mr. James Gordon's nursery garden, and made a drawing of it, which I magnified for my friend Mr. Peter Collinson; since which, Mr. Solander has seen and described it as a new genus, and I suppose sent it to you. You will pardon me when I tell you that people here look on a little mean-looking plant as reflecting no honour on the person whose name is given to it; though I am convinced, as it is a distinct genus, the compliment is equally great with the largest tree.

I sent you the specimen of a shrub, with white flowers, not unlike a *Philadelphus* *, which Dr. Garden described and added to his description of some fishes he sent you. I desired this plant might be called *Schlosseria*. If it is not too late, and you find it a distinct genus, I would rather choose to

* This proved a *Styrax;* either *grandifolium*, or *lævigatum*, of Solander, in *Ait. Hort. Kew.*

change plants with Dr. Schlosser. If this is not convenient, and you have any new genus of a specious plant, that will grow in England, you will do me much more honour; because I may communicate it to the gardens of my friends here, to put them in mind of me.

With regard to the conduct of our mutual friend Mr. Solander, no man bears a better character here. He is constantly employed in the business of natural history, and, I am persuaded, has made many discoveries of new genus's in both the animal and vegetable world. His friends are considering of getting him employed in something that may be for his advantage, and they are in hopes of succeeding. He is exceedingly sober, well behaved, and very diligent, no way expensive; so that I hope he will do very well. I can assure you, the more he is known, the more he is liked; and now peace is near settled, he has a greater probability of succeeding, than when we were engaged in the hurry of a troublesome, though victorious war.

I am going to publish a short memoir on some few of the Coccus's, especially to describe the male fly of the Cochineal. I shewed Mr. Solander two sorts, which I had observed at my friend Mr. Webb's garden, one on the *Ananas*, and one on the *Ribes*; which he says you have not yet described: but I have not been able to discover the males. We have a sort of *Anser* from America, that is not yet described. As we shall be in the country, at Mr. Webb's, during the holidays, we shall send you a description of it; it is not unlike the Barnicle of Willoughby in size. I have not met with any thing new from the sea since I described the *Encrinus*. Though the sea sponges are certainly of the animal kingdom, yet we could not get them so recent as to describe the nature of them properly.

The seeds you sent me of the Rhubarb, which I sent to the Governors of the Provinces in North America, I am afraid were lost; for though it is now a twelvemonth ago, I have not heard of their arrival. I must, therefore, intreat you to send me half an ounce inclosed in a letter, and direct the cover of your letter to Anthony Todd, Esq. at the General Post-office, London. I shall be sure to receive it safe; for that gentleman is the Secretary to the Postmaster General, and my friend. You may send any other seeds, not exceeding an ounce, the same way.

I have not heard from Dr. Garden this long time. I fear he is ill, I expected to hear of his receiving your obliging letter to him.

Now we are to have a peace, I shall extend my correspondence, and hope to meet with many new discoveries; which, if I do, shall immediately communicate them to you. I have still an inclination to enquire into the nature of sponges. I am persuaded the *Fibræ intertextæ* of sponges are only the tendons that inclose a gelatinous substance, which is the flesh of the sponge. Mr. Solander and I have seen the holes or sphincters in some of our sponges, taken out of the sea, open and shut while they were kept in sea-water; but discovered no animal like a Polype, as in the *Alcyonium Manus mortui.* In a *Priapus,* which Dr. Bierken dissected, I made a drawing of one of the Sphincters, which has a great deal of the appearance of one of the openings of the *Spongia medullam panis referens.* This drawing and animal I sent you by Dr. Bierken.

Mr. Solander is now with me, and joins me in our best wishes for your health and prosperity.

I am, dear Sir, with the greatest regard, your much obliged friend, JOHN ELLIS.

SIR, London, Jan. 1, 1765.

In answer to your very kind letter, letting me know that you had, after many attempts, at last got a true Tea plant, which you wished to dispose of to some of our nobility, I wrote you that I had mentioned it to some of our Magnates, and they desired to know what would be agreeable to you in return, as it was impossible for me to put a proper value on it. I have waited long, expecting an answer to my last letter, which I wrote in May last; but, hearing from your disciple, and my very good friend, Mr. Adam Kuhn, that you had expressed a desire to hear from me, and that you wondered at my silence, I now assure you that I do, and always did, esteem it the greatest honour of my life to be a correspondent of Professor Linnæus, always considering you as the great ornament to science, especially Natural

History, which I so much esteem. Mr. Kuhn has fully explained the affair between Solander and you, which I was totally ignorant of before, and for which I am extremely sorry. I wish to see you in the same friendship as before. What friends I made him, were intirely owing to your warm recommendation. I have lately introduced him to our Lord Chancellor, who is, in rank, the first man in dignity, except the Archbishop, to the Royal Family: but if I thought he had acted dishonourably by you, I should never esteem him more. I sent a dissertation to you, and one to your son, upon the *Pennatula*, to the care of Mr. Kuhn, who promised me it should go by the first ship. In that I have observed, that there does not appear any perforation in the base, which not only from my own observation, but that of Bohadsch, I am thoroughly convinced of. Your letter to our friend Peter Collinson, I have heard the contents of. I rejoice to find poor Forshall discovered it, and I observe likewise that you think the *Rheum compactum* is the true officinal Rhubarb. I wish for a few seeds by the post. I am now (thank God) in a place as King's Agent to West Florida, that entitles me to the correspondence of many gentlemen that are gone to reside there, and are curious in Natural History. I expect to hear every day from them; and you may depend on it, no man of the Royal Society of London shall sooner be acquainted with what I receive than you. I shall go to the sea-side to search for Marine Insects next summer. I attended last summer in pursuit of the animals in sponges, but believe me there are none; but the whole is an animal, and the water passes in a stream through the holes, too and fro, in each *papilla:* and whatever has been wrote by Peysonell and others, is not true. Adieu, my dear friend.

I am most sincerely yours,

JOHN ELLIS.

LINNÆUS TO ELLIS. (Latin.)

MY OLD FRIEND, Upsal, Feb. 12, 1765.

I have just received, with the greatest pleasure, a long-wished-for letter from you*. I rejoice with all my soul, that you are in good health and spirits, and that you have now under your command the province of West Florida. Much advantage and increase must hence accrue to our most delightful science, which you have ever promoted, and which you have enriched by laying open a new submarine world, as it were, to the admirers of Nature.

Your *Pennæ marinæ*, I have not myself received, though I have seen and studied them attentively in the hands of Dr. Bæck. The essay on the subject is excellent, and worthy of you. I have, however, received your beautiful drawing of *Pennatula Cynomorium* from Gibraltar.

I have been employed these two months in exa-

* Dated Jan. 1, 1765. Some letters appear to be wanting, on both sides, between 1762 and 1765.

mining East Indian and Cape plants, of both which I have received a great quantity, with their flowers, and I find about 30 new genera among them.

Possibly you have not received the letters I wrote to you, during my stay in the country. In May last I laboured under a very severe and dangerous attack of the pleurisy, and spent some time out of town, to recover my strength. I suspect that, through the fault of my servants, my letters were not duly forwarded.

It gives me much pleasure to perceive, by your last, that Mr. Kuhn is still in London. I should be glad to know whether he received my answer to his only letter. Not knowing how to direct to him, I sent it, Oct. 8th last, under cover to Mr. Lee. Kuhn is one of the most worthy and industrious young men I ever knew. Give my kindest regards to him, and ask him to let me hear what country he means to visit after quitting London. Is he about returning home*? Mr. Alstrœmer was with me a few days since; and I heard with great delight his account of your good health, and indefatigable application.

If you have Ginanni's posthumous works, published at Venice in 1755, in folio, vol. 1st, pray tell me which among his figures of *Sertulariæ* or Corallines, certainly not among the best, answer to yours.

I have published a dissertation on *Opobalsamum*, which is a species of *Amyris*.

* To Pennsylvania.

Ask Mr. Lee whether he received my letter of the 8th of Oct. containing seeds of *Alstrœmeria.*

I learn from Mr. Alstrœmer, that your English cultivators have, in their gardens, that beautiful and most remarkable plant the *Sarracenia*, even both species; which I should scarcely credit unless he had told me.

My Tea-plant is still in health, but has not yet flowered.

By your account of the *Penna marina*, I perceive it to come but too near my genus *Alcyonium.*

In the beginning of spring, I shall begin to print a new edition of the *Systema Naturæ.* If you can point out any thing to be amended, I beseech you to inform me.

There is in America, about Carthagena, a singular bird, allied to Marcgraave's *Jacana*, which, if you can introduce it into Europe, and, in the first place, into England, will do you great credit. It is known at Carthagena by the name of *Chavaria* *. This bird follows the chickens, geese, and other domestic fowls, all day long, as our dogs keep sheep and oxen; and drives away all birds of prey, being formidable to them on account of a sharp claw, placed in the middle of each wing. It might be fed, during the voyage, with bread soaked in water; its natural food being plants.

Is Dr. Garden still in Carolina, alive and well?

Gouan is now publishing a *Flora* of Montpellier; a naked performance, having synonyms of Magnol only, without remarks.

* *Parra Chavaria.* Linn. Syst. Nat. ed. 12. v. I. 260.

Oeder, the Danish Professor, has published Elements of Botany, entirely taken from my *Philosophia Botanica.*

I know of no news in this country. I wish you health and happiness, and hope to be ever numbered among your friends.

ELLIS TO LINNÆUS.

London, July 19, 1765.

It has been some concern to me that I have not answered sooner the kindest letter of 12 Feb. from my worthy and most respectable of friends the Right Hon. Sir Charles a Linné.

I have been studying matter to fill my letter, for meer business to Philosophers is of no account. You flatter me much, when you tell me, that you approve of my little dissertation on the *Pennatulæ.* I am sorry those I sent by my friend A. Kuhn's directions did not arrive.

I have sent to Dr. Schlosser, of Amsterdam, for Ginanni's *Opera posthuma.* He has promised to send the book to me; and when I receive it, I shall make such observations as you desire.

The *Sarracenia* that you mention, is sent over every year from North America, and both species flower with us.

I rejoice that you have obtained a true account of the *Opobalsamum.* I believe Mr. Lee has lately received some seeds from you of the *Alstrœmeria.*

I hope your *Thea* will succeed and flower. None has yet come to England in my memory.

I shall write to a friend of mine, who has been at the Havanna, to ask if there is such a bird as the *Chavaria* at Carthagena, as you describe, and whether there is a possibility of getting one to Europe.

I am now come to tell you I have got a letter from Dr. Garden, which I now inclose you, and shall send the specimen of the very rare animal by the first opportunity, together with the description not only of that, but of several new plants which he has lately discovered. I have not heard from West Florida since I have been agent, I mean from people of science. There has been a ship lately lost, that was bringing over many curious things to me; but I still am in hopes of receiving full accounts of the natural history of that country, from a particular friend, who is now going as Lieutenant Governor to that province. I use every means to promote that happy and endearing part of human knowledge, Natural History. Nothing gives me more real pleasure than when a new scene opens in it, of which I was totally ignorant before. I have been engaged all this winter and spring in business; but hope to get to the sea-side, that I may give you a further account of the kingdom of Neptune.

I have recommended to your correspondence Doctor David Skene, M. D. of Aberdeen, in Scotland, as a very ingenious Natural Historian. He proposes to write to you very soon.

I am, dear Sir, most truly your affectionate friend,

JOHN ELLIS.

Pray let the address of your letters be to John Ellis, Esq. in Gray's Inn, London.

LINNÆUS TO ELLIS. (Latin.)

Upsal, Aug. 15, 1765.

Your letter of July 19 has come to my hands in due course, and affords a joyful testimony of your health and happiness, as well as of your regard for me.

Accept my sincere thanks for Dr. Garden's letter, which mentions some dried plants. If these have reached your hands, I trust that you will not fail to forward them to me the first opportunity.

So extensive a country as Florida cannot fail, under your auspices, to yield a rich harvest to the learned world. Its lot is peculiarly fortunate, in being subject to your controul, and *Florida* may now truly answer to its name. We know but few of its vegetable productions, and scarcely any thing of its animals. Fate has reserved them for you. May God grant you life and happiness, till you have laid open many of these treasures of science!

My Tea-tree is thriving, but still without flowers; nor have I yet dared to expose it, in the open air, to the cold of our winters, having only a single plant.

Your *Pennæ marinæ*, for which I am indebted to you, have afforded me much pleasure. You have well explained their characters; and the work is no less elegant than satisfactory.

If your leisure will permit, I wish you would undertake to describe the marine kinds of *Mollusca*, which can neither be understood nor explained by any one but a person, like yourself, competent to examine them, and to make drawings and descriptions from the living subjects.

Pray enquire of Mr. Lee, whether he received my letter directed to Mr. Solander.

Farewell.

ELLIS TO LINNÆUS. [Without a date.]
It appears to have been written about Aug. 1765.

About a month ago, just before I had the honour of transmitting you Dr. Garden's letter, with the account of his new animal, as he calls it, I was shewn a printed book of the Lives of the Professors at Gottingen in Germany. Among the rest, to my astonishment, I saw that Professor Buttner has had the assurance to write, or give the following account of himself, relating to my discoveries of Corallines. He says, "besides a considerable number of sea plants, he discovered also the manner of generation of many habitations of Polypes, taken for plants."

"These discoveries he laid, 1750, at London, after a voyage thither, before the Society of Sciences there; when John Ellis from Ireland, a member of it, communicated them afterwards to the world, under the title of an Essay towards Corallines, with copper-plates, which he got done at his own expence, besides some additions of his own."

In the first place, you may from me assure the world, that he never laid any observations at all, of any kind, before the Royal Society of London; as the certificates of the Secretaries will soon convince the world of his integrity. In the next place, I most solemnly declare I never received any information from him of the nature of the animals that belong to the *Sertulariæ*, or any of the kinds I have treated of in my book; for he never went with me to the sea side, nor never saw them (I really believe) till he saw the figures of them in my plates. He never cared to talk on the subject to me, and only botany was our conversation when we met. Your friend Peter Collinson can tell you what a vile fellow he is, and how ill he has treated me, after the civilities I have shown him.

I have been told, and I am now convinced, that if it had not been for your determination to find out the true author of my book, this man would have privately robbed me of all my expences, labour, and merit. To you I owe my fame; and the honour you have done me, by electing me a member of your Royal Society of Upsal, I shall gratefully remember; and hand down to posterity your kindness and this honour done me, in my next volume, which I am now about. I shall write to Baron Haller, and shew him the iniquity and unworthiness of such a professor; and Baron Munchhausen shall certainly be informed of it. You and all the world may plainly see that there are no discoveries worth notice but those I made at the sea side, and at these I always mention who was by, and what company went to the sea side with me.

As to the collection of Sertularias, I had them by means of the Secretary of the General Post Office; and my sister, who is curious, and was then at Dublin; so that I do not know that I am indebted to him even for one single specimen, though he had hundreds of me, both of the animal and vegetable world; but I shall trouble you no more with such an unworthy creature.

I received the skin of the *Mud Iguana* of Dr. Garden. This animal is certainly a *larva*. Those that were small, of about seven or eight inches long, have three appendages on each side, in the room of fins, but they are not opened sufficiently, as in the large skin, which is 30½ inches long. Here they look like the fins of the *Lacerta vulgaris* in its *larva* state, and are pinnated on both sides; but the reason why he takes it for a new genus is, that there is no appearance of the hind legs in the least, but the fore ones are very distinct and plain, with bones and nails. It has a remarkable palate, full of rows of teeth. I believe you will find, when I send you a specimen of it, that there is nothing new in it.

I long to have your opinion about sponges. I am intirely against Peysonell's account, in the Philosophical Transactions, vol. 50, p. 592. I am persuaded, whatever animal appearance these bodies have, they are never detached from the body itself; but I never could meet with polype-like suckers in sponges, only *papillæ*.

There are so few people go abroad that are capable of judging, that I have but one friend I can depend on, who went to Guadaloupe. He declares the sponges, which are remarkably fine there, have no appearance of protruding any animal like heads, as suckers. I shall be glad to have your sentiments. My business has kept me close prisoner, so that I have not been able to get to the sea side yet; but I live in hopes to be able very soon to obtain that pleasure. Poor Kuhn has been very ill of a Pleuritic fever, but is now crawling about. The loss of the pacquet from Pensacola has deprived me of many fine and rare curiosities, but I live in hopes of my friends sending me more. My best and sincerest wishes attend my honourable and worthy Friend. I am most truly yours, &c. &c.

JOHN ELLIS.

MY DEAR FRIEND, London, September 10, 1765.

Your very kind letter of the 15th of August I received yesterday, inclosing one to Dr. Garden, which I shall forward, with one of my own, to him by the first opportunity, which I daily expect. The Provinces of the two Floridas afford most certainly an ample field for the wonderful productions of nature; and if one of your pupils were to travel through them, they might be properly described; but, to my misfortune, the people that are there are totally ignorant of Natural History. However I do not despair, as there is now going out a Lieutenant Governor that seems to have some taste, and has promised me faithfully to send me what curiosities he can. I had a letter very lately from thence, from a person who lived long in Georgia, who informs me, that the plants about Pensacola and Mobille are much the same, but that the climate is worse, and that they do not come to so great perfection. I have the greatest hopes from the Banks of the Missisippi, or from the Southern part of East Florida, which extends to 25 degrees, and is more Southward.

I have just now received a letter from Doctor Garden. He has left off his business to take care of his health, and promises to be a very good correspondent for the future. I am sorry to tell you he sent me no specimens of the new genera of plants, only the descriptions, which I shall take the liberty to tell him will not answer your end. I find he has sent them to an ignorant fellow here in Botany, one Robertson, gardener to our Queen; so that I must desire him to send you other specimens with the descriptions, if he expects you to take notice of them. His new animal, which he calls "*Mud Ignana,*" is no more than a larva of a large kind of *Lacerta.* He has sent me a dried specimen that is 30 inches and half long, and has yet but two feet; his other specimens, which are preserved in spirits, are about nine inches long. Having last year met with some larvas of the *Lacerta vulgaris,* I observed little fins like wings come out on each side, a little above the forefeet; and that after I had kept them some time in water, they cast their skins, and then crawled out of the water and became terrestrial animals, the fins soon disappearing. The fins of the large dried specimen exactly agreeing in shape and situation, confirms my opinion, that they will not leave the water till they get their hind legs. As soon as I can get a drawing made of this animal, I shall send you this specimen. However, when I send you the one preserved in spirits, you will plainly discover that it is a larva.

I called just now at our Museum Britannicum on Solander; he desires me to assure you, that he never received those letters, nor Mr. Lee the one that covered them.

I have not yet been at the sea side, but still hope to get there this year before winter comes on. We have had the hottest and driest summer that I remember. We have good wheat, but are scarce in hay, oats, barley, and peas.

There is now growing, at the Museum Garden, a plant of the *Mirabilis*, between the *Ipecacuanha* and the long-tubed one. I leave this to you whether this is a variety or new species: for my part I am puzzled much at these sportings of nature. The leaves have stalks not so long as the *Ipecacuanha*, and the long-tubed ones are sessile.

I have often wished to know from what tree or plant the Russian matts are made, of which commodity an infinite number are used in England. Some say it is from the *Tilia;* some from the *Betula*, or birch; some from the *Salix viminalis*. You can determine it, which will do us pleasure, when we receive your authenticity to the account.

I have got a few seeds of a new genus, No. 7. of Dr. Garden's, which I inclose you in this letter. Solander takes it to be a *Gerardia*. If they are sown now, and kept in shelter all the winter, they may come up in spring.

I rejoice to find that your Tea plant is alive; you are much in the right to protect it from severe weather. I have made another attempt to get the Tea seeds from China, preserved in common bees wax.

I should not have failed the last time, but that I forgot to tell my friend to expose them awhile, to take off the superfluous moisture. This I have done now, and if he can get them, I am sure he will send them by the ships that are expected to arrive here in June next.

I wish you would make experiments this year on oak acorns wrapt in wax. Involve some before they are sweated, others after, and you will find the experiment to succeed infinitely better in the latter manner than the former. I have now an oak growing whose acorn was inclosed in wax in February, and not planted till the February following; whereas those that were inclosed in the beginning of November all rotted, from the moisture that was thrown out on the surface, under the wax.

Your method of the salts saved many curious seeds that were inclosed by your directions; but I believe Mango stones, and the Tea seeds, with many more, may be preserved better in wax. I wish Doctor Russel would direct his friends at Aleppo to send some of their rare seeds in this manner, such as the *Arbutus Andrachne*, and many other curious seeds, which lose all vegetation by the time they come here, as much as East India seeds.

I should be glad to obtain seeds or a plant of the *Ficus sycomorus* for our American plantations; this might be worthy our attention: but few people have the same warmth for extending Natural Philosophy and History as I have.

If Providence had made us so happy as to have placed you here, we should long ago have exceeded all the world. I shall write to our mutual friend Dr. Garden to attend to your correspondence, which I am sure he will do, esteeming it the greatest honour of his life. My best wishes attend you; may you long live, a true light, to point out to us the only and best method to investigate nature.

I am, dear sir, your much obliged and obedient servant, JOHN ELLIS.

LINNÆUS TO ELLIS. [Latin.]

Upsal, September 24, 1765.

I received your letter, without a date, and rejoice in the continuance of your accustomed regard for me. Mr. Buttner's conduct is ridiculous, and more worthy of pity than of anger. You have taken so lofty a rank in science, by your discovery concerning Corallines, that no vicissitude in human affairs can obscure your reputation. He, poor man, destitute of any discoveries of his own, tries to steal that celebrity, which he will be forced, with much ignominy, to resign. Any one must be blind, and a novice in science, who does not perceive, that what you advance is all your own. This is not the first instance of such conduct among germans. It is otherwise with honourable englishmen. I shall write to Baron Munchausen about Buttner's ill behaviour. Those who have met with him, all speak of him as a very singular man, whom nobody can well understand. He is reported to be a misanthropic character, of whom nothing good is known.

I care little about the *larva* of the *Iguana*; but our mutual friend Dr. Garden mentions some dried plants, destined for me. These I shall be very glad to have, whenever they come to your hands.

I rejoice at Mr. Kuhn's recovery from a dangerous pleurisy. Has he been in France? How can I direct to him?

I wonder whether Mr. Lee has received the letter I sent him last summer. If you chance to see him, please to ask him for the *Sanguinaria*, which I know is to be had in England, though I have not received it from any of my correspondents.

We have never known in Sweden a more rainy summer than this last. There were scarcely eight days of serene and warm weather.

Tell Mr. Solander that Mr. Neander, well known to him, died this day. Adieu.

ELLIS TO LINNÆUS.

SIR, London, October 29, 1765.

I have had the pleasure to receive your very kind

letter of the 24th of September, wherein you express the regard and esteem of a true friend. I have received more honour and encouragement from your good opinion of me, to pursue natural history, than from the Royal Society of London, or any person else.

I am greatly obliged to you for supporting my character against that insolent plagiary Buttner, and thank you most kindly for your promise to write to Baron Munchausen. This will be of more service to me, than the certificates I have got from the Secretaries of the Royal Society, *that he did not give in any paper to the Society relating to corallines, or any of those marine bodies which he says he did*; and it is well known to all the members of the Royal Society, lovers of Natural History, that during the time he was in England, nor at any other time whatever, did he ever give in any account whatsoever. Your friend Peter Collinson particularly knows this to be true, and knows that he is a bad man.

I have been confined this month to my chamber by a violent catarrh; but I hope, by the advice of Dr. Fothergill, my good friend, to re-establish my health. I am now looking into the nature of sponges, and think by dissecting and comparing them with what I have seen recent, and with the *Alcyonium Manus mortua*, that I can plainly see how they grow; without trusting to Peysonell's account of them, which is printed in our Philosophical Transactions, wherein he pretends to tell you, that he takes the animal out of them, that forms them;

and that he put it into them, and it crept about through the meanders of the sponge. This kind of insect, which harbours in sponges, I have seen; but sponges have no such animals to give them life, and to form them. Their mouths are open tubes all over their surfaces, not furnished, like the tubes of the *Alcyonium Manus mortua*, with polype-like mouths or suckers. With their mouths they draw in and send out the water; they can contract and dilate them at will, and the Count Marsigli has (though he thought them plants) confirmed me in my opinion, that this is their manner of feeding. If you observe what he has wrote on sponges in his Histoire de la Mer, and the observations he has made on the *Systole* and *Diastole* of these holes in sponges, during the time they are full of water, you will be of my opinion. Take a lobe of the officinal sponge, and cut it through perpendicularly and horizontally, and you will observe how near the disposition of the tubes are to the figure I have given of the sections of the *Alcyonium Manus mortua* in my plate of the Sea-pens.

I am collecting all those zoophytes you call *Corallinæ*. I have met with several species and varieties from the Bahama Islands. It is to shew Dr. Job Baster that he does not know the difference between a Coralline and a Conferva, when he says, he can prove, that all which you call Corallines are no more than Confervas.

Dr. Walker of Edinburgh is now here. He tells me he has collected a great variety, in his voyage round the sea coast of Scotland, of all the submarine productions. I believe he has a view to writing a Natural History of Scotland. The people of that part of this island, seem fonder of it than the English.

I wrote a week ago to Mr. Lee about the letters you sent him, and likewise begged of him to send you the *Sanguinaria*.

Our friend Adam Kuhn is now at Mr. Pitcairn's, a merchant's in Edinburgh, Scotland; I do not doubt but he will promote the study of Natural History there.

By a Swedish ship lately sailed to Stockholm, you will receive a bottle with a *larva* of the *Lacerta Iguana*, and Dr. Garden's account of it.

The Doctor wants much to know your opinion of a specimen of a shrub, which was sent to you about the same time with the specimens of the Fish, in the year 1760. At the latter end of the descriptions of his Fishes, he describes the plant. Dr. Solander and I both saw the specimen, and I have by me now a good drawing of it and some seeds—it is *decandria monogynia*. The seed is a nut. The flowers are white, and grow in spikes like the *Ribes*. The leaves of the Plant are like the *Styrax* *.

I should be glad to know if you continue in the same mind about disposing of your Tea tree, and what I shall call the value that you put upon it. I do not doubt but that there are many persons here who would be glad to purchase it if they knew the price.

* This is *Styrax grandifolium* of Solander in Ait. Hort. Kew.

Pray is the plant of the true *Ipecacuanha* yet known? If it be, I should be glad to know what it is.

Lord Hillsborough received last summer from Madeira, seeds of the Dragon Palm, and they are as they are described, about the size of a pea, quite round, and horny, like the seeds of Palmetto from South Carolina. I shall write for some specimens of the flowers. I am not of Mr. Lœfling's opinion that they belong to the *Asparagus*.

My best wishes attend you; and when you have leisure to write, nothing can be more agreeable than your elegant and instructive letters. I am, dear Sir, with sincere regard, your most affectionate humble servant. JOHN ELLIS.

I have told Dr. Solander what you desired; he presents his compliments to you.

LINNÆUS TO ELLIS. [Latin.]

Upsal, Dec. 27th, 1765.

I received your letter of the 29th of October, my worthy friend, with great pleasure, and return you my best thanks for it, as well as for Dr. Garden's *larva* of a Lizard, which reached me yesterday.

That little twinkling star Buttner is scarcely worth mentioning; for if he had any brilliancy, which he really has not, it would vanish before your meridian splendour. What mortal, when reading or investigating your performances, can think of him?

There is lately come to my hands, from one of

our Swedish lakes, a most beautiful *Spongia*, though in a dried state, in which I can distinctly see some *animalcula* in their transparent vesicles. The branches of this sponge, when brought to the flame of a candle, take fire with a bright effulgence, and the animalcules explode in little fiery globules of a very lively blue.

I wish Mr. Lee would not forget to send me seeds of *Sanguinaria*. I wonder he has not answered my letter. The *larva* of the *Iguana* resembles the *Siren* of the old writers, and is really a great curiosity.

What plant yields the *Ipecacuanha*, I am altogether ignorant. It seems to be a species of *Viola*, and of that tribe whose flowers are erect, to which Lœfling gave the name of *Calceolaria*, and Jacquin that of *Hybanthus*. Possibly Marcgrave mistook the three valves of the capsule for three berries. Barrere, truly a most indolent botanist, in the french West Indies, is the only person who calls this plant a *Viola*. Sometimes a blind hen meets with a grain of corn. We know from the observations of the older botanists, that the calyx and roots of *Viola odorata* have an emetic quality. See on this subject my dissertation on indigenous purgatives, just published.

I have not been able to make out many of Dr. Garden's genera, for want of the plants themselves.

If he had sent specimens with his descriptions they might have afforded much botanical information.

I have received the flowers only of the *Sanguis*

Draconis, which are as like those of *Asparagus*, in size and appearance, as one egg is to another. The seeds are solitary in each cell of the fruit; but in the *Asparagus* there are two together. This is the only mark of distinction I have found.

Dr. Garden's plant with ten stamens did not succeed with me, nor do I remember having seen any thing of it.

Erica Dabeoci was sent by Peter Collinson; a fine specimen which much delighted me. It is truly an *Erica*, though so unlike the rest *.

I have very distinctly seen the flowers of *Salicornia* and *Chara* to be monandrous.

Concerning the *larva* of the *Iguana*, the circumstances which principally excite my surprize are the claws on the toes, which in every other *larva* of a lizard, that I have ever known, are wanting; the *branchiæ* or gills, which do not exist in Water Salamanders, presumed to be the *larvæ* of lizards; the sound or voice which should seem not suitable to a *larva;* and finally the situation of the *anus*.

If this animal does not undergo any metamorphosis, it must unquestionably be ranged with the *Nantes*. Nature is so manifold in all her works, that we can determine nothing certainly *à priori*. Who would not suppose the common Bed Bug a *larva*, about to undergo a metamorphosis, if the contrary had not been determined by observation? Whether the animal in question may or may not be in its final or complete state, it is a very curious production.

May health and happiness long attend you!

* *Menziesia Dabeoci. Sm. Compend. Fl. Brit.*

ELLIS TO LINNÆUS.

DEAR SIR, Grays Inn, London, January 31, 1766.

I received your instructive kind letter of the 27th of December, with one enclosed to our mutual friend Dr. Garden, which shall be forwarded the very first opportunity. Your reasons for thinking the rare animal (Mud *Iguana* as he calls it) to be quite a new genus, are very convincing to me. I forgot to take a copy of his description, which I wish you would send me in your next, for the satisfaction of many curious people here.

I find in Hasselquist, that the *Lacerta Gecko* makes a noise not unlike the croaking of a frog, which is another probable reason you give of our animal being come to its perfect state.

I have desired Mr. Lee's partner in trade, whom I lately saw, to put him in mind of the roots of the *Sanguinaria* for you. I shall write to him to send them in a small box, with earth and moss, for you, to my care.

I have wrote to Dr. Garden about sending you what specimens of his new plants he has left; for he says he sent them to a person in London who I know has no taste. If I can find that person, I will endeavour to get them.

There is now printing in Holland a book on Zoophytes by Dr. Pallas of Berlin, who was two years ago in England, but has since resided at the Hague, where he has access to the Prince of Orange's cabinet. This gentleman, I find by one Mr. Brunnich, a Danish gentleman, sent over here by that court to travel for the sake of Natural History and agriculture, has treated both you and me with a freedom unbecoming so young a man. Dr. Solander has seen the book, and has advised Mr. Brunnich to write to Dr. Pallas to reprint that part, which makes so free with you. As to me, if he does not act as a gentleman, I shall take particular notice of his criticisms in the book I propose to publish, which I hope will travel as far as his. For my part, I know my errors, and should have taken notice of them in my second volume; for I never attempted at being systematical. All I ever aimed at, was to observe those rare things in nature that were before unattended to; and to leave such eminent persons as you to arrange them properly. This behaviour is one of the obstacles to the investigation of nature, and deserves to be taken notice of; for no checks should be given to such laudable pursuits. Dr. Job Baster of Zurichzee and this man are particular friends; and I find Pallas has used me with so much ill-nature, because I exposed the absurdities of Baster's doctrine and experiments, in our Philosophical Transactions.

Dr. Solander was yesterday at our famous anatomist's Dr. Hunter's, to see four specimens of this new animal Mud *Iguana*; but none of them are so large as the dried specimen that I have, which is 31 inches long. He has dissected one of them, and says it has the same kind of teeth, and is, in every respect, like our *Lacerta* (called in english Eft or Newt) in its aquatic or larva state, and differs in nothing from it, but in having only two feet. Dr.

Hunter says that the *larvæ*, both of our *Lacerta*, and the *Iguana*, have gills and lungs. Dr. Solander, Mr. Brunnich, and I dined together this day, and drank your good health. Mr. Brunnich proposes to write to you next post.

I shall forward by the first ship your letter to Dr. Garden inclosed in my letter. He seems uneasy at not having his Diploma from you. I wrote to him to tell him you would give any satisfaction he pleased on that head.

I have some specimens of what were sent here as species of *Corallina* of your system, which I must send you to examine. The limits between the animal and vegetable world are difficult to determine, more particularly in this genus: but I shall be happy for them to undergo your examination. I have likewise some rare *Fuci*, which I have for a long time determined to send you. Now I am resolved you shall have them by the first ship.

My best wishes attend you, assuring you, dear Sir, that I am and shall always acknowledge myself greatly indebted to you, and with great respect your most obedient humble servant, JOHN ELLIS.

SIR, London, Oct. 21, 1766.

By a ship just arrived from South Carolina, I have received a letter, dated August 5th, from our friend Dr. Garden, with one inclosed in it for you. This he has left open for me to read, as it contains the descriptions of nine of those plants he calls new genera. There is nothing new in the letter, only he advises that he has sent you the specimens of those new plants in a pacquet, some few insects, and a bottle containing one of his new animals, your *Siren*, and three fishes, one of which he says is new. In the same letter, and in mine, he says that he has met with the Siren from four inches to three feet and half long, and that he never observed any difference, so that it must be a distinct genus of animals. These things, as soon as I receive them from on board the ship, I shall forward to you by the best method I can think of, together with his letter, which is rather too bulky for the post. He further says he has received your two letters which I inclosed him, viz. one of the 15th of August and one of the 27th of December, 1765; but that your letter of the 19th of May, 1765, which you mention, never came to his hand, and he therefore desires you, if there was any thing new in it, to send him a copy of it.

Since I wrote to you last I have been endeavouring to investigate the nature of your genus of *Corallina*, and have lately added several new species, some of which I have received from the newly ceded Islands in the West Indies.

I am the more solicitous about them, as I find the French and Dr. Baster seem to think them vegetables; particularly Dr. Baster supposes them nothing but *Confervæ* incrustated. At the same time I am endeavouring to shew the nature of some of the *Confervæ*, I have discovered in the *polymorpha* male parts in one plant, and what appears like an *amentum*, and capsules full of seeds, in another. The same I have seen in the *Conferva plumosa*. There is a bad figure of it in Plukenet, Tab. 48. fig. 2. Dillenius has given a better figure, ed. 3rd of Ray's Syn. Tab. 2. fig. 5, and calls it *Fucoides purpureum eleganter plumosum*. I have found the same male and female parts in the *Fucoides rubens variè dissectum*, of R. Syn. p. 37. I have a *Conferva* with fruit like a *Rubus*, when it is seen through a microscope, and this fruit is surrounded with petals, and sits on a footstalk. I expect to have the plate of these curious *Conferva* engraved, before Christmas: as also a figure of the *Coluber Cerastes* from a curious specimen I presented to the Royal Society. It was given me by Dr. Turnbull, who remembers poor Hasselquist. He brought it from Alexandria, where they say it is poisonous.

Mr. Gregg, who formerly sent me the kidney-shaped Sea-pen from Carolina, has lately sent to Lord Hillsborough, a curious collection of specimens of plants, which he gathered himself in the islands of Tobago, St. Vincent, Granada, and Dominica. Among them are the *Brownea* and *Sloanea*. The latter agrees with Lœfling's description exactly, so that what Miller has seen must be a chesnut. Besides these plants, there are some curious birds and submarine animals. When Dr. Solander and I have looked them over, they are intended for the British Museum. There is among the sea productions an *Actinia*, that arises, many together, in a row from wrinkled tubes, which tubes adhere firmly to rocks, shells, &c.; this seems to come near to the *Tubularia.*

I shall send you a specimen of them preserved in spirits. I should be glad to know what you call Bohadsch's *Hydra;* for there are many of them in that part of the world: and likewise what you call his *Tethys.*

In this *Actinia* there are two rows of *tentacula,* that surround the mouth at the top of it. The lines that run up and down are the tendons of the muscles by which it extends and contracts itself.

While I am writing this letter, Solander is examining some of Lord Hillsborough's plants, and among them is the *Jacquinia armillaris,* or *Chrysophyllum Barbano* of Lœfling. He says Lœfling's description is more accurate than any other, especially in the figure of the stigma and insertion of the filaments.

This is called the Dwarf Gum-tree, and is common on the sea coast in all our new islands.

N. B. Your description of the Calyx is most exact of any.

I wish you would continue to write to me often: it will make me spare more time than I have done for the amusement of so agreeable a correspondent as you are, and one whom I honour and regard so much.

Before I conclude I must tell you, I have met with a most remarkable sponge near Chichester, at Emsworth in Sussex, in company with Mr. Brunnich, which I shall soon send you the plate of. Mr.

Brunnich is now in Cornwall; he met with an accident in Wales from a fall from his horse, which has detained him longer than he intended.

My best wishes attend you; and be assured of the love, honour, and esteem of, dear Sir, your obliged humble servant, JOHN ELLIS.

I have a new *Pennatula*, or Sea-pen, the largest and most elegant I have seen, from the East Indies. I have had a drawing made of it; it is about ten inches long. I intend to have it engraved as soon as I can get a proper person to do it.

SIR, London, Dec. 5, 1766.

I am obliged to you for sending me Dr. Garden's account of the *Siren*. I am sorry I could not get the rest of the things he sent you, before the ship sailed, when I sent you the specimens of plants. I have only got the insects, which are of little value, and the skin of a *Siren*. The things in spirits are not yet brought on shore; but I hope to get them; and as soon as I have an opportunity, will send them to you. Peter Collinson spent the evening with me, and shewed me a letter you wrote to him about funguses being alive in the seeds, and swimming about like fish. You mention something of it to me in your last letter *. If you have examined the seeds of them yourself, and found them to be little animals, I should believe it. Pray what time of the year, and what kinds? I suppose they must be

* This letter is missing.

taken while growing and in a vigorous state. I intend to try; I think my glass will discover them, if they have animal life in them. The seeds of the *Equisetum palustre* appear to be alive by their twisting motion, when viewed through the microscope; but that is not animal life.

I have just finished a collection of the *Corallinæ*. I think there are 36 species; but I believe some of them will prove varieties. I have most of the copper plates that represent them finished. They are the most difficult to examine of all the Zoophytes; their pores are so small, and their manner of growing so singular. I have got a copper plate of the *Coluber Cerastes* finished. I send you inclosed a copy of it; I have not yet given it in to the Royal Society. I had two specimens sent me from Egypt. I sent one to the Royal Society, upon condition that it should be drawn and engraved, because the figures in Alpinus are very indifferent. The person who gave it to me says it is accounted very venomous in Egypt. The teeth of these specimens were taken out, for I could find none. It is drawn exactly of its natural size.

I have just received an account that I have a pot of seeds inclosed in wax, from the Northern parts of China, which is now on board an East India ship lately arrived. I hope they will prove curious. They are for a present to her Royal Highness the Princess Dowager of Wales, to be taken care of by Mr. Aiton, who is the best gardener for exotics in England.

As soon as I get a copper plate engraved of my new *Actinia* I will send you a copy. I hope it will be done in two months at farthest. It is the most curious discovery among the Zoophytes (unless your *Fungi)* that has appeared for some years past. I think, from their rising out of this creeping tube, by which they adhere, they join the Actinias to the Sertularias. Dr. Fothergill has sent me a fine specimen of the *Isis Hippuris*, with the true natural fleshy or calcareous whitish covering. The whole surface is full of holes, where the polype-like suckers protrude themselves, each having eight rays.

Pray let me know how your Tea-tree grows. It is very odd that, notwithstanding we have had 15 ships from China this year, we have not had one Tea-tree brought home alive. I have sent a boy to China, whose dependence is on me, to try to bring over several sorts of seeds in wax. I expect him home next summer.

The English are much obliged to you for your good wishes. We every day see a superiority in the Swedes over the other European nations. All your people that appear among us are polite, well-bred, and learned; without the vanity of the French, the heaviness of the Dutch, or the impudence of the Germans. This last nation has intruded on us swarms of their miserable, half-starved people, from the connexion that our Royal Family have had with them.

Let me hear often from you; it will oblige me to write, and find out something new to entertain you.

Sir, I am, with great respect, your obliged and affectionate friend, JOHN ELLIS.

LINNÆUS TO ELLIS. (Latin.)

Upsal, Jan. 1, 1767.

I am every day receiving letters, for the most part, if not altogether, to little purpose; filled with compliments indeed; but when I review the accumulation of the last year, I derive, on the whole, but little real benefit. Yours indeed are always rich in agreeable and learned information, like your last of the 5th of December.

With regard to *Fungi*, you may pick up, in most barns or stacks of corn, spikes of wheat or barley, full of black powder, which we call *ustilago*, or smut. Shake out some of this powder, and put it into tepid water, about the warmth of a pond in summer, for three or four days. This water, though pellucid, when examined in a concave glass under your own microscope, will be found to contain thousands of little worms. These ought first to be observed, to prevent ocular deception. In mould, *Mucor*, you will find the same, but not so easily as in the larger *Fungi*. If, in the course of from 8 to 14 days, the water has been kept up to the same temperature, you may observe how these minute worm-like bodies become fixed, one after the other, and acquire roots. I have just printed a dissertation on the Invisible World, which shall be sent you by the first opportunity. These chaotic worms are nearly akin to the last species of animals which I have placed in my *Systema*, under the genus *Chaos*.

Your *Coluber Cerastes* pleases me much. It differs from that drawn by Petiver. If it really has none of the larger teeth, or fangs, it cannot be venomous. But I beg of you to open its mouth, so as to examine carefully whether the horns be not artificial, formed perhaps of a couple of bird's claws, forced through the *cranium* of the serpent. I wish also to know how many *scuta*, or broad scales, it has under the belly, and how many pair of lesser ones under the tail.

Yesterday I received the plants constituting Dr. Garden's new genera, which you have so kindly transmitted, and for which I return you my best thanks.

I have long been well acquainted with the elastic or jumping seeds of *Equisetum*, described by Staehelin in the Paris Memoirs; and still longer with the elastic seeds (or rather capsules) of Ferns in general, known for above a century. These, having no real vital motion, are totally different from the vermicular bodies of *Fungi*, which are truly alive.

I am afraid your Chinese seeds will not grow. The best method of procuring plants from China would be, to have them sown in pots of earth the day the ships leave Canton. They will then germinate during the voyage, and you will be much more likely to obtain living plants than by any other method; as I have advised in my dissertation on Tea. My tea plant is alive, but has not flowered, nor does it seem to bear our climate so well as heretofore.

I thank you for the *Sertulariæ*, all the species of which I owe to you. These are your pledge of immortality, of which no revolution can deprive your name.

I have seen a small branch of *Isis Hippuris*, with its bark full of star-like pores; but the specimen was a miserable one.

Numerous *Actiniæ* have been sent me from Iceland. No department of natural science has been less successfully cultivated than what relates to the *Mollusca*. I have also, from the same country, the *Conferva plumosa*.

Please to forward Dr. Garden's letter when convenient.

May the new year bring you all possible happiness!

Upsal, April 28, 1767.

By one of our divines going to Pennsylvania I send you a few small dissertations, 57 in number. If any of these are not interesting to you, I beg you will distribute them, as well as the duplicates of the rest, among our friends, especially Dr. Garden, Mr. Collinson, &c.

I am sorry my printer has not yet let me have the first volume of the new edition of my *Systema*, containing the Animal Kingdom. The last sheet of the index alone is not yet finished.

Yours in haste.

ELLIS TO LINNÆUS.

SIR, London, May 29, 1767.

Among the Zoophytes there are none so difficult to investigate as the *Corallinæ;* for this reason, it is so hard to persuade mankind that they are any thing else but plants, and of the same nature with incrustated *Confervæ.*

Few people have either eyes or proper microscopes to examine them; and those who have analyzed them chemically have gone through their experiments in so careless a manner, and with so little attention, that it is no wonder bodies, of so fine a texture, should pass for others, little removed from them in outward appearance.

'Tis owing to this cursory view of them, that Dr. Baster is so positive, that all your *Corallinæ* are *Confervæ.* Vide Phil. Trans. vol. LII. p. 111.

Dr. Pallas, in his article of Corallines (vide Pallas Zoophytes, 418), depending on Count Marsigli's chemical analysis of them, considers them as vegetables. But, if we observe how Pallas has confounded the calcareous crust of Corallines, with the farinaceous covering of vegetables, it will be no longer a matter of surprise: for had he put the true Corallines into an acid menstruum, and the *Fucus pavonius,* which he calls *Corallina pavonia* (vid. Pallas

Zoophytes, p. 419), and the *Lichen fruticulosus*, &c. of Meese Flora Frisica, which he calls *Corallina terrestris* (vid. p. 427), he would have found that the true Corallines would ferment strongly, and the *Fucus* and *Lichen* would not be in the least affected.

It is confessed that the pores of Corallines are very minute, and the difficulty is very great to come at them (while alive) to see their suckers in motion at the sea side: and these are very plausible reasons, which he draws from Dr. Jussieus' observations on them while on the sea coast. But if he compares them with animal bodies next above them in the arrangement of Nature, he will soon alter his opinion.

Let him examine specimens of his *Millepora calcarea*, and *Millepora agariciformis*, Pallas Zooph. p. 263 to 265 (which he admits to be animals), by dividing them longitudinally, and viewing them in the microscope before that he has put them into an acid menstruum, as vinegar, &c. and at the same time let him do the same by the *Corallina officinalis*, and he will find the internal texture exactly the same, and agreeable in form to what he will find in Ellis Corall. plate 27. fig. D. This was the appearance that fragments of the *Corallium pumilum*, &c. and the *Corallium lichenoides* made to me in the microscope. And in plate 24. fig. A. Ellis Cor. he will observe the same manner in the arrangement of the fibres leading to the pores, both of his two *Milleporæ* and the *Corallina officinalis*, after they have been in vinegar. In the same figure A. he will observe the matrices, like seed vessels; some coming out of the end of the branches, and three, like *papillæ*, adhering to the branches. These he thinks another argument to prove them Vegetables. But let him look at his specimens of *Millepora agariciformis*, and he will find many of the very same seed-like vessels on them. On my specimens from Cornwall, of this very same kind, I have seen many of these *papillæ*, which proves to me that they differ only in their manner of growing, some being jointed, and others spread flat. I mean the *Corallina officinalis* and the *Millepora agariciformis*.

Indeed I have observed that the base of the *Corallina officinalis* terminates in a calcareous substance, exactly like the *Millepora agariciformis*, spread upon rocks and shells, to which it adheres. I have many specimens to prove it.

So that there is no more reason why we should think them vegetables from these *papillæ*, than the vesicles being on the *Sertulariæ* (which have a great resemblance to the fructification of land Mosses) prove the *Sertulariæ* to be vegetables, when all the world now knows that they are animals.

About the year 1755, at a meeting of the Premium Society, an accident happened that seemed to give demonstration to the gentlemen then present, of the evident difference between animal and vegetable substances.

A gentleman of Wales, who corresponded with

the Society, had sent up some specimens of the *Lichen tartareus*, as a proper material to answer the purposes of dyeing, instead of *Lichen Roccella*. The Society being desirous of being informed of the nature and appearance of the true Orchell, or *Lichen Roccella*, some gentlemen undertook, against the next meeting, to bring some specimens; which they did, having got them from the Orchell dyers. At the same time Dr. Manningham, willing to be very exact, applied to Mr. Miller, as understanding plants better than most people. Accordingly he presented the Society with a paper, on which was wrote, by Mr. Miller, Orchell, and which contained a large specimen of the *Corallina nervo tenuiori fragiliorique, internodia nectente*, of Sloane, Hist. Jam. vol. I. tab. 20. fig. 4. It being presently controverted which was the true Orchell, as Mr. Miller's specimen differed so widely from the rest in appearance, I proposed to the Society that burning of them both, one after another, in a candle, before the gentlemen present, would convince them of their being of different kingdoms of Nature; for that the *Lichen* as a vegetable would smell like any common weeds burnt, and that the Coralline would give a disagreeable smell, like burnt bones: which experiment was immediately tried, to the entire satisfaction of the Society.

But to carry the proof still further by a regular chemical process: —

I lately procured a parcel of the *Corallina officinalis* from the sea coast near Harwich, and have got it most exactly analysed by my friend Mr. Peter Woulfe, F. R. S. a very ingenious chemist. The process is as follows: — Twelve ounces Troy weight were picked perfectly clean and put into a stone-coated retort; this was set in a reverberatory furnace, and an adopter and quilled receiver luted to it.

The fire was very gentle for the first eight hours; in which time there came over half an ounce and 18 grains of a transparent and most colourless liquor, which was set aside to be examined afterwards. The fire was then increased; and in six hours time there was distilled two drachms and three grains of a turbid liquor, which had some marks of oiliness on its surface. This was likewise set aside. The fire was then increased for six hours longer; and during the last two hours the retort was red hot all over, which ended the distillation.

In this third and last process the portion of liquor that came over was more turbid than the second; and some of it, from the redundancy of its volatile alkali, was crystallized. It also contained better than a drachm of a light empyreumatic oil, very much resembling the smell of hartshorn. In the recipient there were also some small crystals of a volatile alkali. The whole of this last product weighed three drachms and a half. The caput mortuum was quite black, and weighed ten ounces one drachm and one scruple; so that there was a loss of 4 drachms and 49 grains, out of the 12 ounces of Coralline.

The first liquor that was distilled, mixed with spirit of salt, slightly effervesced; and, mixed with syrup of violets, turned green; both proofs of a volatile alkali.

The second and third portions effervesced strongly with spirit of salt; as did also the volatile salt which came over into the receiver; evident proofs of its being a concentrated alkali.

Had this distillation been conducted in a hurry, there would have been no concrete volatile alkali; for then this would have been confounded and dissolved in the first liquor that came over.

In order to examine this Coralline still more minutely, it is proposed to dissolve the calcareous matter of several pounds of it in an acid; and when the remaining membranaceous and fibrous part is washed clean from the acid, till the syrup of violets no longer turns red by being mixed with it, then to distill it. This process would give still stronger evidence of its animal nature.

I am, dear Sir, most truly yours,

JOHN ELLIS.

P. S. I have received your kind favour of the 1st of January, and am ashamed I have not answered it. I have a great deal to write to you, but must defer it to next week.

The horns of the *Cerastes* are genuine, and of the same nature with the scale at the tail, pliable like the small part of a pen.

SIR, London, June 5, 1767.

Dr. Job Baster asserts positively in the Philosophical Transactions vol. 52. p. 111, that all the *Corallinæ*, that you and I have described, are *Confervæ*. As he is not acquainted with the different characters of them, I think it necessary to give you my idea of both.

Conferva est planta, filamentis vel simplicibus, vel ramosis, articulatis; fructificationes, vario modo dispositas, habens.

Corallina. Animal crescens habitu plantæ. Stirps fixa, e tubis capillaribus ramosis, per crustam calcaream porosam sese exserentibus, composita; sæpe articulata, semper ramosa: ramis vel divaricatis, liberis; vel conglutinatis, connexis.

In order to examine more minutely into the nature of some of the fructifications of the marine *Confervæ*, I have had figures drawn from the microscope of several specimens in water. Among the rest I have met with two kinds that appear to me to be *Diœciæ*, having some amentaceous-like spikes on one sort, and capsules with seeds on the other. One kind of these is the *Conferva polymorpha*, represented at fig. d. and e. and the magnified figures at D. and E. The figure E. 1. represents the amentaceous spike higher magnified. The other is the *Conferva plumosa* at fig. b. and c., and the magnified figures at B. and C.

Besides these, to shew the manner of the fructification of three other kinds, I have given the magnified and natural size of a part of the *Con-*

ferva flosculosa, with pedunculated flowers, at fig. a. A. That of the *Conferva geniculata*, with verticillate flowers, at fig. g. and G. That of the *Conferva Plumula* at fig. f. F. and F. 1., and have added another without fructification at fig. h. and H. on account of the singularity of its articulations, which I call *Conferva ciliata*. My plates of Corallines are almost finished; as soon as they are you may depend on copies of them. If you look into our Philosophical Transactions, vol. 47. tab. 5. you will see on the *Mirozoon* or *Pseudo corallium*, &c. which Donati describes (and which Pallas calls *Millepora truncata*), all the pores covered with *opercula*, and likewise when these are open, the polype-like sucker is funnel-shaped. I have observed lately something very like this on the surface of the joints of the *Corallina Opuntia* kinds, where all the cells are covered with convex *opercula*, and in each cell there appears to be the remains of a funnel-like sucker. There is a figure of this, but not well represented, at Pl. 25, fig. a. A. and A. Ellis. Cor. but no figure of the *opercula*, which I have since had an opportunity, by means of recent specimens, of describing more exactly. This will shew you how near the *Millepora* and *Corallina* approach.

The print which I inclose you is the only one I have, but you shall soon have more when your Swedish ships return. I have above 30 species of *Corallina* (I reckon 36); but no doubt many are varieties.

The insolent manner in which Pallas treats us, will make me exert myself to shew him that he is not infallible.

I have been so much employed about public business, that it is with difficulty I can steal this time to write to you, whom I so much love and esteem for your generous manner of treating, not only me, but all mankind.

Dr. Solander (who dined with me to-day) rejoices that I have found, upon dissection of my *Pennatula Cynomorion,* that there is a bone in it; but this bone is not so long as in the other species. This is another objection to the infallibility of Dr. Pallas, who says it is without a bone.

As I find opportunity, I shall examine all his genera, and shew him that he is a mere compiler, and no inventor.

I have by me, to send you, a fine specimen of the Siren from Dr. Garden, and some fishes: also a few specimens of insects from him: but these are in bad order.

If I had received them from on board the ship that brought your specimens of plants from him, you should have had them sent with them; but I did not receive them till the ship was sailed to Stockholm.

I have likewise a specimen of the *Actinia radicans,* which I mentioned before. This and the plate of it, you shall have the first opportunity, with some other plates that I hope will please you. I am, dear Sir, your most assured friend, JOHN ELLIS.

I shall present a memoir to our Royal Society addressed to you, shewing the probability that the *Corallinæ* are animals, and entirely differ from *Fuci* and *Confervæ*, in answer to the assertions of Dr. Baster and Dr. Pallas.

Sir, London, July 3, 1767.

I have received from Dr. Schlosser of Amsterdam, a small specimen of Meese's *Lichen terrestris articulatus*, &c. or the *Corallina terrestris* of Dr. Pallas. It is a different species from what I have. It has longer cylindrical joints than any of the trichotomous kinds, which are in my collection; but those which he would have to be parts of fructification, are evidently no more than defective lateral branches. They are solid, compressed, and very irregularly shaped; whereas all the tribe of articulated stony Corallines, have hollow subrotund ovaries.

Mr. Meese asserts he found it growing in an *Ericetum* in Friesland, which Dr. Pallas, who should know better, agrees to be true, in order to make Corallines plants. But the world is not so easily imposed on. For my part I should as soon expect to find *Pennatulæ* creeping about the woods, as Corallines growing on land; nay as ever to expect to find a calcareous plant. The place where he found it must have been sea formerly. I have received the dissertations you were so kind to send me by the clergyman that is going to Pennsylvania. I shall send Dr. Garden and Mr. Collinson the duplicates. I return you my hearty thanks for them. I have been so hurried with business, that I have not been able to try the experiments on the Fungi, which indeed are surprising, and demand our strictest attention. This is the time of the year that they are likely to succeed best. I called at Gordon's the other day. The *Ellisia* was in flower and seed; but the fruit is testiculated, not one seed growing over another, but side by side.

The *Alstrœmeria* is in flower at Mr. Lee's. Mr. Lee desires to know whether you have received the plants he sent you by order of Lady Ann Monson; when you write next pray let me know.

James Gordon intends you a plant of the Ginkgo of Kæmpfer. I shall send your specimen of the *Siren lacertina* at the same time.

This dissertation on the Corallines employs my thoughts when I have leisure from business. It is the most difficult part of all the Zoophytes to explain, as a great deal depends on examining the structure of them in the microscope. My plates of them are now correcting.

I have several of the *Fucus* and *Ulva* tribe, that amaze me when I come to examine them with glasses. I cannot expect to live long enough to go through so difficult a task, as the attempt to describe them. I will do what I can when I want amusement; as I find the study of nature the greatest happiness I enjoy in this world.

My best wishes attend you. I am, dear Sir, your most obedient and much obliged friend and humble servant, JOHN ELLIS.

I have made an alteration in my character of a *Corallina.*

ANIMAL crescens habitu plantæ.

Stirps fixa, e tubis capillaribus ramosis, per crustam calcaream porosam, sese exserentibus composita. Rami (sæpe articulati) semper ramulosi, vel divaricati, liberi, vel conglutinati, connexi.

LINNÆUS TO ELLIS. (Latin.)

MY VERY KIND FRIEND, Upsal, July 20, 1767.

You have favoured me with several letters, while I passed the spring of the year in sickness and infirmity, being only now recovering. This cause alone has prevented my acknowledging your friendship.

That Corallines belong to the Animal Kingdom, I never had any doubt, on account of their calcareous crust ; being well convinced that lime is never produced by vegetables, but by animals only. Hence I have asserted in *Syst. Nat.* page 1304, that " the calcareous substance of Corallines evinces their belonging to the animal kingdom, all calcareous matter being the produce of animals."

Having never seen the *Corallina* of Pallas's *Zoophyta* 427, which Meese discovered, I have placed a cross against it in *Syst. Nat.* ed. 12. p. 1306. I have a thousand times wondered at this production of nature; considering whether it could have been produced like the *Ulva intestinalis*, a marine plant which grows on house tops in Scania, as recorded in my Travels in Scania, June 20, p. 217.

I shall correct the character of *Ellisia* in the second volume of the *Systema Naturæ* *, now in the press, according to your observations. I took my character of the fruit from Ehret's figure, in which the seeds are so represented that one seems above the other.

I have written twice to Mr. Lee, thanking him for the plants he sent me, and I wonder my letters have not come to hand. They were directed according to Mr. Alstroemer's instructions.

It would be singular enough if *Confervæ* should prove diœcious. I have no opportunity of examining them.

All the characters of the new genera of plants, as well as of fishes, sent by our worthy friend Garden, are already printed in the *Systema Naturæ*, where I have duly cited his name, in every instance without exception. When you write to this most amiable man, I beg of you to present my best compliments.

Have you any botanical news from your kingdom of Florida? Pray send me some dried plants when convenient.

Farewell, and do not forget your friendship for me.

* See that volume, p. 735.

ELLIS TO LINNÆUS.

DEAR SIR, London Aug. 26, 1767.

I have received your kind epistle dated 20 July, and am much concerned to find that you have not been well; every body lamented; but no one can rejoice more sincerely than I, that you are recovered.

My letter to you relating to the nature of Corallines was read at our Royal Society here, and seemed to give satisfaction. Meese has sent me a specimen of his *Lichen articulatus*, or Pallas's *Corallina terrestris*, which he had collected on the sea coast; so that it is evidently no more than what I have always thought, a sea production, or true Coralline, carried into the land by some accident. As to vegetable substances, such as the *Ulva intestinalis*, growing on the thatch or straw covering of houses, exposed to the sea, and often moistened by sea water, I am not at all surprised; but to consider, as the ingenious Dr. Pallas does, that a calcareous animal substance, such as a Coralline, should grow in common earth, above 30 miles from the sea, is absurd, unnatural, and contrary to all experience. I wrote to you fully on this subject the 3d of July, which I hope you received.

The evidence of the truth of some of the *Confervæ* being *Diœciæ* I have submitted to Solander, and many other friends, who are satisfied it is so; as also that there are several *Fuci*, male and female, in different plants.

I have lately received some fragments of *Encrinus*, recent, from the Straits of Malacca in the East

Indies. This kind is found fossil in England. In the centre is a five-sided or pentagonal figure. From these five sides proceed dichotomous arms. There were many small cylindrical jointed arms, that issue out under the five principal ones, but the main stem was broken off, so that no perfect figure could be given of it. I have sent you some of the fragments by captain Robenius: you have likewise a fine specimen of the *Siren* in spirits, with some fishes of Dr. Garden's, that were to go last year, and also his insects, which are not in good order; they were very indifferent when I received them. I have likewise sent you a specimen of the *Actinia radicans*, in a small bottle in spirits. If you dissect one of them, they appear a different animal from the common *Actinia*. I shall write to Dr. Garden how polite you have been. I have been greatly disappointed as to receiving rare and curious natural productions from West Florida, owing to a bad governor, who is recalled, but another is appointed, who has promised me to attend to Natural History. I did not receive any thing new except the *Illicium anisatum*, or Skimmi, from West Florida; and the specimens were so indifferent, I sent them back, that they might know the plant, with strict orders to gather the seeds to be sent to the Princess of Wales's garden at Kew, which I hope to receive about January next; if I receive any specimens, I will certainly send you some.

I have plainly discovered the figure of the mouths or Polype-like suckers, of the *Antipathes spiralis:*

and am fully persuaded that Dr. Pallas is mistaken in the ovaries or cups, which he pretends to have discovered. They are nothing but the covering or skin of the *Antipathes,* which it sends over all such bodies as adhere to it. The form of them is irregular, as is their situation. I have some *Antipathes* that send a covering over a small kind of the *Lepas Concha-anatifera,* in the same manner as the *Gorgonia* does over the *Lepas Calceolus.* There are six small *tentacula* that surround a cup. The *tentacula* are shaped like bull's horns with wrinkles across, and full of a gelatinous matter: the cup in the centre when soaked in water and magnified, is of a most elegant figure. I have examined several small ones that have had the gelatinous substance uncorrupted on them, and they appear to have much the same figured mouths, but extremely minute.

It is a mistake that the *Antipathes spiralis* is found in Norway; Mr. Brunnich assured me it does not grow there.

I have not had time to try yours, and Baron Munchausen's experiments, on the animalcules in the origin of mushrooms and smut in corn; but have recommended it to the public to try the experiments. As soon as I do, I shall communicate my thoughts to you on the subject.

I have so much trouble with my engraver, that my microscopical observations on the marine productions, go on but slowly.

I shall write to Dr. Garden very soon: he has got one of his correspondents here to bespeak a profile of you, taken from a medal of Count Tessin's, which I lent him for that purpose; it is a small one, that Solander gave me.

My best wishes attend you! May heaven preserve your life for the good of mankind, is the hearty prayer of, dear Sir, your most affectionate friend,

JOHN ELLIS.

DEAR SIR, London, Sept. 8, 1767.

I wrote to you the 26th of August, wherein I mentioned that I had sent you some fragments of an *Encrinus*, or Star-fish, from the Straights of Malacca; but this animal I have since found is well described in Linkius, on the *Stellæ marinæ*, and appears to have no stem, only a few claws under the principal arms.

I have lately been trying experiments on the seeds of the Fungus, called by you *Agaricus campestris;* and also on those called the *Agaricus fimetarius.* The minuteness of these bodies, obliged me to make use of the first magnifying glasses in the double microscope. This plainly shewed to me, that these seeds, though put into water according to your directions, have no animal life of their own, and are only moved about by the *animalcula infusoria,* which give them such a variety of directions, both circular, as well as backward and forward, that they appear as if alive.

The *animalcula* are so numerous, and at the same time so pellucid, that without good glasses the most accurate observer may be mistaken. I wait for an opportunity to try the seeds of the *Lycoperda*, and the dust of the *Ustilago* in corn.

Our friend Solander has been ill of a slow fever for these ten days past: the physicians here give it the name of the influenza. I hope he will get the better of it; but he grows very weak, notwithstanding he goes out every day, and has the best advice.

I am, dear Sir, your obliged humble servant,

JOHN ELLIS.

LINNÆUS TO ELLIS. (Latin.)

MY DEAR FRIEND, Upsal, Oct. 1767.

I received yours, in which you speak of the living seeds of *Fungi*, asserting that you have only seen the *animalcula infusoria*, moving the powder of these vegetables.

I am not able rightly to understand whether you have actually seen the *animalcula* or not. If really so, they ought, at the end of 14 days, to begin to attach themselves to the bottom of the glass, first a solitary one, then several more adjoining to it, till almost all of them are thus become fixed; after which they grow up into *Fungi*.

With respect to the *animalcula infusoria* themselves, unless I am totally mistaken, I think I have seen these to be the living seeds of Mould, *Mucor*. But before I venture to put forth such an opinion, I beg of you to lend me your lynx-like eyes; and you will see in the vessel or glass, where there is so little water that it may soon evaporate, whether these bodies do not change to plants of *Mucor*. This point is of the greatest importance, and if my ideas be correct, we shall no longer be surprized at the quantity of such *animalcula* in common water, any more than at the mould itself on decaying food, &c. I beg and intreat of you not to slight my request. You will find it worth your while to look closely into the nature of these minute beings, as they are related, though remotely, to your own marine *animalcula*. Every body wonders at the *animalcula infusoria* being produced by an infusion of pepper, and such substances; whereas the difficulty vanishes if they belong to *Mucor;* for pepper if long kept moist, is as liable to grow mouldy as any thing else.

Having once discovered the little worms in the *Ustilago*, by the help of the microscope, I can now see them with my naked eyes, though less distinctly; and I showed them a fortnight ago to some of my pupils.

What you mention in your last, as having sent me, are not yet come to hand.

I am anxious to hear of Mr. Solander's recovery.

The 2d volume of my *Systema*, which treats of plants, will appear soon.

Cannot you procure for me some seeds of *Ellisia?* I have corrected its character, according to your kind instructions.

Farewell! May you long live, to the honour and benefit of our most lovely science!

ELLIS TO LINNÆUS.

My dear Friend, London, Oct. 30, 1767.

I have received your obliging letter about the seeds of *Fungi* being animated. By your letter, you seem to think, that the seeds of the *Fungi* are animated, or have animal life, and move about; my experiments convince me of the contrary. I must first let you know, that I am convinced that in almost all standing, or even river, water, there are the eggs, and often the perfect animals, of those you call *animalcula infusoria*. As soon as these meet with their proper *pabulum*, they grow and increase in numbers, equal to the *Musca vomitoria*. I often have examined river water and pond water, and scarce ever found it without some species of these *animalcula*, especially in summer and autumn: besides, the same *animalcula* that attack, eat, and move about the *farina*, or seeds of the *Fungi*, do the same with other vegetables, as I have lately been convinced of by a fair experiment. I have tried at your request my experiments over again, and shewed them to D. C. Solander. I will keep these infusions according to your desire 14 days, and examine the particulars you desire of the *animalcula* fixing themselves, first one, then many more, to the bottom of the glass; and will endeavour to find out what you mean by their "growing up into *Fungi.*" If you mean that *animalia infusoria*, when they are dead, are a proper *pabulum* for *Mucor*, I agree with you; for I have many animal substances that are covered with *Mucor*, even between the Muscovy Talcs (or *Glacies Mariæ*) used on purpose for microscopic animals, in the microscope. But what appears to me most difficult to comprehend is, for instance, I have now a *Lycoperdon Bovista*, which I received from our good friend P. Collinson four days ago. I put part of it into river water, and in two days' time I perceived the seeds or *farina* of it moving about distinctly. The fourth day I perceived the figure of the *animalcula* that moved them. Are these seeds, or these *animalcula*, (for they are evidently distinct bodies), to turn into *Fungi*, *Mucores*, or *Lycoperda?* This is what I do not comprehend in this new discovery. If the *animalcula*, that moved the seeds of the *Lycoperdon*, it would be amazing; and again, it would be as surprising that the seeds of one genus should produce another; for instance, that the seeds of *Lycoperda* should produce *Mucores*. However, I have determined to go through these experiments with precision, and to call in witnesses of the several appearances.

I have not yet got any of the *Ustilago*. If you will be so good as to send me a Spike of Corn infected with it, proper for trial, you may depend on me in carefully going through the experiments properly.

I have made some observations lately on a small

kind of *Lumbricus,* which serves as food for the *Hydra,* or fresh water polype of Trembley. I cut these small worms into three or four pieces, and they all have grown into perfect animals. Dr. Solander (who is obliged to you for your kind enquiry after his health) has examined them, at my chambers to day, in the microscope, and was surprised at their structure. They differ quite from the *Lumbricus terrestris,* and approach near to the *Tœnia.* I have wrote to James Gordon for some seeds of the *Ellisia* in their capsules. As soon as I get them, will send them to you.

I hope you have got the *Actinia radicans,* and the things I sent you by Captain Robenius. I long to have your thoughts on this *Actinia;* it differs very little from the *Sertularia.*

I find by your last edition of the *Syst. Anim.* that you look on the stem of the Gorgonias to be vegetable. I differ in opinion, and intend to write a letter particularly on that subject, as I have had many opportunities, from seeing different specimens, together with chemical and microscopical enquiries, to prove them as different almost as the *Dendrites,* or the crystallization of Sal Ammoniac, from vegetables. Pray let your young gentlemen examine the vesicles, or ovaries (as Pallas would have them), of the *Antipathes;* for to me they approach nearer to the *Gorgonia,* as you have placed them, than to the *Sertularia;* and what I have seen are no more than some extraneous bodies, which have adhered to them, being partly covered with their spiny skin, and have formed little irregular turbinated cups, here and there, on the branches, not in that regular shape and manner that the *Sertulariæ* have their vesicles.

You may depend upon it I shall always attend to your commands, and shall not make light of these enquiries; you shall always find me your most affectionate friend, and that I shall always acknowledge myself indebted to you, for the many useful observations that you have been so kind to communicate to me, by your letters as well as your writings. My best wishes attend you. I am, with great respect, your most obedient servant, JOHN ELLIS.

I wish you would send your son next spring to England. I will do him all the service in my power.

LINNÆUS TO ELLIS. (Latin.)

Upsal, Dec. 8, 1767.

TO THE MOST CANDID AND FRIENDLY OF MEN,

Your most welcome letter of the 30th of October reached me a few weeks since; but I deferred my answer till I had received your communications by Captain Robenius. These indeed are safely arrived at Stockholm; but as yet there has not been a fit opportunity of forwarding them hither. I hope they will soon come to hand; but in the mean time I must not keep silence any longer.

I have been seeking for branded spikes of corn with all diligence, but in vain, the wheat being all threshed out long ago. If I live, I will send you some next summer.

I am beyond measure delighted with your observations upon the *Lycoperdon* in river water; that its powder moved about, and was transformed into that species of *Mucor* which I have named *Mucedo.* I have long suspected this *Mucedo* to belong to *Lycoperdon;* but my suspicion has never before been confirmed.

Your observation is very curious respecting the little worm, *Lumbricus,* whose transverse fragments grew to complete animals. Is this the species published in Trembley's Polypes, *tab.* 6. *f.* 1.; *Roes. Ins. Polyp. t.* 79. *f.* 1, and *t.* 78. *f.* 16, 17; *Schæff. Monogr. Regensb.* 1755, *t.* 3. *f.* 1, 2, 3?

I am very anxious to see your *Actinia radicata,* which is to prove akin to *Sertularia.*

You puzzle me by saying I have removed the *Gorgoniæ* to vegetables in the last edition of my *Systema.* Previous to the year 1740, I, like all the rest of the world, considered Corals and Zoophytes as plants. But from that period, when the smaller tribes of animals came to be rightly understood, I have never done so. I cannot comprehend whence you have taken up this opinion, of my having removed the *Gorgoniæ* to the vegetable kingdom. You will find the case to be otherwise, if you look at the tenth or the twelfth edition of my *Systema,* where the genus *Gorgonia* stands among the *Zoophyta.* Though furnished with a stem, branches, occasional roots, a bark, and annual internal circles, in which it resembles a vegetable, it is perfectly animal, as every one must perceive who investigates its metamorphosis, or flower, which is not of a vegetable but animated nature. I have in my possession a number of very beautiful specimens of this genus, from the sea of Norway.

There is lately come to my hands an elegant little (Italian) treatise, entitled *Discorso dell' Irritabilità d'alcuni Fiori nuovamenta scoperta*, Florence, 1764, which is new to me. This shews how the florets of certain capitate compound flowers move, when touched, as if they were alive.

I have just seen mentioned, in the Memoirs of the Academy of Turin, a new species of worm, called *Hirudo alpina*, which is the real cause of the cholic or heartburn of the Laplanders, and others, who drink alpine water. If you should meet with this animal alive, I beg of you to examine accurately whether it be not a species of *Fasciola*; perhaps nearly related to the *Fasciola hepatica*, Faun. Suec. 2075, though a distinct species.

Tell Dr. Solander, as he has received Bergius's Descriptions of Cape Plants, published this year, in octavo, that the following (supposed new) genera of that author are as follow:

His *Dilatris* is my *Wachendorfia*.
Stilbe - - - *Selago*.
Aulax - - *Leucadendrum*.
Nectandra - *Struthiola*.
Nemia - - *Manulea*.
Melasma - - *Nigrina*.

His *Nothria* is my *Frankenia.*
Cyphia - - *Lobelia.*
Lidbeckia - *Cotula.*
Laurembergia *Serpicula.*
Thamnochortus *Restio.*

Pray persuade Solander to write to his excellent mother, who has not received a letter from her beloved son for several years, which she much laments. Her residence is now at Pithoa.

I learn from the newspapers that Tea is already planted in Carolina. Is this true? By the same authority I was told that the Tea plant had borne flowers and seed in a garden in Denmark. On inquiry through a friend, this proved nothing but *Veronica maritima.*

The second volume of my *Systema,* comprehending the Vegetable Kingdom, is just published. It contains 50 new genera of plants. — May you long live, the ornament and the polar star of science!

Dear Sir, London, Jan. 15, 1768.

I have received with great pleasure your kind letters of the 8th and 18th of December, with one inclosed for Dr. Garden, which I shall send him immediately. We have had a most severe frost for three weeks past, which has interrupted the navigation of our river; but now the weather is warm, and Farenheit's thermometer in the open air at 50.

In your letter of the 8th December, you seem to misapprehend the meaning of the letter which I wrote to you the 30th October. I find, on looking over the copy which I have of it, it runs thus, or much to this purpose: — "I have just now received from our friend Peter Collinson a *Lycoperdon Bovista*, nine inches diameter, perfectly ripe for experiments on this subject. I have put part of it into river or soft water, and find in 24 hours the same kind of *animalia infusoria* moving all the seeds about as if they were alive. I shall follow your directions, and keep this infusion for 14 days, and observe exactly the result of the experiment; but I cannot think that these seeds of the *Lycoperdon* ought to produce any thing else but *Lycoperdons*. I am well persuaded that these *animalia infusoria*, when they die, afford a proper nutriment for the several species of *Mucores*, as much as the dung of animals or any other animal substances, such as cheese in a putrescent state, provided they are kept in a warm close place, or where the air is stagnant, moist, and warm. Further, that upon putting small animals between muscovy talcs in sliders for the microscope, I have observed that in some time a *Mucor* will proceed from them."

Thus far I thought it necessary to quote from my former letter of 30th October, as my real opinion. have kept a regular journal of my observations in making my experiments on the seeds of the *Fungi*, which I have shewn often to Dr. Solander, to prevent any mistake; and do assure you I have convinced him that they do not move of themselves when kept in water; but it appeared evidently to him, and many more gentlemen who saw my experiments, that the motion which they had, proceeded from *animalia infusoria,* whose shape we plainly saw, and observed distinctly the particular motion, with some attention, which these little creatures had while they were eating the seeds of the *Fungi,* and which they communicated to the seeds of the *Fungi,* so as to make them appear alive.

The *Lumbricus,* which I cut into several pieces that became distinct animals, I believe is mentioned by Trembley; for it was he that taught our people to feed the common fresh-water *Hydra* with them.

I shall send you by the first opportunity a magnified drawing of it.

If Rœsel has drawn a figure of it, he has made a mistake in the appearance of the head, which, if I remember right, he has made like a snake, which is not so. There is a waved gut runs through the whole, with several divisions, one at every joint; and for about two-thirds of the whole, from the head, are two bristles on each side of each joint, which it can draw in or extend.

The animal which I send you in spirits I have called *Actinia socialis.* I believe it is quite different from your *Priapus humanus.* If you dissect it lengthways, with a fine lancet, you will find the structure very different from the common *Actinia* :

besides, you do not mention that this, like mine, propagates itself from radical tubes, which adhere to marine substances, like the *Sertulariæ,* many of the same animals arising near one another out of the same common adhering wrinkled tube.

It is looked upon here, by Dr. Solander and many of our curious people, as the rarest sea animal that has been lately discovered. It is found in our Southern American Islands. I have laid an account of it before our Royal Society, which will be published, with a print, in our next volume of the Philosophical Transactions.

What I mean by your making the Gorgonias vegetables is, in your description you call a *Gorgonia,* "Planta radicata more fuci excrescit in caulem ramosum, cortice indutum, deponente librum, indurandum in lignum secundum annotinos annulos concentricos, intra quos medulla animata, quæ prodit in animalcula florida." No man who reads this but will conclude that they are at least half vegetable and half animal: but I am sure there is no communication between the *Medulla* and the *Flores,* as in vegetables; and as to the concentric rings, they are not produced after the same manner with those of the wood in trees, there being no visible communication between them. I have lately carefully examined several, that were taken out of the sea, and put into spirits. Besides, how is the calcareous crust or *cortex* (for it is the same) to be accounted for, that is so often found between these circles? Dr. Pallas's description is wrote with great art; but Natural Historians and Lawyers are very different people. I hope to shew my objections in a memoir very soon; for which purpose I have some plates engraving, to explain this abstruse point more clearly to our senses. Artful people may puzzle the vulgar, and tell us that the more hairy a man is, and the longer his nails grow, he is more of a vegetable than a man who shaves his hair or cuts his nails; that frogs bud like trees, when they are tadpoles; and caterpillars blossom into butterflies. These are pretty rhapsodies for a Bonnet. Though there are different manners of growth in the different parts of the same animal, which the world has long been acquainted with, why should we endeavour to confound the ideas of vegetable and animal substances, in the minds of people that we would willingly instruct in these matters?

I have shewed your letters to Solander, and have made him promise to write to his mother. I have many things more to write, but must defer them.

I heartily wish you many happy years; and am, most truly, your affectionate friend, and obliged humble servant, JOHN ELLIS.

I have got some seeds of the *Ellisia* for you. I hope to have some curious new seeds from the Musquito shore, on the Spanish Main. They are lately arrived at my friend the Earl of Hillsborough's, who is now made our third Secretary of State for America, Africa, and the East Indies.

DEAR SIR, London, March 15, 1768.

I have lately received a letter from Dr. Schlosser of Amsterdam, advising me that he had received a letter from Dr. Pallas, dated 20th November 1767, O. S. from Petersburgh; wherein he confesses that he is now convinced your genus of *Corallina* are animals, and that Mr. Meese has imposed on him; for Meese had found the same *Corallina* on the sea shore; and it is possible this might have been blown to Bergummer heath by the winds.

The Acorns which were inclosed in wax by a new method have succeeded, which is the inclosing each Acorn first in soft pliable bees' wax, and when the whole number you intend to inclose are rolled up separately, and quite cold and hard, some melted wax must be put into a thin deal box, about half full; and when this wax is cool enough to bear your finger in it, the Acorns must be placed in rows till the box is full; afterwards pour more wax on them when it is barely fluid, so that they may be covered quite over; and let all cracks, after the wax cools, be filled up with very soft wax. In this manner they will keep very well for a twelvemonth. Care must be taken to choose the Acorns sound, and neither too fresh from the tree nor too old; and they must be wiped very clean, that no condensed perspiration may remain on their surface. You will see an account of those I preserved for the Royal Society in the public papers; every body here seems pleased with the success of this experiment.

I have now discovered why putrid vegetable substances yield volatile alkaline salts, the same as animal substances. I put a rotten potatoe into water, about ten days ago, in a glass, and covered the top with a card, having a weight on it to keep it close, and placed it on a shelf in a room over the fireplace. In four days the water was full of small animalcules, so that, I believe, for every particle of the potatoe there were ten minute animals, as in the experiment with the *Fungi.* In order to know for certain whether these animals came from the potatoe or the water, I boiled a small potatoe till it was ready to fall to pieces. I likewise boiled some water for half an hour, and then put the mashed potatoe and the boiled water together in a glass, and they were placed in the same warm situation, with a card over it. I examined a drop of the liquor four days after, and could perceive millions of animalcules, of a tadpole shape, turning about the crystalline round particles of the potatoe in all directions, just as I had observed them turning the seeds of the *Fungi* last autumn. I must inform you I used the largest magnifiers I had, which were the first and second of Wilson's microscope. I wish you would try the same experiments; you will find many new scenes in nature will be discovered by this hint. I have shewed the experiment to Solander and a very few friends, but have not yet made it public.

I wish Baron Munchausen would try the same.

You mention, in one of your letters, that you see, in our newspapers, that they have got the true Tea tree in America; but you must not mind such accounts; they are false.

I have discovered the eggs of a *Gorgonia* in the polype-like suckers of the *Gorgonia muricata*. I have had a specimen of it, preserved in spirits, from the West Indies, which has given me new lights into the nature of the *Gorgoniæ*. I have likewise had some Corallines in the same spirits, which are full of slime, and not hollow, but filled with irregular tubes, I mean the Corallina nervo tenuiore fragiliorique internodia nectente, of Sloane, Hist. Jamaica, vol. I. tab. 20, fig. 4. This is what Mr. P. Miller gave to the Premium Society, for the *Lichen Roccella*.

We have got a new pine from China; it was raised last summer at the Princess of Wales's Garden at Kew by Mr. Aiton.

I have many things to say, but have not time at present, only to assure you that I am most sincerely yours, JOHN ELLIS.

DEAR SIR, London, Aug. 19, 1768.

Not hearing from you since I wrote to you last, which was 15th March, I have concluded that you must have been prevented by sickness.

I take this opportunity of Capt. Trobenius, to send you a specimen of a *Siren* from Dr. Garden, and likewise some proof plates of what I intend for my second volume, as well those that are already printed in our Philosophical Transactions, as those that are entirely new, and have not yet been given to any body; depending on you, my dear friend, that no copies may be taken of them, but to remain with you for your own information. I have not as yet finished the description of them, but shall now soon begin to arrange what I intend on the subject; and am only detained by five plates more, which I must get executed, as I have several *Mollusca*, Corals, and particularly some *Gorgonias* and *Alcyoniums*, which I have received from our West Indies, preserved in spirits, with all their suckers, or *tentacula*, extended as when alive. But it is impossible to describe to you how difficult it is to get drawings and engravings well executed, and finished in any reasonable time, besides the extravagant expenses that attend them.

I must now inform you, that Joseph Banks, Esq. a gentleman of £.6000 per annum estate, has prevailed on your pupil, Dr. Solander, to accompany him in the ship that carries the english astronomers to the new-discovered country in the South sea, Lat. about 20° South, and Long. between 130° and 150° West from London, where they are to collect all the natural curiosities of the place, and, after the astronomers have finished their observations on the transit of Venus, they are to proceed under the direction of Mr. Banks, by order of the Lords of the Admiralty, on further discoveries of the great Southern continent, and from thence proceed to England by the Cape of good Hope. They are to leave Plymouth about this time, and from thence proceed to the island of Madeira; from thence to Rio Janeiro, afterwards to the Falkland Islands and the entrance of the Straits of Magellan; from thence round Cape Horn, and so to George's land. No people ever went to sea better fitted out for the purpose of Natural History, nor more elegantly. They have got a fine library of Natural History; they have all sorts of machines for catching and preserving insects; all kind of nets, trawls, drags, and hooks for coral fishing; they have even a curious contrivance of a telescope by which, put into the water, you can see the bottom to a great depth, where it is clear. They have many cases of bottles with ground stoppers, of several sizes, to preserve animals in spirits. They have the several sorts of salts to surround the seeds; and wax, both bees-wax and that of the *Myrica*; besides, there are many people whose sole business it is to attend them for this very purpose. They have two painters and draughtsmen, several volunteers who have a tolerable notion of Natural History; in short, Solander assured me this expedition would cost Mr. Banks ten thousand pounds. All this is owing to you and your writings.

About three days ago I took my leave of Solander, when he assured me he would write to you and to all his family, and acquaint them with the particulars of this expedition. I must observe to you, that his places are secured to him, and he has promises from persons in power of much better preferment on his return.

Every body here parted from him with reluctance; for no man was ever more beloved, and in so great esteem with the public from his affable and polite behaviour.

The reason he was not introduced to the King was owing to Dr. Hill's being so great a favourite with Lord Bute; for if Solander's merit had been known to the King, Dr. Hill must have sunk in the opinion of all persons that attend the court, as much as he is fallen in the opinion of all lovers of Botany and Natural History, on account of the little credit that is to be given to what he advances; for though he certainly has great abilities, the want of that fairness and exactness makes the world suspect him; and his writings are a mere drug. Mr. Fitzhugh, Factor to the East India Company for many years in China, who has been in the tea country, is just arrived, and has brought a Tea-tree home alive. As he is very curious and intelligent, I asked him particularly about the species of tea, green and bohea. He declares it is but one and the same plant, and that if you take the green tea and transplant it into the country where bohea is made, it will produce bohea tea; and so the contrary. That it has but six petals, and that you must have been imposed on by Dr. Hill. Dr. Solander and I examined him minutely about it but a week ago. He knows the captain that brought over your Tea-trees very well; and confesses that the sowing the seeds in the earth, as they are going to leave the coast, is the only way. Mr. Fitzhugh's tea plants came up from seed, and, when at St. Helena, were destroyed by rats. The tree he has brought over is an old one; it is under the care of J. Gordon. I hear there is another Tea-tree just arrived in fine order.

I am in hopes of specimens of the *Illicium anisatum* very soon, and seeds in January. The specimen I lately received has 30 *stamina*, and 13 *pistilla*; about 27 petals, and 4 parts in the *perianthium*. The flower is of a red colour. There are no *nectaria*. It seems to come next to the *Dillenia*. I have lately had several fair seed-vessels of the *Ellisia*, and I find that there are four seeds in each, two above the other two. The calyx or *perianthium* is *monophyllum, quinque-partitum*. The seeds when magnified are not muricated, but like the seeds of *Papaver*, only black. I have lately received a *Boletus* from the Mosquito shore, whose upper and smooth surface is of a fine sky-blue colour, the under part, where the pores are, of a fine yellow. It seems to be a parasitical one; but the stalk, by which it adhered to the tree, grows to the middle of the upper or blue part.

I have likewise received from the North part of South Carolina a new kind of *Fungus*. I shewed it to Solander, and he had never seen any thing like it; it seems to approach the *Phallus*. I compared it with Michelius's figures, p. 83, but it wants the *pileus*. The inside, when it is dissected, is like a honey-comb. The person who gave it to me collected it, and says when it was fresh it was of the richest scarlet colour he ever saw, but that the smell was intolerably fetid. I should be glad to know whether you have seen any thing like it, or whether it is worth making a drawing of.

If you have made any new experiments on animal and vegetable infusions, pray let me know. I am determined to go through Mr. Needham's, and have now some mutton gravy sealed up like his in a bottle. I intend to examine it in a fortnight. I have seen something surprising on a dead fly, that was drowned in water where some flowers were. I begin to think that there are several zoophytes in the fresh water, that appear to grow when animal and vegetable substances are putrefying. I will do justice to these experiments; for truth is all I want to pursue. The zoophyte I mean is that which is represented in the Phil. Trans. Vol. 45, No. 490. Tab. 5.

The one that I have figured is highly magnified, and when the minute globules came out at the top, they seemed to be alive and float about. After a time this kind extended itself into small ramifications, like roots, and then died away.

I have no more room now, but wish to hear from you, and that you may get what I send you safe.

I assure you of my best wishes for your health and long life, dear Sir, your much obliged friend,

JOHN ELLIS.

I shall now endeavour to get some of the *Ustilago* to try that experiment fairly.

Poor Collinson, our friend, is dead.

My Lord Hillsborough has sent me some specimens of the wild nutmeg, with small fruit on them;

they are the same that is taken notice of in the *Herb. Amboinense* to grow at Surinam. I have wrote to Tobago to get specimens of the flowers, to fix the genus.

LINNÆUS TO ELLIS. [Latin.]

DEAR SIR, Upsal, Oct. 16, 1768.

I yesterday received your welcome letter, accompanying the description, character, and figure of that most rare and singular plant the *Dionœa,* than which, certainly, nothing more interesting was ever seen. I laid this communication before our Royal Academy of Sciences to day, nor was it received without high admiration and astonishment. I am charged by that learned body to transmit to you its best thanks, for one of the most valuable communications it ever received, and which is ordered to be the first article in the ensuing volume of our Transactions.

For my own part, though I have doubtless seen and examined no small number of plants, I must confess I never met with so wonderful a phænomenon. Your history of the plant, and its botanical characters, are so complete, that nothing can be added to either.

Letters received yesterday from Stockholm, inform me, that your packages sent by Capt. Robenius are safe arrived. They will soon reach Upsal, and I will write to you soon after they come to hand.

May God preserve you to illustrate still further the works of his divine wisdom!

ELLIS TO LINNÆUS.

SIR, London, Nov. 1. 1768.

I received your very kind letter of the 12th of September *, and now can inform you that I have received a letter from our good friend Dr. Solander from Madeira, dated the 18th of September: he was then sailing out of harbour of Funchal upon their voyage to the South seas. He says they have met with a great many curious *Mollusca*, of which they have made drawings and descriptions; he adds they will send what they have got home by way of Lisbon, and promises me a large packet by that way, as no more ships are expected this year to sail for England from Madeira. He has sent me a letter inclosed for his mother, with orders to deliver it to Mr. Lindegren to be forwarded to her; he begs his kind respects to you and all his friends. I have lately received from my friend, just returned from China, such a parcel of flowers of tea as you mention you formerly received; and upon examining them, I find they exactly agree with your description in your *Genera Plantarum*. My friend says they were sent him from the country where the bohea grows. The Tea-plant here succeeds well from cuttings, under the care of Mr. James Gordon.

* This letter is wanting.

I hope you have received the print of the *Dionœa muscipula.* I wish it may stand our winters; as it is the most extraordinary plant that ever I have met with.

I have lately met with the same kind of zoophyte-like appearance from bad wheat bruised and put into a watch glass with river water, and covered with another watch glass. I am afraid the season of the year is grown too cold to make observations; this is the same that Needham describes in the *Act. Anglica.* I put some more common house-flies to putrefy, but they afforded no such appearance as the former did.

I acknowledge these *animalia infusoria* are the most surprising things in nature, but I cannot agree with Needham, that the *animalcula in semine masculino* are of the same nature. The minute animalcules that appear in a boiled potatoe, one or two days after, amaze me. I find that putting grass and other vegetables into phials of water, and corking them up close, prevented the appearance of any animalcules, though kept from June to September; but in a few days the corks were taken out of the bottles, and they were covered only with white paper tied close down. The animalcules then began to appear, both in the *Poa annua* and the *Rumex Acetosella* infusions. The boiled potatoe, corked up with the water it was boiled in, afforded no animalcules. In the gravy of lamb, though corked up close, there were a few very minute linear animalcules; but I never could see the various sorts and sizes that

Needham says he met with: indeed our summer has been very wet, and but a few warm days. Dr. Hill tells the world that he has found out that the *Fungus Boletus*, or what used to be called *Agaricus*, is a zoophyte as much as any of the rest that are allowed to be true animals. His friend Dr. Watson magnifies him and his discovery. We wait for demonstration. I cannot discover the animality of them; but I have seen the seeds in the cells as plain as Micheli has described them. I am sure in burning they have no animal smell.

I rejoice to hear that your third volume of your System of Nature, or the *Regnum Lapideum*, is finished: we long to see it here. I expect very soon to get the seed-vessels of the *Illicium anisatum*. They write that they have thirteen pods in the form of a star, agreeable to the *Pistillum* in the flower; the blossom is red. I enquired of Dr. Fothergill about Mr. Collinson's collection: he says his son keeps them, and does not intend to dispose of them.

I hope you received the specimen of the *Actinia sociata*, which I sent you above a year ago. If you have, pray give me your opinion; though I have joined it with the *Actiniæ* from its outward appearance, I believe from the inward part or what we discover in the longitudinal section (which is exactly drawn from nature) it may be another genus: but that I submit to you. I am now drawing up an account of the difference between the growth of trees and of zoophytes, such as the *Gorgonia*, *Isis*, and the *Antipathes* of Pallas.

By comparing the upright and cross sections of both, the difference is easily discovered.

I have removed the *Isis nobilis* to the *Gorgoniæ*, and called it *Gorgonia æstimabilis*. In my opinion the articulations of the *Isis* are the only and proper distinguishing character of this genus, otherwise it would belong to *Gorgonia*. I know you will object that it wants the *Medulla*. All the *Antipathes* that I have seen, and I have seen many species, and dissected them carefully, have spines on their surface, and even in their inner circles. I should not have been at the pains to do this, though it is very curious, but that I think Dr. Pallas seems to treat me too freely, without giving a full answer to my remarks on the *Gorgonia Flabellum Veneris*. He has seen the specimen in my possession, and he cannot account for it, especially for that part of the top where it grows back upon itself, contrary to what has been observed in plants.

While I was writing this part of my letter, I received your very kind and extremely complaisant letter of Oct. 16, acknowledging the receipt of my letter of the 23d of September last*, with the print and description of the *Dionæa*. I am glad it has given you and the illustrious Royal Society of Upsal so much satisfaction. Be so good as to return them my most hearty thanks, and assure

* This letter is missing.

them I esteem it a particular honour done to, dear Sir, your most affectionate and much obliged humble servant, JOHN ELLIS.

When the King of Denmark was here, there was not a person of any curiosity in Natural History among all his attendants.

LINNÆUS TO ELLIS. (Latin.)

Upsal, Nov. 8, 1769.

It is long since I had any of your letters, and I cannot help being anxious for your welfare, especially as the fates seem always adverse to the undertakings and the discoveries of eminent men. Whenever I look at those beautiful figures of Zoophytes which you have sent me, an anxious wish arises that I may live to see their descriptions from your pen. Your discoveries may be said to vie with those of Columbus. He found out America, or a new India in the west: you have laid open hitherto unknown Indies in the depths of the ocean.

What American news from our excellent friend Garden? Is he in good health? I beg of you to send him my best wishes.

Have you heard any thing of Solander? Perhaps he is so far removed from us, that Fame herself cannot convey a whisper concerning him.

My whole summer has been devoted to the plants of southern Africa, a rich harvest of which has been brought to me from the Cape of good Hope;

and among them I have met with various singularities.

I have received, from whom I know not, seeds of the *Ellisia*, which must have been sent by yourself, and I return you my best thanks. They had all the appearance of being sound and perfect. I therefore committed them to the ground immediately, and I hope they will next year yield a progeny to remind me every day of my excellent friend.

Can you, my friend, who take your daily walks and contemplations among the plantations of Neptune in the deep, like a botanist in the fields, tell me any thing about those numerous phosphorescent *Mollusca*, so luminous at night? What is your opinion of the *Noctiluca marina;* the *Scolopendra lucida* of Grisley, or the *Lucionella* of Vianelli? Do these animalcules render sea water luminous? Or is that appearance caused by the salt of the sea only, as the french think?

Have you learnt any thing further respecting the character of *Illicium,* so as to make it complete?

Is your Tea-tree in health, that came last year? The king of France has received one, of which a branch was sent me by his command; but this proved a species of *Camellia,* and not the genuine *Thea.*

Farewell, my dear friend; do not forget me.

ELLIS TO LINNÆUS.

DEAR SIR, London, Nov. 27, 1769.

I have been prevented, by many unforeseen accidents, from paying you the regard I owe you for your most polite letter, of the 8th of November 1768 *. I forwarded you a letter from our worthy friend Solander, which you have received before now. How inhuman were the Portuguese at Rio Janeiro, to hinder Banks and Solander from making observations in Natural History! I have published Solander's letter to me, in our newspapers, to expose the governor for his impolite behaviour to our philosophers.

I send you inclosed some seeds of the *Illicium anisatum* fresh gathered, which I believe will grow. I have sent the king some with a specimen of the plant, and intend to have it drawn and dissected ; it is a different species, I think, from Kæmpfer's.

I am in hopes it will bear the frost, as they have smart frosts both in East and West Florida, where it grows. John Bartram says it stood a severe frost, in his account of his journey up the river St. John, in East Florida. The smell of this is more aromatic, not rank, with the anise-seed smell of the oriental species. They drink it with *Cassine,* as tea, in Florida.

I have raised a Tea plant from a seed that happened to lie in the bottom of a tin canister from China, which I received this time twelve month. I put it into a pot of fresh loamy earth, under a glass cover, without heat, and in March put it into a gen-

* This letter is missing. There is one of Nov. 8th, 1769, which arrived in London Dec. 8.

tle heat, and by July it was six inches high. There have been a great many raised that came over inclosed in wax, and many have been sown at St. Helena, and were brought here in tubs growing; so that I make no doubt, but by this time twelvemonth we shall have many hundred plants of the true tea growing in England.

I only wait for the engraver to publish my explanation of the plates I have already done. I wish to have it finished, as life is very precarious, and I would willingly shew how far I differ in opinion from Dr. Pallas: and how little Corallines have (besides form) of vegetables in them. My amusement of late has been observing the effect of various vegetable infusions, as that study can be carried on in the house. I have seen many new animalcules, such as wheel animals; one particularly, like a duck without its head, and a wheel in its breast.

Mr. De Saussure from Geneva, says he has seen one sort divide into two parts, and become two separate animals. I tried the experiment on many, and found six sorts divide. I have ranged them under your genus of *Volvox*. The most curious is the discovering of their fins, by an infusion of the *Geranium zonale*, which is death to them. Those that die without, or for want of, water, never shew their bristles or fins. I have given the Royal Society a complete skin of the *Felis*, which Buffon calls *Coquar*. Perhaps it is your *Onca*: but this has no stripes nor spots. It is the colour of a rabbit, or common fallow deer; a fine smooth skin. The breast and belly are whitish, and the hairs a little longer; it is seven feet six inches long from the nose to the tail. The tail has black hairs at the end. From the middle of the back to the ground, as it stood when alive, is two feet six inches. They are often much larger, and do mischief, killing oxen and buffaloes, wolves, &c. This was shot near Pensacola. There is one now alive in London, but not so large. Inclosed I send you the print.

I have discovered salts of the same kind in various infusions besides the Hempseed, as in *Sesamum, Linum, Vicia, Dolichos, Triticum;* and this day a new salt in the infusion of the *Agaricus fimetarius.* These are surrounded by animalcules which swim about them, and are not dissolved by an addition of water; they are about the size of the fine salt used at table. Adieu, my dear friend, yours most sincerely,

JOHN ELLIS.

DEAR SIR, Gray's Inn, London, January 16, 1770.

I received your favour of the 8th of November, and wrote you a letter about the same time, which I hope you have received before this: it inclosed some fresh seeds of the *Illicium anisatum.* About a month before that time I forwarded a letter to you, which came into my hands from Dr. Solander, from Rio Janeiro. I hope we shall have the pleasure to see him next summer, returned with the spoils of the new-discovered world. I have not heard from

Dr. Garden this year past, but expect to hear from him every day, as I have some curious seeds sent from West Florida, through his hands. When I write to him, shall be sure to present your kind respects.

You enquire after the Tea plant which I raised from seed. It grows very well, being about seven inches high, and had a side branch which is cut off and planted, and will become another plant. It is under the care of Mr. William Aiton, botanic gardener to her Royal Highness the Princess Dowager of Wales. He keeps it at present in a green-house where the Camphor tree, Oranges, and Myrtles stand in the winter, though I do not doubt but it would stand the frost of this climate. Is it a deciduous plant? I suspect it is, as the leaves begin to fall. We have now a great many plants of Tea, near 100, in England. Of some the seeds were sown at St. Helena; many brought over in wax, in the manner which I have directed for the Oak Acorns in the Philosophical Transactions, besides many raised from cuttings. I believe we have got a *Camellia* from China as well as the king of France.

The *Illicium anisatum* of Florida is a new species; the number of the *radii* of the seed-vessel is almost constantly 13. The petals of the flower are red, whereas Kæmpfer says the Japanese is yellow. The smell of the eastern seed-vessels is like aniseed; ours are much more fragrant, and the leaves of the Florida tree are more lanceolate. I shall give an exact figure of this plant very soon, with as good a description of it as I can; Mr. Ehret has promised to make me a drawing of it.

You desire to know whether I have seen any of the *Mollusca* that shine in the dark. I have seen a sort like a *Nereis*, but I never particularly examined them, nor the sea water; as I took it for granted, that, according to Vianelli and Baster, the luminous quality of the sea was owing to minute insects; but this last month a memoir of Mr. Canton's, F. R. S. was read on that subject, wherein he proves from experiments, that putrid fish, in fresh water made as salt as the sea water, and kept in a vessel, will in 24 hours give a luminous appearance to the water. It will do the same in sea water, but not in fresh water where no salt has been put. I have now a Whiting, which I put into river water made as salt as the sea; and in 24 hours the water was very luminous when stirred; the fish was bright and shining, but the water, though luminous, like milk or white paper, did not sparkle or shine like the fish. I shall try more experiments on the subject. I have viewed a drop taken off the luminous bright part of the fish, and found, when it was put on a slip of glass, it did not shine; and when it was put into the microscope, it appeared full of globules of a very bright clear oil. In the drop or small portion that I placed on the same slip of glass, which was taken from the water stirred up, I could perceive but very few of these bright oily globules. In both there were many minute *animalcula* moving to and fro, as in the infusion of flesh. At present it appears that this oil gives the phosphorical light. I must own I do not think the appearance of the water like what I have seen at sea; for to me the sea appeared as bright as the putrid fish does now. I mean to try chalk, coral, poundéd shells, and such like, in fresh water and putrid fish; for as it is said that fresh water will not become luminous with putrid fish, it must be owing then to the salt. I first supposed it might be owing to animalcules preying on the fish, that occasioned this light; but as yet, these that we see are too small, and yet the light is very strong on the fish. There is a brightness in the oil, when magnified, that surpasses any other oil that I have ever seen. Other salts likewise should be tried in fresh water with putrid fish.

The sea fish is at first quite fresh, when the experiments are begun; you have frequent opportunities of trying these experiments. I am persuaded you will have great pleasure in them, and that you will find out something new.

I am now going to try experiments on Turnip seed, which when sown in summer, and a dry season follows, are destroyed by a kind of *Curculio* or *Scarabæus*. From some late observations that I have made on seeds kept close and warm, these insects, which are in them, must be the same that destroy their seeds—lobe leaves, after they come up out of the earth, when the season is very dry, and not others that are accidentally in the earth. This is a matter of great consequence to this country, where, by means of Turnips, we can afford to keep so many more sheep: and therefore it would be a great discovery to be able to destroy insects in the seed, without hurting the vegetation of the plant. I propose to fumigate the seeds (when soaked in a liquor) with sulphur, before they are sown. I recommend it to you to try what will destroy these animals without impairing vegetation. There is a large cargo of the *Dionæa Muscipula* come to Mr. James Gordon the Nurseryman near Mile End. I have seen some of the leaves, or that part which supports the lobes, serrated; this is a variety. The anatomy of the leaves, with their lobes, would be curious; to find out how the lobes close with so much force as to kill large flies.

My best wishes for your health and long life, for the general good of mankind.

I am, dear Sir, your much obliged and obedient humble servant. JOHN ELLIS.

LINNÆUS TO ELLIS. [Latin.]

MY DEAR FRIEND, Upsal, Sept. 4, 1770.

I can no longer bear to be deprived of your most agreeable correspondence. It seems an age since I heard from you. Pray inform me speedily of your welfare.

How are you going on with your immortal work? I long for it more anxiously than a little bird looks for the dawn of day. I wish I may live to see it.

Have you heard any thing of Solander. or when we may expect his return?

If you have by this time published, in the Philosophical Transactions, the character of *Illicium*, I beg the favour of you to communicate it correctly to me.

The *Ellisia* has flowered well with me this summer, and is still living. It is a pretty plant, which I never saw before.

Does Dr. Garden enjoy his health in his own Carolina? and has he, as usual, sent us any thing new? Whenever you write to him, pray offer my compliments.

I have made a description of the fructification of *Cycas*, hitherto referred to the Palm tribe; but it proves a genuine species of Fern.

I beseech you to indulge me with some news. The whole creation is laid open to you; nor can any thing escape those lyncean eyes, which have penetrated into the invisible world.

Will you be so kind as to forward the inclosed, after sealing it, to your able cultivator Mr. Gordon?

Does the Japanese tree with two-lobed leaves (Ginkgo) ever blossom with you, or do you know its botanical character?

I have now in flower the *Dalechampia*; as well as the *Calceolaria* of Fewillée.

ELLIS TO LINNÆUS.

Dear Sir, Gray's Inn, London, Sept. 25, 1770.

Our mutual friend Dr. Garden sent, in August

last, a bottle with some animals in spirits, and also a letter, directed to you. I sent the bottle to Mr. Lindegren, who promised to forward it to you by Capt. Robenius (I think that was his name), and the letter by the same ship, or some other conveyance. I hope you have received them safe.

I received, a few days since, your very kind letter of the 4th of September, with one inclosed for James Gordon, our famous gardener, and one for Dr. Garden; the first I sent immediately, the latter shall go by the first ship.

I have not yet inserted the characters of the *Illicium Floridense* in the Acts of the Royal Society, but propose to do it immediately. It differs in smell from the eastern one, being more agreeably aromatic; the Petals of a fine red; the *Stamina* about 30; the *Capsulæ* always 13; the Calyx from 4 to 6 leaves, deciduous; the linear Petals in 3 arrangements; the outward 9, the broadest; the next 9, between them, smaller; and the innermost 9, still smaller; in all 27. They are not *Nectaria* but *Petala*. I think it comes next to the *Dillenia*. As soon as I have finished it, I will send you my description.

Solander and Banks are daily expected, but no news of them yet.

The Ginkgo of Kæmpfer has not yet flowered, but grows freely, and we are in hopes will soon flower.

The *Hypericum Lasianthus* (Anglicè, Loblolly Bay) is a new genus; the Petals join in a tube at the base; besides the five ciliated subrotund *foliola* of the Calyx, there are constantly four bracteal leaves, like a calyx, under it, but not regularly placed, one of them being always lower than the rest. It is now going into flower; and I shall have an exact drawing made of it, with a description, which I shall send you, proposing, with my friend Dr. Garden, to have it called *Gordonia*, after our friend James Gordon.

I have seen many *animalia infusoria*, which I shall make drawings of; but have not time from public business to attend to this, or my System of Zoophytes, as I could wish. I live in hopes to print it in time; at least I will leave my figures and descriptions to the world, and wish they were done by abler hands; but am fully convinced that they are in no degree vegetables, but branched animals.

The Tea plant thrives with Mr. Lee and Mr. Aiton. No new seeds were sent this year from China. The wax inclosure is the best method. There are many of these plants in England. The *Gardenia* thrives well, both single and double: a peaty earth, with very little sand in it, is the best soil; this I lately discovered.

I wish your ambassador here would take charge of any seeds I may give him for you. I think you may procure an order from your Court for this purpose, that he may forward you any seeds when he sends expresses to Stockholm.

We every day receive new seeds from different parts of the world. Mr. Aiton, the Princess of

Wales's gardener at Kew, is the person to whom I send mine. I think him worth your correspondence, for he has it greatly in his power to serve you.

My best wishes attend you.

I am, dear Sir, your most affectionate friend,

JOHN ELLIS.

DEAR SIR, London, Dec. 28, 1770.

I received your kind favour of the 24th of November * yesterday. I now send you the characters in English of the *Illicium* and *Gordonia*, as presented to the Royal Society last month. I have not time to translate or contract them: that I will leave to you.

ILLICIUM *Floridanum.* Starry aniseed tree from Florida. *Polyand*[a] *Polygyn*[a] next to the *Magnolia.*

Calyx. The *Perianthium,* or flower-cup, consists commonly of 6 little concave membranaceous leaves, that soon fall off; the outward 3 are of an oblong-oval form, but the inner 3 are narrower, and not unlike petals.

N. B. The Calyx seems sometimes to have but 3 little leaves, so that the others may probably be changed into petals.

Corolla. The Flower consists of many Petals (26 or 27), which are lanceolate; they are of three sizes; the outward 9 are long (about an

* Missing.

inch), concave, obtuse, and spreading open *(patentia)*; the next 9 are shorter and narrower; the innermost 9 are still shorter, narrower, and very sharp-pointed.

N. B. Dr. Linnæus (from Kæmpfer) supposes the inner leaves to be *nectaria;* but they are true petals, and give the blossom the appearance of a double flower.

Stamina. The Filaments are many (about 30), very short and flat, placed close over one another *(imbricatim)*, surrounding the *germina*, or embryo seed-vessel. These support as many *Antheræ* or Summits, which are erect, oblong, obtuse, and emarginated; they have a cell on each side, full of farina of a globular form when magnified.

Pistillum. The *Germina* or Embryo seed-vessels are 20 or more in number, placed in a circular order above the receptacle of the Flower; they are erect, compressed, and end in so many sharp-pointed styles, bending outwards at the top. The *Stigmata*, or openings on the top of the styles, are downy, and placed lengthways along the upper part of each style, as in the *Magnolia*.

Pericarpium. The seed-vessel consists of 12, but much oftener of 13, little pods or *Capsules*, that come to perfection. These are of a compressed oval shape, and of a hard, leather-like substance, with two valves to each, containing one seed, and are disposed edgeways in a circular order, like so many rays of a star.

Semina. The Seeds are smooth and shining, of an oval shape, a little compressed, and appear obliquely truncated at the base.

I am sorry I cannot oblige you in changing the name of *Gordonia* to *Lasianthus*, as it has been presented to the Royal Society, and my worthy friend James Gordon has accepted this compliment from me: but I shall retain the trivial name of *Lasianthus.* For my part I do not know a greater proficient in the knowledge of plants, particularly their culture, nor so warm a friend of yours; for he always toasts your health, as the King of Botany, by the name of My Lord Linnæus, and that before he drinks the King's health. His son has your books in his hands oftener than the Bible, and is now assisting a person here in translating your *Genera* into English. Notwithstanding the efforts of Dr. Hill, and his fine plates in his grand work called the Vegetable System, your Vegetable System will bear many editions when this magnificent work will scarcely be known: for my part I know very little of it, and do not think it worth purchasing. The Royal Society, and the Society of Arts and Manufactures, will never admit him among them; he has been often refused at both. I know he is patronized by some great people. which I think does them no credit. The botanical people of my acquaintance say the figures are bad, and do not admire the work. What you read in the news-papers about Tea growing in Carolina is void of truth; as much as the *Rhus* that bears the true China Varnish, though you have let it continue in your Species. Poor Miller, through his obstinacy and impertinence to the Society of Apothecaries, is turned out of the Botanical Garden of Chelsea. I am sorry for it, as he is now 79 years of age; they will allow him his stipend, but have chosen another gardener. His vanity was so raised by his voluminous publications, that he considered no man to know any thing but himself; though Gordon, Aiton, and Lee have been long infinitely superior to him in the nicer and more delicate part of gardening. Our booksellers have made fortunes by their imposition of new editions of Miller's voluminous Dictionary, the whole or useful parts of which might have been comprehended in a book of the size of your *Genera Plantarum.*

I suppose you know Ehret is dead! We have nobody to supply his place in point of elegance. We have a young man, one Taylor, who draws all the rare plants of Kew garden for Lord Bute; he does it tolerably well; I shall employ him very soon. Besides, there is a valuable work now carrying on upon your system by Mr. John Miller, a German painter and engraver, under the direction of Dr. Gowan Knight, of the British Museum. This will make your system of botany familiar to the ladies, being in english as well as latin. The figures are well drawn, and very systematically dissected and described. I have desired that he may send to your ambassador for you the two first numbers, to know your opinion of it; and if you approve, you may get him subscriptions.

I cannot say much of Hudson. I wish you had published a *Flora Anglica*, like your *Flora Suecica*; it would have sold well here. The English names might have been taken from Lee's last edition of his Botany.

I shall certainly speak to Mr. Aiton about sending you what he has that is new. He has a most profound respect for you, and follows your system, and no other.

Dr. Kuhn is one of those American chiefs that despise us Englishmen. I sent him some seeds of the *Rheum palmatum* by a friend, and he had not the decency to thank me; but his German pride will do him no service; for, thank God, we shall now humble those American revolters. He is, to my knowledge, infinitely obliged to you; without your care in cultivating his mind he would have been a mere savage.

I have received no Tea seeds last year, but hope for some next summer. I will speak to Aiton for a leaf of the true Tea for you. We protect it in winter under a glass frame, as we do the *Illicium*; but hope both will stand our winter, when we have sufficient to try the experiment.

Near to the *Stuartia*.

GORDONIA *Lasianthus*. Anglicè, Loblolly Bay
Monadelphia Polyandria.

Calyx. The *Perianthium*, or Flower-cup, consists of 5 concave roundish downy stiff leaves, hairy on their margins, embracing the *Germen* very strongly at the base, and permanent.

Obs. About the stalk, under the Flower-cup, are four *bracteæ*, or floral leaves, placed at unequal distances, of an oblong form, concave and roundish at top, and truncated at bottom, where each seems to embrace a part of the stalk. These cover the flower, and its proper calyx or flower-cup, in its younger state; but as it grows up and expands, they begin to appear more distant from each other, and soon decay; otherwise they have the appearance of an exterior flower-cup.

Corolla. The Flower consists of 5 large, fleshy, concave, inverted-oval petals, united at the narrow part of the base, forming as it were one petal. In the inside of this narrow part, or joining, is a funnel-shaped fleshy substance, like a *Nectarium,* which is united to, and appears to be, a part of the petals: this surrounds the *Germen,* or embryo seed-vessel. The upper or broad part of this *Nectarium* is waved in such a manner, that the rising of each wave answers to the middle of each petal.

Stamina. The Filaments are numerous and linear, or rather awl-shaped; they are inserted all round, on the top, or margin, of this waved part or *Nectarium* (not separated in parcels, and united only in different *Phalanges,* as in the *Polyadelphia*, but equally distant from each other, and connected together at the bottom, by this *Nectarium* or fleshy substance). The *Antheræ*, or Summits, that contain the male dust, are of an oval form, and erect; they have a cell on each side, full of dust, of a globular shape when magnified, different from the *Farina* of the Malvaceous tribe, which is full of points.

Pistillum. The *Germen*, or embryo seed-vessel, is oval, and acuminated at the top, where the Style begins. The Style is very short, and has 5 prominent ridges, as if 5 Styles were united together; the tops of these apparent Styles end in 5 acute *Stigmata*, or openings of the Style, disposed horizontally in a radiated form, a little contorted towards the points. On the upper part of each of these Styles is a longitudinal furrow or cavity, covered with down, and ending with each of them in a point, being the true stigma.

Pericarpium. The Capsule is egg-shaped, and acuminated at the top; it is woody, and splits open at top into 5 valves, with 5 cells or loculaments.

Semina. The Seeds are kidney-shaped, and winged obliquely on one side. There are two seeds in each cell, which adhere to a membranaceous receptacle (one on each side), that proceeds from the *Columella* in the centre.

I have now no more room than to wish you most heartily the compliments of the season, and to assure you that I propose soon to set about my Zoophytes.

I am, dear Sir, most truly yours,

JOHN ELLIS.

N. B. I have got specimens of the *Theobroma Cacao* and *Guazuma*, preserved in spirits. I intend to send them to you. I am surprised at Tournefort and Plumier; how little they know of this genus! Their figures magnified are most elegantly contrived, very different from dried specimens.

Q. In what part of Sweden, or the North, do Woodcocks breed? I mean in numbers, because they by accident breed here?—Has it ever been observed that they carry off their young in their beaks?

Q. Whether the *Turdus Iliacus*, or Red-winged Thrush, sings well in Sweden? They come in flocks in winter, but have no song here.

Q. What is the title of the third part of your *Fundamenta Botanica*?

DEAR SIR, Gray's Inn, London, May 10, 1771.

It is with pleasure I inform you that we have certain accounts from the East Indies, that our worthy friend Dr. Solander, and Mr. Banks, with the Astronomers, were safe arrived at Batavia the 10th of October 1770, and that they proposed to sail for England in a month's time; so that we have great hopes of seeing them here next month.

This news gives universal joy to the learned and curious in England; but it must give the highest satisfaction to you, who esteem him so deservedly; and to all his countrymen.

I was much disappointed in not seeing your present King, as I hear he proposed coming to England if his father had not died.

I hear that Dr. Solander will, as soon as he arrives, be introduced to the Royal Family. He has been kept from the knowledge of our King by the interposition of some great men, who favour quacks; but I hear his Majesty is determined to see him when he arrives, and probably then will reward his merits.

I have had two letters lately from our friend Dr. Garden of Carolina. He has sent over a new Tortoise, or *Testudo,* and proposes sending you a specimen. I have desired Mr. Pennant to give an account of it in our Transactions, as he is now writing a *Fauna Americana.* Dr. Garden is much obliged to you for your last letter, which he received in December, and says he has more things to send you soon.

I have had so severe a fit of sickness in March last, that I expected never to have lived to have finished my account of Zoophytes.

I recollected what you desired me to do often, and have now set about it in earnest.

I cannot reconcile myself to vegetating animals: the introduction of the doctrine of this mixed kind of life will only confuse our ideas of Nature. We have not proof sufficient to determine it; and I am averse to hypotheses.

I am resolved not to mention Dr. Pallas in my book by name.

I shall adopt very few but your Synonyms.

I have made some discoveries in the *Gorgoniæ*, which I hope will please you. I have been very particular in this genus, by the advice of my friend Solander, as he thought it had not yet been sufficiently cleared up. Nothing retards my work so much as the engravers, who are not to be depended on; my best workman is dead.

I have introduced the *Actinia* among the Zoophytes; but am doubtful, though I have added to that genus the *Actinia sociata*, whether it is not a new genus. I sent you over a specimen some time ago, but never received your opinion of it. It is represented in one of the plates I sent you, with a dissection, shewing the internal part.

Look at Pallas's *Sertularia Gorgonia*, and you will find a *Sertularia* overrun by an *Alcyonium*. I have a curious plate of this, to shew the absurdity of his making this a new species. I have detected many errors in him; so that I think he had no reason to find so much fault with you.

In all probability the new and many curious observations that must have been made by Solander and Banks on the Submarine world will retard me some time longer. I am not satisfied about the ovaries of the *Antipathes*. Marsigli's account puzzles me. Solander, no doubt, has been attentive to this genus, which is in great plenty in the East Indian sea.

I have discovered many new *Animalia infusoria;* the viewing of these amuses me while I am at breakfast.

Last Autumn I had some of the common *Hydra* of the ditches, which I had preserved some time, and fed with minute worms, with which the river Thames abounds. Some of these were infested with a kind of lice, but differently shaped from those of Trembley. In order to kill these *animalcula,* I put into a watch-glass, half full of water, in which the Polype with its lice were placed, to be viewed by the microscope, two drops of a decoction of the *Geranium zonale.* The lice were killed immediately, and the Polype rendered insensible for four hours, after which, it recovered, and was perfectly free from lice. I observed the ends of two of the claws were rotted off; but this might happen from the *animalcula* corroding them.

I have mentioned to you before that the *animalcula infusoria,* which arise from the decaying stalk of the *Geranium zonale,* are killed by the fresh juice of the leaves of the top of the same specimen, that has been put into the water on purpose to produce these animals.

I always rejoice to hear from you; pray oblige your most affectionate friend,

John Ellis.

Dear Sir, London, July 16, 1771.

I could no longer defer sending you the agreeable news of the arrival of Dr. Solander and Mr. Banks, from their voyage round the World, laden with spoils, particularly of the vegetable world, some few rare ones of the animal kingdom; but I do not hear much of the mineral kingdom. They were all very healthy till they got to Batavia, where a violent fever carried off almost half their ship's crew. Dr. Solander has been very ill, but is now very well.

They have made great discoveries in Geography: the account of their voyage will soon be published, but as to their Natural History, I fear I shall not live to see it. They have sufficient for one thousand folio plates. They unfortunately lost both their painters, but the last died between Batavia and the Cape; so that most of their rarest things were drawn, but not completely finished.

They are so very busy getting their things on shore, and seeing their friends, after an absence of three years, that they have scarce time to tell us of any thing but the many narrow escapes they have had from imminent danger. I long to hear from you, and fear you have been ill, otherwise you would have wrote to me.

Be so good to inform Dr. Solander's friends of the success he has had in returning safe after so many perils, laden with the greatest treasure of Natural History that ever was brought into any country at one time by two persons.

They came last from St. Helena, with 11 East

Indiamen, who have, I hear, brought a great many rare plants from China.

I hope Dr. Solander will write to you soon himself; I shall beg of him not to defer it.

I am, my dear friend, most sincerely your zealous well wisher and humble servant,

JOHN ELLIS.

LINNÆUS TO ELLIS. [Latin.]

Upsal, Aug. 8, 1771.

I received about an hour ago, my ever valued friend, yours of the 16th of July, nor did I ever receive a more welcome letter, as it conveys the welcome news of my dear Solander's safe return. Thanks and glory to God, who has protected him through the dangers of such a voyage! If I were not bound fast here by 64 years of age, and a worn-out body, I would this very day set out for London, to see this great hero in botany. Moses was not permitted to enter Palestine, but only to view it from a distance: so I conceive an idea in my mind of the acquisitions and treasures of those who have visited every part of the globe.

I should long ago have written to you, my most esteemed friend and correspondent, had I not been detained a great while at Stockholm, in consequence of my deputation from the academy to wait on our new King.

I have received the collection of rare and valuable plants which Mr. Gordon, at your recommendation, has sent me. I know not which of you most merits my thanks, but I am very sure I should never have had these rarities without your interference. I know not how to express my gratitude sufficiently to you both. The plants all came alive, except No. 5 *Dionœa*, 8 *Lonicera marilandica*, and 11 *Andromeda*. I know too well the nature of the *Dionœa*, ever to hope for it in a living condition. I must rely on you to explain to Mr. Gordon, with due force, the high gratification I derive from the possession of what I had never hoped, in the course of my life, to behold. Whenever I look at these plants, I shall think I have you both, as it were, in my company.

The publication of my 2d *Mantissa* has been delayed by my absence. It will certainly come out before the end of next month.

But when, my good friend, shall we behold that work of yours, whose plates you sent me so long ago? Oh that I may live to see the day of its appearance! I know nothing in this world that I desire more. Do not, my friend, procrastinate. Nothing is more precarious than days to come. Pray do not trust to them.

Fate has robbed the world of many valuable things. It is commonly hostile to all the greatest of human undertakings. You alone have walked forth into the deep abyss, unexplored by all before you, nor will you ever have a rival.

If the *Cycas*, which I have been examining, should ever blossom in your collection, or any other near London, I shall be obliged to you to examine its *pollen* microscopically at the bursting of the *antheræ*. Please to delineate the figure of the said *pollen*, which I have not been able to ascertain. This point will unquestionably prove of the greatest botanical importance.

Somebody in Germany has lately written on the subject of Mosses*. He says, the round heads of these plants are their anthers, or male organs of impregnation; but he denies that Mosses have any female flowers; asserting that they are all propagated by deciduous buds, *gemmæ*. It seems absurd to me that males should be created without females, and therefore in vain. I wish you had time and leisure to examine this tribe of vegetables.

I cannot sufficiently admire Mr. Banks, who has exposed himself to so many dangers, and has bestowed more money in the service of natural science than any other mortal. Surely none but an Englishman would have the spirit to do what he has done. You are of all people the most liberal, as I experienced in Lord Baltimore.

May health and happiness attend you! Please to forward the enclosed letters.

* Necker is here alluded to. His opinion, though adopted by Haller, is now not worth refuting. There is no fact nor experiment in its favour, and the time of the perfecting of each organ in question, is decisive against this theory.

FROM THE SAME TO THE SAME. [Latin.]

MY DEAR FRIEND, Upsal, Oct. 22, 1771.

I have just read, in some foreign newspapers, that our friend Solander intends to revisit those new countries, discovered by Mr. Banks and himself, in the ensuing spring. This report has affected me so much, as almost entirely to deprive me of sleep. How vain are the hopes of man! Whilst the whole botanical world, like myself, has been looking for the most transcendent benefits to our science, from the unrivalled exertions of your countrymen, all their matchless and truly astonishing collection, such as has never been seen before, nor may ever be seen again, is to be put aside untouched, to be thrust into some corner, to become perhaps the prey of insects and of destruction.

I have every day been figuring to myself the occupations of my pupil Solander, now putting his collection in order, having first arranged and numbered his plants, in parcels, according to the places where they were gathered, and then written upon each specimen its native country, and appropriate number. I then fancied him throwing the whole into classes; putting aside, and naming, such as were already known; ranging others under known genera, with specific differences; and distinguishing by new names and definitions such as formed new genera, with their species. Thus, thought I, the world will be delighted and benefited by all these discoveries; and the foundations of true science will be strengthened, so as to endure through all generations!

I am under great apprehension, that if this collection should remain untouched till Solander's return, it might share the same lot as Forskall's Arabian specimens at Copenhagen. Thus shall I be only more and more confirmed in my opinion, that the Fates are ever adverse to the greatest undertakings of mankind.

Solander promised long ago, while detained off the coast of Brazil, in the early part of his voyage, that he would visit me after his return; of which I have been in expectation. If he had brought some of his specimens with him, I could at once have told him what were new; and we might have turned over books together, and he might have been informed or satisfied upon many subjects, which after my death will not be so easily explained.

I have no answer from him to the letter I enclosed to you, which I cannot but wonder at. You, yourself, know how much I have esteemed him, and how strongly I recommended him to you.

By all that is great and good, I intreat you, who know so well the value of science, to do all that in you lies for the publication of these new acquisitions, that the learned world may not be deprived of them. They will afford a fresh proof that the English nation promotes science more than the French, or all other people together. At the same time, let me earnestly beg of you to publish, as soon as possible, your own work, explaining those elegant plates of rare Zoophytes, &c. which you last sent me. I can no longer restrain my impatience. Allow me to remind you that "nothing is so uncertain, nothing so deceitful, as human life; nothing so frail, or surrounded with so many diseases and dangers, as man."

Again the plants of Solander and Banks recur to my imagination. When I turn over Feuillée's figures, I meet with more extraordinary things among them than any where else. I cannot but presume, therefore, as Peru and Chili are so rich, that in the South-sea islands, as great an abundance of rarities have remained in concealment, from the beginning of the world, to reward the labours of our illustrious voyagers. I see these things now but afar off. If our travellers should take another trip, I shall have seen them as Moses saw Canaan.

When I ponder upon the insects they have brought, I am overwhelmed at the reported number of new species. Are there many new genera? Amongst all the insects sent from the Cape, I have met with no new genus; which is remarkable. And yet, except four European ones, they were all new species.

Pray make use of your interest with Solander, to inform me to what class and order the Nutmeg belongs. I shall not take advantage of this information, without making honourable mention of my authority.

When I think of their *Mollusca*, I conceive the new ones must be very numerous. These animals cannot be investigated after death, as they contract in dying. Without doubt, as there were draughtsmen on board, they would not fail to afford ample materials for drawings.

Do but consider, my friend, if these treasures are kept back, what may happen to them. They may be devoured by vermin of all kinds. The house where they are lodged may be burnt. Those destined to describe them may die. Even you, the promoter of every scientific undertaking in your country, may be taken from us. All sublunary things are uncertain, nor ought any thing to be trusted to treacherous futurity. I therefore once more beg, nay I earnestly beseech you, to urge the publication of these new discoveries. I confess it to be my most ardent wish to see this done before I die. To whom can I urge my anxious wishes but to you, who are so devoted to me and to science?

Remember me to the immortal Banks and Solander.

P. S. I can never sufficiently thank you and Mr. Gordon, for the beautiful and precious trees of *Magnolia*, both the *Gardeniæ*, both the *Kalmiæ*, and the *Rhododendrum*; all now in excellent health. But the *Calycanthus*, and a tree of a new genus allied to *Hamamelis* *, I am sorry to say, are no more. They were very sickly when they came, nor did they put forth any new roots. *Dionæa* died, as might be expected, in the voyage.

My Lord Baltimore passed a day with me about a year ago, at my country house. I read over to

* This subsequently obtained the name of *Fothergilla*.

him whatever he desired. After his departure, he sent me a most elegant vase of silver gilt, certainly worth more than 150 guineas. I never received so splendid a present before. No Frenchman, nor perhaps any other person, was ever so bountiful. The English are, doubtless, the most generous of all men.

My second *Mantissa* is, at length, published. After it was finished, I received from Surinam what I call *Hypericum Lasianthus,* so similar to your *Gordonia,* that at first I thought them the same. The flower is, in like manner, internally hairy; the stem is shrubby, and the leaves similar. But the stamens are in five sets, separated by five hairy nectaries. On a careful examination, I conclude your *Gordonia Lasianthus* to be really a different plant, agreeing with that of Plukenet, in having winged seeds, as you rightly describe it. The synonym of Plukenet, therefore, does not belong to my *Lasianthus,* which, however like it, is truly a species of *Hypericum;* but that synonym must be referred to your plant.

ELLIS TO LINNÆUS.

DEAR SIR, London, Nov. 19, 1771.

I received both your kind letters of the 8th of August, and the 22d of October, from Upsal. I delivered the two letters inclosed in that of the 8th of August to Mr. Banks and Dr. Solander, and they both promised me that they would write to you;

but they have been so hurried with company that they have very little time. They are now at Oxford, where they went to receive the honorary degrees of Doctors, and are soon expected in town, when I shall lay your last letter before them.

I assure you it greatly distresses me to think of loosing Solander for ever, for I cannot expect to see him more, should he return; but I fear he never will return alive.

I shall do what I can to persuade him to print the botanical account before he goes, as it is all ready. The specimens are to be deposited in the British Museum till they return. I shall write to you soon again. I then will tell you what they determine to do, and I will beg them to write to you.

I thank you for the account you sent me of the *Gordonia*, and that you approve of it as a new genus. Mr. Gordon returns you thanks, and says you are heartily welcome to any thing in his garden.

I now inclose you a specimen of the *Polemonium rubrum**, which I think comes near to the *Ipomœa*, only the stigma is trifid. This plant I had the seeds of from Pensacola, West Florida. Gordon propagates it from cuttings; it is now two years old from seed, and only blossomed the other day. It requires heat to make it flower, but stands all weather but frost out of doors.

I shall soon set about printing the plates already engraved, and long as much as you do to finish this

* *Ipomopsis elegans. Sm. Exot. Bot. t.* 13.

work. I am satisfied that I run great risque at this time of life, of leaving it unfinished.

I sent to Baron Nolken several packages for you, from Dr. Garden; for the future shall commit them to the care of Mr. Lindegren, who has promised to take care of every thing for you. The world should know how genteel Lord Baltimore has been to you; it shews the esteem we have of your great merit.

We have had many young Tea trees brought over this year, and several undescribed trees, from China.

My best wishes attend you.

I am, my dear friend, with the utmost respect, your faithful friend and humble servant,

JOHN ELLIS.

LINNÆUS TO ELLIS. [Latin.]

MY DEAR FRIEND, Upsal, Dec. 20, 1771.

I yesterday received, with great pleasure your letter of the 19th of November.

I beseech you, by your warm regard for me, and your sense of what is just and fair, to persuade Solander to send me some specimens of plants from *Banksia*, or *Terra australis*, that I may have some idea of the vegetable productions of that hitherto unknown region. You may ask this, on the ground of his long-established friendship for me, and of my attachment to him; of his honourable character, and his botanical zeal. You may remind him, that it was I who obtained his father's consent that he should study Botany; that I have cherished him as a son, under my own roof; that I advised his visiting England; that I introduced him to you, and consequently to all your friends; that I procured him the Petersburgh professorship. If he slights my request, I scarcely think he can answer it to himself.

Polemonium rubrum is known to me by the work of Dillenius alone *(Hortus Elthamensis)*. I am greatly obliged to you for sending the plant. I perceive it is no *Polemonium*. It might, perhaps, be referred to *Ipomœa*, notwithstanding the three-cleft stigma. In order to judge of this matter with certainty, I want to know the following particulars: 1. Is the plant milky? 2. Has the fruit two or three cells? 3. Does the fruit burst at the summit? 4. Are the seeds few and large; or numerous and very minute?

I much approve of your bellows for clearing plants of their lice.

You are entitled to my best thanks for undertaking to persuade Solander to publish his first botanical discoveries, before he sets out on another expedition. Otherwise his collection may long remain in the British Museum, a prey to moths and other insects, and the fruit of so much care, labour, expense, and hazard, may share the lot of but too many human projects, to the grief of the whole world. Have the Banksian plants any great affinity to the Peruvian discoveries of Feuillée? Do any of them resemble the productions of Europe, or the Cape, or do they very widely differ? Are they akin to the plants of America? Are any new genera of Insects brought home by these travellers?

The new-found country ought to be named BANKSIA, from its discoverer, as America was from Americus.

I sincerely rejoice that you mean shortly to publish your account of the wonders of the deep, that I may behold them before I quit this mortal life.

I beg the favour of you, whenever you indulge me with another letter, to give me some particulars of Solander's discoveries in Natural History, for in my distant arctic situation, I have as yet heard nothing.

My pupil Sparrmann is just gone to visit the Cape of Good Hope. Another of my scholars, Thunberg, is to accompany the Dutch embassy to Japan. Both of them are well skilled in natural science.

The younger Gmelin still remains in Persia. My friend Falk is in Tartary. Mutis is making splendid botanical discoveries in Mexico. Koenig has found many new things in Tranquebar. Professor Früs Rottböll, of Copenhagen, is publishing the plants found in Surinam by Rolander. The Arabian discoveries of Forskall will soon be sent to the press at Copenhagen.

Pray favour me with an answer to my preceding letter, as soon as you have had an opportunity of conferring with Solander.

May you long enjoy life and prosperity, for the benefit of science!

ELLIS TO LINNÆUS.

My dear Sir, London, Jan. 14, 1772.

Upon the receipt of your obliging letter of the 20th of December, I wrote to Dr. Solander to call and spend a day with me; accordingly he came a few days ago, and I shewed him all your letters to me, and begged the favour of him, that he would send you an assortment of all the plants, which he and Mr. Banks had collected on the voyage. He assured me that he would do it, and also would send you as many of the descriptions as he could get copied; but he is doubtful whether he can send you those of the animal kingdom. I desired he might send them to the care of Charles Lindegren, who has promised to send them to you by the first ship. I shall not let him depart this kingdom in peace, unless he pays that respect to you which I know he owes you, for the pains you have taken to qualify him for the company of our King, and the greatest people in this kingdom. Mr. Banks is taking great pains to preserve the animals in spirits, at a very great expence; but I fear we shall not live to see them described. This voyage towards the South Pole is more to please the head of our Admiralty, Lord Sandwich, than is consistent with prudence. They would have had employment enough for seven years, to have finished completely what they have discovered.

Many of their seeds are destroyed by the boxes being obliged to be exposed on the shore to the heat of the sun, and bad weather, when they had like to be lost on the coast of New Holland. I hope I shall be able, in the spring, to raise the seeds of a most valuable plant, which they call *Chlamydia,* from the people of New Zealand, in the latitude of 40 deg. South, making themselves cloaks of it. It seems to have leaves like an *Aloe* or *Yucca;* the flowers are hexapetalous and ringent, and the seed-vessel is above the receptacle. Solander says, it comes nearest to the *Hemerocallis.* From the leaves of this the natives prepare a fibrous substance, like the best flax; these fibres they connect together in a very new and curious manner with a needle and thread, so as to form very durable and convenient garments.

As the seeds of it were all destroyed by the sunshine and bad weather, I begged Solander to let me look at the specimens which they had preserved in papers, and was so fortunate as to find several seed vessels perfectly sound and full of ripe seeds. I have got some of them from him, and given them to our best gardeners, and have examined them in the microscope, and find the germ in the seed perfectly sound. I suppose you will have some sent you in your specimens. I find seed vessels will preserve their seeds a very long time.

You ask me some questions about the *Ipomœa;* in answer, it is not lactescent. I do not remember whether it is *bilocularis* or *trilocularis.* I think the fruit opens at top. The seeds are not many in proportion to the size of the loculament; they are angular in respect to form.

I have received specimens of the *Brownœa,* from

Tobago, preserved in spirits, and some seeds in pods. I shewed them to Solander, and desired him to compare Lœfling, and Jacquin's descriptions. He thinks Jacquin's interior *corolla* is properly a nectarium. The seed-vessel is very strong and coriaceous. It is a *legumen* of five inches long, with two large seeds like Tamarind seeds, but much larger and rough; many of them are shorter, and contain but one seed. When the seed-vessel bursts open, it is with great force, to discharge the seeds; for the valves immediately curl up in a spiral manner, and there is no bringing them back to their form without breaking them. I think Lœfling's description better than Jacquin's. The French inhabitants at Tobago and Grenada call it bastard Courbaril, and Rosa de Monte. It is a most beautiful flower. I mentioned, on the other side, that seeds were a long time preserved sound in their seed-vessels, an instance of which I lately tried. When Doctor Alexander Russel came from Aleppo, he brought with him about the year 1755, some seeds of the *Convolvulus Scammonia*, saved in the year 1754. In looking over my seeds last year, I found some few seed-vessels full of seeds of this *Convolvulus* which he had given me. I immediately gave them to a gardener in my neighbourhood, and I can assure you several of them came up, and continued to grow during the last Summer, 1771; so that they were out of ground near 17 years. I kept them in a paper in the drawer of my bureau, in the room where I generally sit.

The most curious animal that I think they have brought home is the skin of a four-footed beast, like the *Mus Jaculus*, or *Jerboa*, described in Edwards. It seems to be about three feet and a half high, standing on its hind legs. It weighed (if I remember right) 80 pounds, and was too swift for their greyhounds, so that they were obliged to shoot it. I think they say it differs from the *Mus Jaculus*, but I shall enquire of Solander more particularly about it*. They have added very little to my Zoophytes; and could not distinguish the animal of the *Tubipora*, though the rocks were covered with it. It was covered over with a mucilage; but they at that time had their lives to think of, by endeavouring to get the ship off the rocks.

My best wishes attend you.

I am, dear Sir, your truly affectionate friend,

JOHN ELLIS.

LINNÆUS TO ELLIS. [Latin.]

Upsal, Jan. 20, 1772.

I have just received the valuable communications of our friend Garden, through your hands, and return you my best thanks. You are still the main support of Natural History in England, for your attention is ever given to all that serves to increase or promote this study. Without your aid, the rest of the world would know little of the acquisitions made by your intelligent countrymen, in all parts of the

* This was the now well-known Kangaroo.

world. You are the portal through which the lovers of Nature are conducted to these discoveries.

For my own part, I acknowledge myself to have derived more information, through your various assistance, than from any other person. I rely on your usual kind attention to convey the annexed letter to our mutual friend Dr. Garden, the happy illustrator of Nature in his own region of Carolina. If your intercession does not procure me a few plants from my friend Solander, I shall despair of ever seeing any vegetable production of the antarctic world. I chiefly should have wished to see whether these Southern plants differ in aspect from what we know in other countries, as the Cape ones do from those of Europe.

I trouble you so often with my letters, that I fear I shall exhaust your patience; but I could not defer writing to the worthy Dr. Garden.

Farewell, bright star of Natural History! Do not forget your affection for me.

FROM THE SAME TO THE SAME. [Latin.]

MY DEAR FRIEND, Upsal, Aug. 13, 1772.

At the meeting of our Academy of Sciences of Upsal yesterday, you formed the chief subject of conversation. I explained the nature and the quantity of your discoveries; exhibiting your book, along with some specimens there delineated. I laid before the assembly your unpublished plates, and gave an account of what you now have in contemplation to communicate to the world. All the naturalists who heard me were delighted. The rest of the party, both divines and philosophers, were astonished. They are lynxes abroad, but moles at home. All with one accord requested me to express their respect for you, and their anxiety that this still greater work of yours should not long be delayed. This I likewise wish, with all my heart, that I may behold it before I die. There surely never was a bolder man than you, who being wise enough to be fully aware of the brevity and frailty of human life, nevertheless delay the publication of a work which is to last for ever.

I again trouble you with a letter to Dr. Garden, for whose acquaintance, and for the fine plants he has sent me, I am entirely indebted to your kindness. Be so good as to express my regard for him whenever you write.

Pray let me know whether you have begun to print your book.

I have two pupils now at the Cape of Good Hope. If they send me any thing in your department, you may depend on having it.

I will, if I can, get you that portrait of myself which you are desirous of having. I wish I had a likeness of you in my museum.

Adieu, may you long live the Phœnix of our science!

Biographical Memoir

OF

ALEXANDER GARDEN, M. D. F. R. S.

AND HIS

CORRESPONDENCE WITH LINNÆUS AND ELLIS.

ALEXANDER GARDEN, M. D. F. R. S. a native of Scotland, and educated at Edinburgh, resided at Charlestown, South Carolina, where he was extensively engaged in the practice of Physic for near 30 years. He married there, on Christmas-eve, 1755, as appears from one of his letters to Mr. Ellis, with whom he maintained a frequent scientific and friendly intercourse, and by whom he was introduced to the correspondence of Linnæus. Botany, and some of the more obscure departments of Zoology, especially fishes and reptiles, were his constant resources for amusement and health, amid the sometimes overwhelming duties of his profession, and the inconveniences of a delicate constitution. In Natural History he was, throughout, a zealous and classical Linnæan. No one welcomed the publications of the Swedish luminary, from time to time, with more enthusiasm, or was better able to appreciate them; for he had felt by experience the insufficiency of preceding systems of Botany, and had been, in consequence, near giving up the science in despair.

When the political disturbances of America came on, Dr. Garden took part with the British Government, and, like many others, suffered a very considerable loss of property. He returned to Europe about the end of the war, with his wife and two daughters, residing for some years in Cecil street in the Strand. A pulmonary consumption, confirmed by the effects of sea-sickness, terminated his life April 15, 1791, in the 62d year of his age. His son conformed to the new American Government, and remained in Carolina.

The cheerful, benevolent character of Dr. Garden is conspicuous in his letters. His person and manners were peculiarly pleasing; and he was a most welcome addition to the scientific circles in London, as long as his declining health would permit.

The botanical history of that elegant and delightful shrub, the *Gardenia florida*, makes a conspicuous figure in the correspondence of Ellis, who established the genus, and of Linnæus. So many species have since been found to belong to it, that this is now one of the most extensive, as it is certainly one of the most beautiful and fragrant genera, in the whole vegetable kingdom.—See the botanical as well as biographical articles, GARDEN and GARDENIA, in Rees's *Cyclopædia*.

Correspondence.

ALEXANDER GARDEN, M.D. TO LINNÆUS. [Latin.]

Charlestown, South Carolina, March 15, 1755.

LEARNED SIR,

A year ago your *Fundamenta Botanica, Classes Plantarum*, and *Flora Virginica Gronovii* came to my hands. I have read over and over again, with the greatest pleasure, your *Fundamenta Botanica*, and, if I am not deceived, have greatly increased my knowledge. From that time I have sedulously devoted myself to the study of your sexual system, by which I have, most certainly, made greater progress in the space of a year than in the three preceding ones, following the method of Tournefort. The reading of that inimitable little collection of aphorisms engaged and delighted my mind so powerfully, that, for one whole summer, scarcely a week passed away, without my re-perusing it with the greatest attention. Such neatness! such regularity! so clear and supremely ingenious a system, undoubtedly never appeared before in the Botanical world. Nothing can be more finished than your works, which will be read with avidity, by those most deeply versed in such studies, for ages to come. I may say I have myself bestowed some little labour and time in these studies; but I freely acknowledge that I, when I read your works, learn from you, not only things of which I was previously ignorant, but even what I thought I had already learned from other teachers. Botany never was placed before in so clear a light. It is not only more easy for beginners, but more perspicuous to the learned, appearing in so new and elegant a form, as to be much more attractive. How much you have deserved of all lovers of Botany is strikingly evident; and I therefore earnestly intreat you to accept this testimony of my gratitude, for the benefit which I have received from your writings.

Meanwhile I have greatly lamented that none of your publications, except the abovementioned, has reached me, though I ordered all your works, hitherto published, to be purchased for me last year in Holland. But how egregiously was I disappointed! It was written, in answer, that no booksellers would have them for sale, *Bibliotheca Botanica* excepted, till new editions came out. An opportunity perhaps might never have occurred to me of seeing your most excellent and elaborate characters of plants, your *Systema Naturæ*, or *Critica Botanica*, had I not been compelled, last summer, on account of my health, to visit the northern parts of America.

But whilst I travelled over these regions, I not only profited by reading your books, but also by the company of several most learned botanists, while I enjoyed the discovery and examination of many plants not known where I reside. I could scarcely regret the indisposition that had brought me hither. What botanist, for the sake of such enjoyment, would not even submit to a temporary diminution of health?

When I came to New York I immediately enquired for Coldenhamia, the seat of that most eminent botanist Mr. Colden. Here, by good fortune, I first met with John Bartram, returning from the Blue Mountains, as they are called. How grateful was such a meeting to me! and how unusual in this part of the world! What congratulations and salutations passed between us! How happy should I be to pass my life with men so distinguished by genius, acuteness, and liberality, as well as by eminent botanical learning and experience!—men in whom the greatest knowledge and skill are united to the most amiable candour.

—— Animæ, quales neque candidiores
Terra tulit. ——

Whilst I was passing my time most delightfully with these gentlemen, they were both so obliging as to shew me your letters to them; which has induced me, Sir, to take the liberty of writing to you, in order to begin a correspondence, for which I have long wished, but never before found the means of beginning. In the mean time I am sufficiently conscious of my inability to repay such a favour as I ought, or as these eminent men are competent to. Still I will not yield to them in my ardent desire to imbibe true science from the same source, and to quench my thirst at so pure a spring. My anxiety is not inferior to theirs, that I may meet with any thing in my neighbourhood worthy of your observation; nor shall I be less ready to send whatever may be likely to prove acceptable. We have, I perceive, many plants wild in the more southern parts of North America, which cannot bear the climate of the northern provinces. If, Sir, any of these would be interesting to you, I would either, with all possible attention, make descriptions from living plants, or I would collect dried specimens or seeds, and send you living plants or roots in pots, according as may be most useful.

I have passed nearly three years in America; the first, and also a great part of the second, I have almost entirely thrown away (pardon the expression, for I speak feelingly,) in following Tournefort's system, in which I was first instructed by Dr. Charles Alston in the Edinburgh garden. Having been invited to Carolina (where from that time I have, with tolerable success, practised medicine), being furnished with the *Institutiones Rei Herbariæ* and the writings of Ray, I made daily excursions into the country. But the immense labour of reducing the plants I collected into proper orders, to say nothing of the uncertainty attending the investigation of genera and species, and, still more, of determining varieties, according to the Tournefortian system, was all so very tiresome, that at length my patience was exhausted; and had I not, by good fortune, met with your most excellent works above-mentioned, I might have been stopped in my progress, and have altogether given up this most pleasing of pursuits. I had already almost yielded to this uncertainty and want of satisfaction, when my friend Mr. Bull put your writings into my hands. These have first introduced into Botany the science of demonstration, founded upon mathematical principles; and at the same time a sound, as well as easy mode of classification.

What has been passing in the literary world, is entirely unknown to us, in these regions, wholly, alas! separated from Europe. If therefore any thing new has appeared, either in medicine, botany, or any other part of natural history, you will render me a great favour by informing me of the titles of such publications. Pray send me those of all your own works; and inform me how my letters may best reach you. I have inclosed this in one for Mr. Gronovius, earnestly entreating him to send it you by the first opportunity. If I am not mistaken, your answer, if you should deign to write one, may easily be conveyed; for every year four or five ships from our part of the world go to Rotterdam, laden with rice, and soon after discharging their cargoes, return to Carolina. I have now many things in readiness, which I would willingly communicate to you; but lest I should give you trouble rather than pleasure, I think it better to postpone sending them till I find, by your letters, whether my correspondence be agreeable to you or not.

I request your friendly correction of the description * inclosed herein; and I rely on your advice and kind encouragement to confirm and strengthen me in this delightful study, which I beseech you not to deny me. May you, my dear Sir, long continue to be, as you already are, the pride and ornament of botanical science! Such is the sincere wish of,

your most devoted admirer,

ALEXANDER GARDEN.

If you honour me with an answer, please to direct to Dr. Alexander Garden, Physician, Charlestown, South Carolina.

FROM THE SAME. [Latin.]

April 2, 1755.

About three weeks since I sent you a letter †, to the care of Dr. Gronovius. But, lest the captain of the ship, who took charge of it, should be negligent, or any other accident should prevent its reaching you, I have thought best to transmit a duplicate herewith by another captain. I should be very sorry if any thing were to impede our correspondence. Since that letter was dispatched, I have received from England the *Critica Botanica, Characteres Plantarum, Philosophia Botanica, Amœnitates Academicæ*, and *Systema Naturæ*. Furnished with these arms, I am preparing to make war upon the Vegetable kingdom, and to submit the lofty ho-

* This description does not appear.

† The foregoing.

nours of the forests to the rule and authority of Botanic Science.

A correspondent informs me that you have lately favoured the publick with a most valuable book, which has long been wanted, entitled *Species Plantarum.*

May success ever attend you! and may you not forget your devoted disciple! Once more farewell!

FROM THE SAME. [Latin.]

SIR, Charlestown, Nov. 30, 1758.

Three years ago I troubled you with a letter, by way of Holland, of which I sent also a duplicate; but I fear they have both accidentally miscarried.

From that period I have often thought of soliciting afresh your friendship and correspondence, but shame has deterred me. I am well aware that your time must be fully occupied with more valuable correspondents, and that I am likely to be more troublesome than useful, having nothing worthy to repay such an indulgence. I do however stand in great need of your advice and assistance in the prosecution of the most delightful of studies; and such is my conviction of the benevolence of your character, that I cannot refrain from writing you another letter. I earnestly beseech you to take this in good part, and not to refuse me the favour of your friendship. Mr. Ellis, in a recent letter, encourages me to believe that my correspondence may not be unwelcome to you, which, you may well suppose, has greatly delighted me; and it has induced me to hope you will pardon this intrusion. I learn from him that you have already written to me; and it has given me no small concern that your letter has never come to hand. I flattered myself, as long as I possibly could, with the prospect of its arrival; but I have now given up all hopes, and am only sensible of my loss and mortification.

Had it not been for the repeated encouragement of Mr. Ellis, I should scarcely ever have ventured to expect that my friendship and correspondence could engage your attention; nor can I now attribute your favour and kindness towards me to any other cause than, probably, to the too partial representations of this friend. I fear that his usual indulgence for me, of which I have had repeated instances, may have prompted him to say more in my recommendation than my abilities deserve, or than truth can justify.

Of this I am very certain, that if you do deign to correspond with me, I can never repay such a favour as it deserves. Nevertheless, I am ready to receive and to obey your wishes and directions; and if this country should afford any thing worthy of your notice, I will, if you please, make descriptions, or send specimens, with all possible care. Your commands will indeed prove most welcome to me. I have only to request that you will inform me of every thing you want, and of the best methods of preserving and forwarding specimens. Every opportunity that you may be so good as to afford me of serving you, I shall esteem an honour; and if at the same time you favour me with your advice, and allow me to drink at the fountain of pure botanical science, from your abundant stores, I shall esteem it the highest honour, as well as gratification, that I can enjoy.

Almost every one of your works is already in my hands, and I trust I have thence greatly improved my knowledge of Botany. Mr. Ellis informs me of your being about printing a new edition of your *Systema Naturæ* and *Genera Plantarum*, both which I have ordered to be sent me as soon as they appear. From the riches and erudition of what you have already published, your whole mind being devoted to this one pursuit, I am at no loss to anticipate the still greater degree of information, elegance, and perfection of your future performances. Nothing indeed more excites my wishes, as a certain source of pleasure and improvement, than to be more deeply conversant with your writings; that I may not only profit by your genius, but, at the same time, have the information of the most eminent and approved writers in Botany always ready at hand.

I am disgusted with the coarse and malicious style in which some carping and slanderous criticks have attacked these works of yours, the delight and ornament of botanical science. But such men are objects of pity rather than anger. Their blind inclination to find fault leads them so far into the mazes of absurdity, that they censure what ought to afford them nothing but instruction. Their futile reasonings indeed fall harmless to the ground, like the dart of Priam from the shield of Pyrrhus. The works they abuse shine brighter the more strictly they are scrutinized, and will certainly be read with delight by men, in every age, who are best qualified to appreciate their value. Your censors, when duly weighed themselves, seem to have acquired what they know by application rather than by any great powers of mind; and they make but a poor figure, with all that they can find to say, when they enter into a controversy with a man whose learning has received its last polish from genius. Nor are you, my excellent friend, unsupported in the contest; for you are surrounded by all who have entered on the same studies at the impulse of genius, or under the auspices of Minerva, and whose industry has gradually improved, sharpened, and given the last finish to the powers of their understanding. These stand ready armed for the battle, in your defence. They will easily put to flight the herd of plodding labourers; for nature can certainly do much more without learning, than learning without nature.

If your adversaries and detractors had candidly pointed out the disputable, inconvenient, or faulty parts of your System, for your better consideration and revision, I have no doubt that they would now have found in you a friend and patron, instead of an enemy and conqueror. But they were excited by an envious malignity, and a depraved appetite for controversy, to write without judgment or genius, and to blame without candour or liberality. Not that I pretend to say, that your System is already brought to the supreme point of perfection. That would indeed be a foolish assertion, which your better judgment would at once reject, as mere flattery. But to give due praise to supreme merit in botanical science, and to recommend, as they deserve, your most ingenious and most useful writings, is a duty incumbent on me, as well as on all who are not destitute of every spark of gratitude, for the immense services which your labour and ingenuity have rendered to the whole world. Nor are you, Sir, so little able to appreciate your own merits as not to be perfectly conscious that the attacks alluded to originate in envy, rather than the commendations you receive, in flattery. Compliments out of the question, we certainly ought to give every one his due.

But it is time to conclude. I venture to inclose, for your opinion, the characters of a very handsome plant, which seems to me a new genus. I am very anxious that it should bear the name of my much-valued friend Mr. Ellis; and if, upon mature examination, you should judge it to be new, I wish you would correct my description wherever it may be necessary, and publish it in the new edition of your *Genera Plantarum*, under the name of *Ellisia.* This plant grows about the bases of the Apalachian mountains, rising annually, from its old roots, to the height of about 12 feet, ornamented with whorls of leaves, at the distance of 18 inches from each other.

It only remains for me, Sir, to beg your pardon for this intrusion. I am well aware how many important labours you have on your hands, and you probably have many more in prospect. Grant me only your friendly assistance in my ardent prosecution of the study of Nature; and may you, at the same time, go on advancing in reputation and success! and after you have given your works to the publick, may you long enjoy the honours which your abilities have acquired!

May God grant you a long life, to investigate the secrets of Nature, as well as to improve the powers of your mind in their contemplation!—and may your valuable exertions benefit the literary world as long as you live! Such is my sincere prayer. Farewell!

ELLISIA.

Cal. Perianth of 4 lanceolate, acute, spreading, permanent leaves, shorter than the Corolla.

Cor. Petals 4, horizontally spreading, deciduous, oval-lanceolate, with convoluted points; the upper side variegated with spots. Nectaries 4, one on the upper side of each petal, in the form of a hollow, callous, gaping tubercle, bearing honey, and fringed with innumerable long, thread-shaped, twisted, converging bristles.

Stam. Filaments 4, inserted between the petals, at their dilated combined bases, spreading, awl-shaped, the length of the calyx. Anthers oblong, incumbent.

Pist. Germen erect, ovate, compressed. Style awl-shaped, round, deciduous, longer than the stamens. Stigma cloven.

Peric. Capsule elliptical, compressed, rough, pointed, of 1 cell and 2 valves, separating when ripe. Receptacles at the junction of the valves.

Seeds numerous, compressed, encompassed with a thin membrane, imbricated, inserted into the edges of the valves.

The *Root* is fibrous, thick, perennial.—*Stem* simple, straight, hollow, rigid, about 12 feet high, annual.—*Leaves* simple, very large, elliptic-oblong, entire, smooth, ribless, succulent, spreading, sessile, 6 or more together, in whorls on the stem 18 inches asunder.—*Flower-stalk* terminal, loosely panicled, consisting of numerous axillary, slightly inflexed, partial stalks, crossing each other. It flowers in June *.

* Linnæus confounded this plant with his *Swertia difformis*, described from the Herbarium of Clayton, of which therefore, having no specimen in his own collection, he forgot the appearance. Mr. Pursh, on examining Clayton's specimen at Sir Joseph Banks's, detected this error, and followed Walter and Michaux in calling Dr. Garden's plant *Frasera*; at the same time adverting to a difference between its fruit and that of *Swertia*, which he does not distinctly define, and which seems to us of no moment. The plant has the true habit and characters of a *Swertia*, and is described, under the appellation of *S. Frasera*, by the writer of this, in Rees's Cyclopædia, v. 34.

[Postscript by Mr. Ellis, through whose hands this letter was sent.]

Sir,—I have inclosed a flower of this plant for your examination. As I may be out of town when you write to me about this new genus, please to direct to Mrs. Nevill's, Park Street, Westminster, and you will oblige your humble servant, J. Ellis.

London, April 27, 1759.

DR. GARDEN TO LINNÆUS. [Latin.]

Charlestown, Jan. 2, 1760.

A few weeks since I received your most welcome letter of May 30, 1759, by which I have the happiness of perceiving that you do not refuse to honour me with your friendship, and that you are pleased to express your good opinion of me in the most open and candid terms, however unworthy I may be. I cannot but congratulate myself on being numbered, through your indulgence, among the friends of so distinguished a man. Be pleased to accept my best thanks for this favour.

Whatever may be my ardent devotion for the study of Nature, and whatever my excellent friend Mr. Ellis may have written to you respecting my proficiency, I must beg you to attribute it to his partiality, rather than to any thing else. For a long time no inability or indolence of mind, no want of botanical books, could hinder my making considerable progress in this study. But although the wonderful works of God were constantly before my eyes, while I was destitute of a true guide I viewed them all to no purpose; and, I confess to my shame, I have trampled some of the most beautiful carelessly under foot.

A few years since, however, chiefly through the advice and assistance of Mr. Ellis, I procured many helps. The same generous friend has just made me a present of your most elaborate and splendid work the *Hortus Cliffortianus*. As I now enjoy the support of your encouragement and counsel, I have no doubt that I shall speedily make more important advances.

The plant I sent you last year, under the name of *Ellisia*, has again come under my careful examination, and I must confess I can scarcely make it a *Swertia*. The *calyx* is certainly of four leaves; and not of one piece, in four or five divisions, like that of *Swertia* *. The *corolla*, moreover, is certainly of four petals, and not in four deep divisions as in *Swertia*; neither do the stamens grow out of the petals, as in every *Swertia* that I have seen, but are quite separate from them, combined at their base, and inserted into the receptacle of the flower, between the corolla and germen. The *Swertiæ* generally have no style; but the *Ellisia*, or whatever

* Most of the differences here pointed out really do not hold good, with respect to the true type of the genus, *Swertia perennis*; whatever anomalies may exist in some American species. The nectaries indeed are two to each petal in *S. perennis*, and less strongly fringed.

you may please to call it, has a style longer than the stamens. The structure and appearance of the nectaries is also altogether singular.

It is two years since I examined the fresh plant; nor do I recollect any thing of its taste. At your request I have now tasted the flowers, leaves, and stem of the dried specimen, in which I find rather the flavour of tobacco than of gentian; for though I took but a very small piece, it produced nearly the same kind of nausea that tobacco usually does. The external appearance, or habit, is likewise more like a *Nicotiana* than a *Gentiana*.

I only suggest my ideas for your consideration. Your experienced eye will at once, as by intuition, perceive the truth. Allow me, however, to request that you will again examine whether this plant be a *Swertia* or not, for which purpose I send you a fragment. I wish I had a better and more complete specimen to offer you. I have freely laid before you my reasons for thinking it a new genus, in which I may, very probably, be mistaken. If so, pray set me right; for nothing can be more grateful or valuable to me, than to borrow a ray from your brilliant light, for my guidance and correction.

I shall not fail to obey your commands in collecting insects, hoping next summer to procure you plenty, which I shall stick into a box, lined at the bottom, as you direct, with wax impregnated with verdegris. I mean also to send you as many serpents as I can; noting at the same time the number of their broad and narrow scales, lest any accident should happen to the specimens.

Before the receipt of your letter, I had scarcely paid any attention to our fishes; but all your wishes are commands to me, so that I wish you never to have occasion to repeat them. Nothing would grieve me more than to disappoint you in any respect, and therefore I immediately set about procuring all the kinds I can. I have caused their skins to be dried, by which I think you will be able to see the true situations of the fins. This will be more satisfactory to you than a bare, and perhaps inaccurate, description of mine. Nevertheless, I will subjoin their characters, however imperfectly described.

I am sorry that but few of those described in Catesby are, at this time of the year, to be had; but I believe what I send will be sufficient to make you aware of the inaccuracy and inability of this writer, in describing and delineating natural objects. Please to observe the *Albula* *, our Mullet; and you will immediately perceive that he has not only forgotten to count and express the rays of the fins, but that he has, which is hardly credible, left out the pectoral fins entirely, and overlooked one of the ventral ones. So he has done in most other instances. It is sufficiently evident that his sole object was to make showy figures of the productions of Nature, rather than to give correct and accurate

* *Mugil. Albula. Linn. Syst. Nat. v.* 1. 520.

representations. This is rather to invent than to describe. It is indulging the fancies of his own brain, instead of contemplating and observing the beautiful works of God.

"*Pictoribus atque poëtis*
*Quidlibet audendi semper fuit æqua potestas**."

Of Catesby s second volume, I have in my possession only a few imperfect sheets, and therefore am unable to determine how many of the fishes now sent may be in that volume. Being unacquainted with the scientific names of this tribe, I have attached to each fish a ticket, bearing its common name in this country, with numbers, referring to my several descriptions, by which you will be enabled easily to compare them together.

I entreat you to send me their true characters, as well as their proper names, and the principles of their scientific description; for although I have never attended to this part of natural history, I cannot sufficiently admire the wisdom of the Creator, as displayed in the finny tribe. Your letter has powerfully excited my curiosity to investigate the characters of fishes, and I return you my best thanks for the taste and ardour you have given me for this new study, in which I must rely on you for your kind assistance, lest the spark you have struck out should be again extinguished.

Allow me to submit to your examination the characters of a very pretty shrub, which I first met

* Painters and Poets have been still allowed
Their pencils and their fancies unconfined. ROSCOMMON.

with, about three years ago, in a marshy and shady situation. I have seen it three or four times since, in the same kind of places, and on more accurate investigation, I have thought it a new genus. For fear of mistake, however, I send a dried specimen for your inspection; and if I am right in this opinion, be so good as to give the plant a suitable name, and refer it to its proper place in your *Systema Naturæ*. Adieu.

New Genus. *Decandria Monogynia.*

Cal. Perianth of one leaf, downy, tubular, gibbous, two-lipped, permanent; both lips torn.

Cor. of one petal, twice the length of the calyx, in five deep segments; tube gibbous; segments spreading, somewhat reflexed, abrupt, minutely pointed.

Stam. Filaments ten, awlshaped, attached to the tube, shorter than the petal. Anthers erect, gibbous.

Pist. Germen globose. Style erect, rather longer than the filaments. Stigma simple.

Peric. Drupa dry, oval, hoary.

Seed. Nut smooth, marked with two or three longitudinal furrows; the kernel sweet and eatable.

(Linnæus has marked this description *Halesia;* but he afterwards, at the suggestion of Ellis, gave that name to a more distinct genus, of the same natural order, as it now remains. This plant of Dr. Garden's is *Styrax grandifolium* of Solander, in *Ait. Hort. Kew. ed.* 1. *v.* 2. 75.)

FROM THE SAME [Latin.]

SIR, Charlestown, April 12, 1761.

You may perhaps wonder at my intruding upon your more important business with another letter, before you have encouraged me to do so by answering my last. Nevertheless, I am confident of obtaining your forgiveness, when you are informed that my chief motive is my anxiety to prove my devotion to you, as well as to fulfil my promise, by sending you some preserved specimens of fishes. I have, indeed, for some time refrained from writing, in daily expectation of hearing from you; but I am afraid to miss the present opportunity of sending by a sloop of war, not knowing when another equally safe conveyance may occur; perhaps not in the whole course of the summer; and in the mean while all the fish I now send might be devoured by ichthyophagous worms and insects.

Since I wrote to you last year, the 1st and 2d volumes of the new edition of your *Systema Naturæ* are come to hand. I have read this work over and over with the highest delight, and shall read it again and again; for when I have, as it were, devoured it, so as to convert it into substantial nourishment, I hope to be so far advanced in my acquaintance with the Creator's works, as to be less dissatisfied with myself, and more on a par with you. I find in this golden book so many things altogether new, so many great and rare discoveries, all classed in so natural and original a manner, and adorned with such luminous and classical descriptions, that I have no doubt of its procuring you, from every quarter, the greatest honour and applause. It has thrown the most brilliant light upon Natural History, and the facility it affords is no less admirable. I confess myself to have derived so much benefit from this excellent book, that I cannot but feel under the highest obligations to its illustrious author. What service could you render me of equal importance with that of furnishing my inexperience with a clue through the intricate study of the wonderful works of God, the love and contemplation of which will ever be, as they are, my chief delight, as long as life remains.

May you, Sir, who are the favoured priest of Nature, and already deeply initiated into her mysteries, go on to inform yourself more and more, to examine and discover every thing that is possible, and to instruct us in your invaluable writings. May your labours serve gradually to dispel every cloud that envelopes the brightness of her glories, and conceals any of them from mortal eyes! I heartily rejoice that there are men, whose superior talents and acquirements lead them kindly to assist and direct less powerful minds, to examine, study, and adore the evident footsteps of the Supreme Creator, so striking and so glorious, even in this world.

I have sent you all the fishes that I have been able to collect, accompanied by as exact descriptions as I could make. I trust you will accept this communication, however small, with indulgence. It is a trifle which, though faulty, is beneath your resentment, and I hope you will rather correct than condemn me. Doubtless, many things will offend your more acute and practised judgment, and these I beg you to expunge, by which, most certainly, you will essentially oblige me, especially if you are not sparing in your remarks and sentiments upon the whole.

Many specimens of fish, preserved in rum, are sent herewith, that you will not find noticed in my descriptions; as well as several specimens of insects, and *Amphibia* of the orders of *Nantes* and *Reptilia*, natives of this country, particularly some Tortoises and Lizards. Among the latter is a young Crocodile, called here the Alligator, and there are a few more things picked up by chance, which seem to me little known.

The following fish I believe you will find to constitute new genera, not yet noticed; viz. No. 11, called here the *Mud Fish**; No. 16, the *Toad Fish*†; and No. 33, which has no English name that I have been able to discover. The two former I have sent you; the latter, being a solitary specimen, I keep by me for the present, till I hear your opinion of the rest. I cannot determine whether No. 23, our *Bermuda Whiting*‡, ought to be a new genus or not. I send two or three specimens, that you may examine them with your own eyes, and give me your opinion.

* *Amia calva, Linn. Syst. Nat. v.* 1. 500.

† *Gadus Tau. ib.* 439.

‡ *Perca Alburnus. ib.* 482.

I conceive that many of the species already sent are either entirely new, or not as yet perfectly well determined. I subjoin a list of such, with their numbers, and the names by which they are known here, till you can hereafter examine them yourself.

The following I think new.

No. 12. *Perca*, here called Yellow-tails. (*P. punctata. Linn. Syst. Nat. ed.* 12. *v.* 1. 482.)

14. *Perca*, here called Black-fish. (*P. atraria. ibid.* 485.)

17. . . . . . here called Silver-fish. (*Argentina carolina. ibid.* 519.)

10. *Silurus*, here called Cat-fish. (*S. Felis. ib.* 503.)

25. *Balistes*, here called Old Wife. (*B. hispidus. ib.* 405.)

27. *Pleuronectes*, here called Taper Flounder. (*P. Plagiusa. ib.* 455?)

38. *Raja.* Is not this in some measure allied to the *Torpedo?*

40. *Labrus*, here called Fresh-water Trout. (*L. auritus? ib.* 475.)

41. *Labrus*, here called Fresh-water red-bellied Perch.

42. *Labrus*, here called Fresh-water speckled Perch. (*L. Hiatula? ib.* 475.)

43. *Perca*, here called Fresh-water Bream.

I have sent you the skins of these, as well as of what are described, except a *Balistes*, No. 23, all carefully taken off and dried, with a slip of paper to each, bearing the numbers and vernacular names, as last year; that you may compare my characters with the specimens, and determine whether they are properly defined. I shall be happy if any of my descriptions meet your approbation; but shall be thankful for your friendly corrections, wherever they may be requisite, being conscious of my own deficiences in this department. Pray be not sparing of your criticisms; for by their means my faultering steps may be encouraged and confirmed.

Whilst I was occupied about these descriptions, the last edition of your *Systema Naturæ*, and the Ichthyology of Artedi, have been always in my hands. I have studied those books over and over again, never without great satisfaction, and, I trust, with considerable profit. I have also consulted Catesby, as it seemed proper to do so, but never without disgust and indignation. I cannot endure to see the perfect works of the Most High, so miserably tortured and mutilated, and so vilely represented. His whole work, but especially his 2d volume, is so imperfect, and so grossly faulty, that to correct its errors, and supply its deficiencies, would be no less laborious, than it is necessary.

The characters of Insects and Serpents have not been attended to by me, nor have I been able to turn my mind to the principles on which they depend. My time and thoughts are occupied with many things unconnected with literature, or at least with Natural History; nor do I feel myself endowed with the peculiar talents requisite for the study of those tribes. Even the time spent in describing these fishes, has necessarily been stolen from the usual hours of sleep. If your most delightful letters had not, from time to time, so powerfully excited me, I doubt whether I might not, before now, have given up these pursuits, interrupted and teized as I am by other avocations.

I have long looked for an answer to my last, and hope speedily to be favoured with hearing from you. Your letters, and those of Mr. Ellis, always serve as a fresh stimulus to my mind, and the only thing I have to request is, the continuance of your kindness and encouragement.

I much wish that my opportunities of sending you any natural curiosities, were more convenient and frequent. I should gladly take advantage of them; for nothing can be more welcome to me than any means of testifying my veneration and attachment, or making some return for the friendship you have shown me.

You will here find the descriptions above mentioned, which, however imperfectly drawn up, may, by your greater skill, be corrected and improved, so as to prove of some utility, provided you think them worth keeping as a pledge of my regard. Farewell.

(These descriptions have afforded many materials for the 12th edition of the *Systema Naturæ*, where Dr. Garden's name is subjoined to numerous new, or little known, species of fish and reptiles in particular.)

FROM THE SAME. [English.]

SIR, South Carolina, Charlestown, June 2, 1763.

Your most agreeable and most acceptable letter, dated at Upsal, 5th Oct. 1761, was very safely conveyed to me by our mutual and obliging friend Mr. Ellis, some time last summer. I can scarcely express the pleasure which I had, to find that my last letter, with the specimens of fishes, insects, serpents, &c. that accompanied it, were agreeable to you. Nothing could, possibly, give me more joy or satisfaction, than the obeying your commands, and my happiness will always be in proportion to the success that I may have in executing them to your liking and expectation. The favourable opinion which you are pleased to express, of the characters which I wrote of the fishes, could not miss to be very agreeable to me, as I am well assured that your approbation not only stamps a value on them, but it likewise convinces me that they were properly executed, the more especially as you intend me the honour to give several of them a place in the new edition of your most excellent and admirable System of Nature.

It is to you, Sir, only, that I am indebted for all that mental pleasure and rational enjoyment, which I have had in examining, determining, contemplating, and admiring this wonderful part of the works and manifestations of the wisdom and power of the Great Author of Nature. Before I received your letters, I had never turned my thoughts to that branch which I now find so full and replete with innumerable marks of Divine wisdom and goodness, the study of which I will pursue with great care and attention, as soon as my business of the practice of medicine will permit me. And, as I intend to withdraw myself from that in two or three years at most, I shall then wholly devote my time and thoughts to the cultivation of the delightful and engaging study of Nature. A few stolen moments are all that my present hurry of business will permit me to spend in these enquiries; but when that passes over, I hope I shall be able greatly to extend my annual collection, which will be wholly dedicated to you.

In the mean time, you may be assured, that I will send every curious specimen that I can procure; and as the general peace will let our correspondence be more regular, I will take care to apply oftener to you for your assistance and advice, and I shall always have the greatest satisfaction when you write fully on each article, because it is only from your observations and remarks that I can improve or make advances in the science.

I shall pay particular attention to the several articles which you mention that you want from America, but I must observe, that I never saw nor had the 3d volume of Catesby, so that I know not what these things are to which you refer there.

Your conjecture about the amphibious nature of the *Diodon*, &c. is perfectly right. The bones are cartilaginous in part, and they have all lungs, but they have *Branchiæ* or gills likewise; which, I suppose they use alternately, or occasionally, as they happen to be on land or in water. The lungs are rather smaller than in those that have no *branchiæ* at all. The prickly belly may, probably, serve them when on dry-land, to promote and forward their motion or progression, as the abdominal scales on the snake's belly serve them.

I cannot conceive any other use of the prickles or roughness of their bellies. This is another mark of the wonderfull art and contrivance of the All Wise Author of Nature, who so curiously adapts and fits every make and structure of creatures to answer all the ends and purposes of life, and their creation. I dissected the fish No. 12, of which I send you the specimen (which by the by is not very exactly preserved), and found lungs in it, which I cut out and preserved in spirits, and now send you in the phial No. 1, for your own examination. Let me beg your remarks at length on this matter.

I come now to give you an account of the several specimens and their characters, which I have sent you; and first, you will please to observe, that No. 1, 4, 5, 6, 9, 10, 12, 13, 14, 15, 16, 17, 18, 19, 20, 21, 22, 23, 25, 27, are entirely new; I mean entirely different from any that I have formerly sent you.— Some appear to me to be new genera, and several new species, but I beg leave to submit the determination of them to your own sagacity and discerning judgment.

No. 1, 3, 4, 6, 8, 9, 11, 23, are all fresh-water fishes, catched commonly in dams, ponds, lakes, or rivers of fresh water. No. 23, or fresh-water rock fish, is by far reckoned the finest fish; nay, indeed, we esteem it before any fish that we have, either in fresh or salt water. It abounds in almost all the American rivers, south of New York, above where the salt water or tides reach; and it is caught from two or three to thirty pound weight and upwards. I never had but only one to examine, and the company who permitted me to make out the description, insisted on their having the pleasure of eating it, otherwise I would have preserved the specimen for you.

Nos. 2 and 5, are probably the same species, called here *Yellow Tails*, or *Scomber caudâ luteâ.* By Nos. 2, 5, and 16, and by the *Scomber* of a former year, you will observe that we have various species of the *Scomber*, all which we call by different names, from little trifling appearances. Nos. 2 and 5 have abdominal *fossulæ* for the concealment or lodgment of the ventral fins, so that they may or may not be used, and of consequence cannot be necessarily and specifically useful. This species may connect the *abdominales* with the *apodes?*

No. 4 is, as I imagine, a new genus. It is always met with in fresh water, and is an ugly, ill shaped and ill looked and sluggish fish; of little use for the table.

No. 7 is the same as No. 33 of my last parcel, which I sent you; the specimen of which I did not send at that time. You will please to observe that this is not the specimen from which I drew these characters, though the difference is trifling. The vermin destroyed that specimen, and had nigh robbed me of this, which I now send for your examination.

No. 9, is a very remarkable species of the *Esox*.

No. 10, is often found in our harbour.

No. 13. When I first saw this fish, I imagined it to be one of the Angel fish, or *Chœtodons*, but, on examination, I found it very different. We have no name by which it is generally known, though by some few these fish are called Leather-coats.

No. 14, is rare and curious with us.

No. 15, is the only specimen of this fish that I ever saw, and the gentleman who was kind enough to let me have it, had unluckily ordered it to be dressed for supper, so that the scales were taken off before he thought of me; and hence you will observe that I could not see the natural appearance of the fish, nor make the characters complete.

No. 17, is very rare with us.

Nos. 19 and 20, are both most beautiful fishes when they are first catched alive; but the last, or No. 20, exceeds in beauty every fish that I ever saw; but alas! the lustre, glory, and beauty of their colours and hues are of short duration, and cannot be preserved; nay, indeed, a great part of their beauty dies with themselves.

Nos. 21 and 22, are species of the *Squalus*, of which I have now the specimens by me, and if you desire to have them, they shall be sent to you immediately on your signifying your desire. They can easily be preserved, and on that account I have kept them, as they are uncommon with us. I was much mistaken last year in sending you No. 34, or the *Canicula*, for the *Zygæna*; but I afterwards got the specimen which I now have of *Zygæna*, and the sight of that convinced me of my error, which your instructive letter afterwards confirmed.

No. 26, is our *Toad-fish*, or No. 16 of the former parcel. As my characters of that fish were then made out from a dried specimen, I have on that account now sent you characters taken from a fish newly catched from the sea; and as I think them much fuller and exacter, I shall beg that you will only pay regard to them. I think that in your letter you have considered it as a species of the *Gadus*, but when you further examine it, you will undoubtedly find it different. The branchiæ are open only by a small hole, and it seems to me to be a fish that connects two classes together, and is probably the more distinguishable, middle, or connecting, link in the scale.

No. 27, I got this curious fish lately, and as it was found among some things belonging to a deceased person, who had several African and Indian curiosities, I cannot positively assert whether it is a production of our American seas or not, having never met with any other of the kind.

I have likewise sent you many of our insects, of which I think many are very curious; but as I have never yet attempted this part of natural history, I have not ventured to digest, class, or number them;

but I must beg leave to assure you that it is want of opportunity and time, not want of inclination, that prevents me; being well assured that "*Natura nusquam magis tota, quamin minimis*," and that the most minute parts of the works of our great, all-wise, and good God, have many marks of his wisdom and goodness impressed every where on them.

You will likewise find several other unclassed articles in the box, which I flatter myself will not be unacceptable to, or unworthy of, your notice.

I am now to return my best thanks for the very distinguishing and unsolicited mark of respect you conferred on me, in having me elected a member of your Royal Society of Arts and Sciences at Upsal. The high title of honour which this election confers on me, I must beg leave to attribute wholly to your friendship, as I must be unknown to the generality of the honourable and learned members who compose that august society; and while this gives me the liveliest and most delicate sense of your favourable opinion of, and friendship towards me, so it deservedly claims my highest gratitude to you in particular, and my warmest thanks, which I now beg leave to offer to all the learned and honourable members who concurred in the election. My sincere aim shall always be to promote the laudable purposes of the society, as far as the narrow sphere of my abilities will permit, and I shall always, as I do now, heartily pray for the prosperity of this society.

By the first vessels, I fancy I may expect to be

favoured with a diploma, or ticket, of my election and admission into the society.

Ever since the present year began, I have been exceedingly hurried by a very epidemic small-pox which has raged amongst us, and besides the vast numbers which I have attended in the natural way, I have inoculated several hundreds.

Some years ago, Dr. Adam Thompson began, the first in America, to use, with great success, Boerhaave's proposed specific of mercury and antimony, and since that time the practice has extended all over the continent. I have differed a little from him in my method, though I had the hint from him; since I used it, I have inoculated about 800 whites and blacks, out of which number I have only lost one white child and one negro boy. I intend soon, when the hurry is over, to publish an account of the method, for the benefit of mankind, and to subjoin any remarkable observation that occurred to me. I think it will from thence appear that the variolous disease is not a disease of the blood originally, and that the fever is *ex se* a simple nervous intermittent. You need not doubt but I shall immediately order a copy to be sent to you as soon as ever it is published.

I had a letter, very lately, from John Bartram of Pennsylvania, in which he informs me that he was just setting out to visit the banks of the Ohio and Mississippi, as far as Louisiana. He will return loaded with spoils. It is a great pity but he knew the method of characterising the plants which he must meet with that are entirely new.

I am now to apologize for the length of this, and likewise for writing to you in English; but I shall only say that I trust to your goodness to excuse both, after telling you that our good friend and mutual labourer Mr. Ellis informed me, you was equal master of the English as of any other language. My warmest wishes for your health and prosperity are the daily offerings of my mind to heaven, and after begging a continuance of your correspondence, I beg leave to subscribe myself

Sir,

Your most obliged and very humble servant,

ALEXANDER GARDEN.

DR. GARDEN TO LINNÆUS. [English.]

SIR, Charlestown, South Carolina, May 18, 1765.

I should scarcely venture to break in on your time, by giving you the trouble of a few lines, if I was not again induced by some paragraphs of Mr. Ellis's letters to me, to believe that you still remember me, notwithstanding the great space of time that has elapsed since I had the honour of hearing from you. I have, besides, much reason to think, that some of my letters to you have miscarried, and that a collection of specimens of fishes, and many other things, which I sent to you in 1763, and which were put into Dr. Solander's care in London, never were delivered to you, notwithstanding I took every precaution to have them carefully forwarded.

Indeed the warm and friendly expressions, and the extreme politeness, which you use in your letter to me, 5th October, 1761, dated at Upsal, which is the last I have had the honour of receiving from you, will scarcely permit me to think that my disappointment has been owing wholly and alone to accident or miscarriage of letters.

Finding, as I apprehended, that my collection of specimens of fishes, insects, &c. above mentioned, had miscarried, I lately ordered a fresh copy of my characters of the fishes to be made out, and sent them to London with a copy of my original letter to you, to be forwarded to you by Mr. Ellis, thinking by this not only to shew you that I had been diligent in collecting and observing, but likewise that I had profited much by your advice and kind directions. I hoped also to furnish you with some new genera which I think the last collection contained. I trust these are already safely delivered to you, as I know I may expect every piece of friendship and politeness from Mr. Ellis.

By the present opportunity I now send you the characters of what appear to me to be new genera of plants. They are seven in number, and I hope that after an examination of them, if you should then find them new, you will be pleased to give them a place in your new genera, and I shall take the liberty of leaving the naming of them to you, only begging leave to reserve one, but any one that he and you shall agree on, for my most valuable friend Mr. Ellis. Nothing will give me more joy than to have the honour of being the person that discovered and described his namesake plant. He merits every thing at my hand, and I hope you will do me the honour of allowing me the above request.

Indeed, if it was agreeable to you, I would likewise beg leave to mention Dr. Hope, professor of botany at Edinburgh, as well meriting to have his name conveyed down on a plant. He is my worthy friend, and has a fine genius for the improvement of the science, and will, no doubt, soon make the same known to the world; but this I beg leave to submit entirely to your own judgment and determination, only begging that, when you are pleased to honour me with an answer, you would then please to mention the names you give them, or the genera under which you class them. While on this subject, permit me to ask, to what genus you associated the last plant which I sent you; it was a beautiful shrub, and to me it appeared to be a nondescript?

The first genus of these now sent is a bushy, low shrub, with a large, contorted, woody root *.

The second is a very beautiful evergreen shrub, about the height of two or three feet †.

The third is a pretty shrub with beautiful but deciduous leaves ‡.

The fourth is a beautiful, tall shrub §, much like

* *Stillingia sylvatica. Linn. Mant.* 126.

† *Prinos glaber. Linn. Sp. Pl.* 471.

‡ *Fothergilla alnifolia. Linn. Suppl.* 267.

§ *Cyrilla racemiflora. Linn. Mant.* 50.

the *Itea*, but much nobler and prettier in its appearance.

The fifth is a tree with deciduous leaves *.

The sixth is a tree with evergreen, laurel leaves, decussated †, and is one of the finest ornaments of our gardens.

The seventh is the most extraordinary plant that ever I saw, and till lately I never could get all its characters ‡. It is very difficult to propagate from the root, but bears great quantities of seeds, and is, I think, of the shrub kind, notwithstanding its low stature.

Besides these, I have sent you the characters of a very extraordinary animal §, which I have never seen till lately, though they are very common here. I have likewise sent you two specimens of it, though they are much less than the one from which I made out the characters, which was between two and three feet in length. I have sent it to Mr. Ellis for the museum. It appears to me to be a middle link between the *Lacerta* and *Muræna*, having some things in common with both, and yet differing from both.

Happy shall I be if these few things reach you, and if you will be pleased to accept them as some testimony of my esteem and regard for you. If you find a spare hour to favour me with a line, pray

* *Hopea tinctoria. Linn. Mant.* 105.

† *Olea americana. ib.* 24.

‡ *Orobanche americana. ib.* 88.

§ *Siren lacertina. Linn. Syst. Nat. ed.* 12. *v.* 1. at the end.

mention any new book or performance that may have appeared on natural history, or any advances it makes. Allow me now only to say, that I esteem it my greatest honour to subscribe myself, Sir, your most obedient servant, ALEXANDER GARDEN.

DR. GARDEN TO LINNÆUS.

SIR, South Carolina, August 4, 1766.

Since my last letter to you I have had the honour of receiving two from you, the first dated 15th August, and the last 27th December 1765, both from Upsal. Your letter to me of the 19th May, 1765, which you say you did me the honour to write, never came to my hands, so that I lost the pleasure of hearing from you, or of answering any thing mentioned in it. I must therefore still beg that you would just take the trouble to hint at what you then wrote, as I reckon it a particular misfortune to have lost any thing that you was pleased to communicate. Every one of your letters, besides the light which they throw on the subjects on which they treat, give me a new stimulus and incitement in the study and pursuit of natural history, and for that reason, as well as many others, I stand indebted to you for the greatest obligations.

It gives me much satisfaction that you had some pleasure in viewing and examining the extraordinary two-legged animal which I sent you last year; we have numbers of them in this province. They live in dams and ponds of fresh water, and in low marshy grounds all over the province. I have seen them of all sizes, from four inches in length, to three feet or three and half. They always appeared to me to be the same animal in every thing but magnitude, and we have none of the *Lacertæ* in this country (excepting the Alligator) that ever grow above six or seven inches in length at most; and all these are land animals, never using or living in water but when drove into it. I think you may depend on this being no *larva* of any animal, but that it is a real distinct genus by itself. My confinement to the practice of physic in the town, prevented me from having so good an opportunity to examine the anatomical structure of this animal as I could have wished this last spring; but in one that I opened, I think I saw the male sperm, or what we call the sperm of fishes.

I had an entire animal sent me, which I put into spirits and now send to you; and the skin of a pretty large one with the head and *vertebræ* of the neck of one of them, which I have put into the box among the insects.

In the former year I desired my servant to dash the head of one against a stone to kill it, as I intended to open and examine it. But he, misapprehending me, dashed the whole animal forcibly against the ground; when, to my surprise, it broke short off in three or four pieces, just in the same manner as the Glass Snake does when struck. From this I judged it was not of the Eel kind, whose skin and muscles are remarkably tough. Though I never yet have been able to know ocularly whether they propagate their species in this form, yet I have great reason to believe it, for we should certainly meet with them in their perfect state if these were only *larvæ*, but we positively have no animal that re-resembles them *.

Agreeable to your request I have sent you specimens of all the genera of which I sent you the characters last year, and I shall beg leave to make a few observations on each. N. B. I retain the same numbers as last year.

No. 1. you say in your letter to me, " *non est Plukenetii citata, quæ Ptelea*"—I cannot help thinking still, that the plant fig. 1. tab. 141, of Plukenett's *Phytographia* has a great resemblance with this, and none or little resemblance with the *Ptelea*, which is never *tricocca* as this is. The fruit or *pericarpium* of this plant is very like the *pericarpium* of the *Ricinus*, as Plukenett justly observes; and the seeds are with difficulty to be distinguished from those of the *Ricinus*; but you will best determine by the specimen. (*Stillingia*—see preceding letter.)

No. 2. I have sent you a specimen of this, though I entirely agree with you it is the *Prinos*.

No. 3. I think has some affinity with the *Hamamelis* in the habit, but none in the characters; therefore I would still submit it to you whether it be not a new genus. (*Fothergilla.*)

* This, the *Siren* of Linnæus, is *Muræna Siren* of Camper, and of *Gmel. Syst. Nat. v.* 1. 1136.

No. 4. I have sent this with great pleasure, in order that you may see the *Capsula multiccularis*. *(Cyrilla,* now *Itea Cyrilla, Ait. Hort. Kew. ed.* 2. *v.* 2. 37.)

No. 5. This you call the *Hopea*. I have my most sincere and hearty thanks to present you on this head, and I have wrote my friend Dr. Hope what you was pleased to write to me on this matter.

No. 6. you say, seems to be an *Olea,* but I am persuaded, when you examine the specimens, you will alter your sentiments. When I wrote to you last year I did not know that it belonged to the *Polygamia Dioecia* class, (See *Mant.* 24.) but I have since fully satisfied myself that there are male flowers only on one tree and hermaphrodite flowers on another, in which it resembles the *Diospyros* and *Nyssa*. The characters of the flowers, which I sent you, were made out from the male tree which grows in my own garden; and though I saw that there was scarcely any style, yet not suspecting that it belonged to the *Polygamia,* I took the bifid appearance of the centre of the receptacle for a *stigma bifidum*.

The character of the *pericarpium* and *semen* were made out from seeds which I gathered in the woods, from a tree that I have since seen in flower, and found the flowers hermaphrodite, so that it is certainly a new genus.

No. 7. I still think is a new genus; but if, on examination of the specimen, you determine it to be a *Gesneria* or *Orobanche,* you will deliver me from a great mistake.

The following are two genera which I now send you.

No. 8. is the specimen of a tree which I have often met with of late, and I cannot well determine whether it is a new genus or not; but I judged it best to send you a specimen, with the characters, in order that you may determine it, when I hope you will be so good as to inform me *.

No. 9.† is a copy of the characters of a plant which I had the honour of sending you in the year 1760, with a specimen and draught of it; but, as you never mentioned it in any of your letters since, I imagined that you had overlooked it. Since that time I met with what I take to be another and lesser species ‡ of the same genus, and as they appear to me to be new, I thought I would send them to you with the others for your examination. I have sent specimens of each along with the characters—they are both among our prettiest flowering shrubs. Along with them I have sent a few insects, which are all that I have by me at present.

I hope these things will arrive safe, and that you will be pleased to receive them in good part. You may depend that I will always be ready to pick up any thing curious, and will most cheerfully communicate every thing of that kind to you. I remain, with the greatest respect and esteem, Sir, your most obedient and most humble servant,

ALEXANDER GARDEN.

* *Sideroxylon tenax.* *Linn. Mant.* 48.

† *Styrax grandifolium.* *Ait. Hort. Kew. ed.* 2. *v.* 3. 59.

‡ *S. lævigatum.* *Ibid.* *Curt. Mag. t.* 921.

List of things sent.

No. 1. is a large packet of specimens. No. 2. is the insects. No. 3. is the bottle containing one of the new animal *bipes* * and three fishes, one of which you will find new; it is the West India Parrot-fish †, and the other is the Parrot or Sparrow-fish ‡, of which I sent you a specimen and character before.

DR. GARDEN TO LINNÆUS. [Latin.]

SIR, Charlestown, May 14, 1770

Although I am at present provided with few things worth offering you, I am happy to take this opportunity of writing, if merely to acknowledge, with my best thanks, the honour you do me in your letters. I cannot but attribute to your friendship alone the place you have given in your works to the new genera sent last year. To labour successfully in the same vineyard with you, though with a feebler hand, I esteem a great honour.

Since your last letter reached me, I have examined the shrub No. 3. *(Fothergilla)* with the naked eye, as well as under a magnifier; and I find it a new genus, undoubtedly distinct from *Hamamelis*, in the number, figure, situation, and proportion of all the parts of the flower, however similar, at first sight, in habit and appearance. The characters of *Hamamelis*, which is common here, agree exactly with

* *Siren lacertina*—see this and the preceding letter.

† *Mormyra n.* 1. Browne's Jamaica, 446.

‡ *Coryphæna Psittacus. Linn. Syst. Nat. ed.* 12. *v.* 1. 448.

what you have published, having uniformly, without exception, four stamens, and the same number of petals and nectaries. Whether this new shrub ever bears separated flowers, as you suspect, I greatly doubt, having never met with any such, though many specimens have passed under my careful examination.

No. 6. has the character of the class and order *Polygamia Dioecia*, and yet you make it an *Olea*, which to me appears very paradoxical *.

No. 8. I quite concur with you in making this a *Sideroxylon*, for the fruit perfects but one seed, and therefore it cannot be referred to *Chrysophyllum*. It will constitute a new and handsome species of *Sideroxylon; (tenax. Linn. Mant.* 48*)*.

I have taken many opportunities of examining whether the *Siren* undergoes any metamorphosis or not, and though I have often seen it in various stages, from its smallest to its largest size, I have never perceived any variation in form or other respects. This letter is accompanied by a fuller and more exact detail of the characters of the singular animal in question, than I have before sent you, made from numerous dissections, which I trust may prove worthy of your attention †.

* The excellent writer had not sufficiently attended to the limitations of the artificial system of Linnæus, and still less to the principles of natural orders and genera.

† These have been made use of in the dissertation entitled *Siren lacertina*, published under the presidency of Linnæus, June 21, 1766; see *Amœn. Acad.* v. 7. 311.

I send you likewise the characters of a plant, known here by the name of Indian Pink Root. I formerly communicated to our friend Dr. Hope, an account of the qualities, and proper doses, of this plant, and of the best manner of giving it, requesting him to lay my letter before the Edinburgh Society. This is soon to appear in the third volume of their *Essays and Observations*, and perhaps you may see that volume before my present letter reaches you. If otherwise, I shall be particularly obliged to you to examine my description, and to write to Dr. Hope your opinion on the subject. The plant seems to me to differ generically, as well as specifically, from *Spigelia Anthelmia*, for which I have given my reasons, but I shall be glad to have your opinion on my side. It is a very powerful anthelmintic, and is entitled to the first rank among medicines for killing and expelling worms.

You may depend on my utmost diligence in all inquiries to satisfy your curiosity, that I may imitate in some degree, if I cannot equal, your kindness towards me. I therefore extremely regret that it is impossible to procure here the *Pulex*, Catesb. Car. 3. t. 10. f. 3, about which you ask for information. This insect never appears in our province, as lying too far north. It only inhabits the southern part of this country, and the West Indies.

We have, however, an extremely minute venomous insect, of a reddish hue, generally occupying old rotten trunks of trees, and known by the name of *Potatoe Louse*. These insects though so very minute as to be hardly visible to the naked eye, are capable of doing great mischief. For in summer time, when they are most abundant and active, they attack persons in the country, especially while incautiously sitting upon fallen trees, or lying down on the ground. These highly troublesome visitors come in great bodies, and no sooner attack any person, than they disperse themselves all over his skin, which they quickly penetrate, not only instilling their poison, but at the same time effecting a lodgment for their bodies, under the epidermis. Hence a violent itching is produced, troublesome at first, and soon becoming intolerable; then follow inflammation and swelling, with a pungent burning pain. Fever not unusually supervenes, which for some days increases, terminating in little itching ulcers. The animal I am describing differs altogether from the abovementioned flea of Catesby, called Chigo; but what its genus may be, I know not. This summer I will enquire into the matter, and if specimens of so diminutive a creature can be procured, you shall have them.

I hope the serpents, fishes, lizards, insects, &c. now sent, though not so numerous or valuable as I could have wished, will not be totally unacceptable. I entrust them to our common friend Ellis, and shall long to hear what you think of them; for your letters not only afford me instruction, but they powerfully excite me, more and more, to pursue these studies.

May you long live, to the honour and benefit of

our science; and in the hope of your continued favour, I beg leave to subscribe myself, &c. &c.

ALEXANDER GARDEN.

The above specimens, preserved in spirits of wine, in a bottle well stopped, will be forwarded by Mr. Ellis, in the first vessel for Stockholm.

FROM THE SAME. [Latin.]

SIR, Charlestown, June 20, 1771.

Since I wrote to you in the course of last year, I have duly received two of your letters, by means of our common friend. I need not say how much pleasure they afforded me, for they assure me of your health, and your continued friendship for me, than which nothing can be more welcome. I rejoice also that my letter, and all the things which accompanied it, came safe to hand; and especially that you found some rarities among them, as well as some entire novelties, all which you are so good as to receive favourably.

Having many things about which I want to consult you, I shall now mention them as they come into my mind, without any regular order.

The fleas described in Catesby by the name of Chigoes, have at length fallen in my way, and I send you some of them preserved in spirits of wine. You will find them in all their different states; some full-grown, others younger and smaller, others in the little bags in which they conceal themselves, while they wound the skin of the feet, or other parts of the human body, for the sake of depositing their eggs there. I have not examined any of these specimens, being unwilling to diminish their number, perhaps to no purpose; and therefore I send you my whole stock, that you may have the more ample means of investigation.

To procure these, and other natural curiosities, I sent a black servant last summer to the island of Providence. During his stay there, he collected and preserved some fishes amongst other things; but, meeting with tempestuous weather in his return, and being, for several days together, in dread of immediate shipwreck, he neglected all his specimens, many of which perished. Some were fit only to be thrown away, and others were greatly damaged. What remain, such as they are, I shall, by this opportunity, send for your examination. Some fishes among them, whether found in our sea, or in that of the Bahama islands, you may perhaps find to be new. The following is a list, with their vernacular names.

No. 1. Goat Fish. (*Mullus* Linn. MSS.)
2. Scots Porgee. (*Umbla* Linn. MSS.)
3. Marget Fish.
4. Yellow Grunt.
5. Blue Fish. (*Umbla* Linn. MSS.)
6. Chub. (*Sparus* Linn. MSS.)
7. Spanish Hog Fish. (*Labrus* Linn. MSS.)
8. Hind. (*Perca* Linn. MSS.)
9. Small White Grunt.

10. Pork Fish.
11. Trumpet Fish. (*Fistularia* Linn. MSS.)
12. Schoolmaster.
13. Parrot Fish.
14. Leather coat.

You will easily reduce these to their proper genera, and your remarks upon them would be highly welcome to me.

You will find in phial No. 2, a few of those venomous insects called the Potatoe Louse, wrapped up by themselves in a little piece of paper, on account of their smallness. Their appearance under a microscope is like that of an *Acarus*, and they may possibly belong to the *A. Batatas* of Rolander, in your *Syst. Nat. ed.* 12. *v.* 1. 1026. In our country however they are met with, not only among the Potatoes, but likewise most abundantly on old trees, and all kinds of rotten wood.

In the phial No. 3, are some Fire-flies. Many thousands of these are seen flying about here every night in summer, darting their phosphorescent light, and affording an amusing spectacle to travellers, though causing considerable alarm to those who are not accustomed to them. (*Lampyris* Linn. MSS.)

The phial, No. 4, contains four other Fire-flies, brought from the West Indies, which I send for comparison. (*Elater noctilucus* Linn. MSS.)

The coleopterous insect No. 5, in the box, is one of the most beautiful things I ever saw, in the elegance of its body and the colour of its limbs.

When first caught, the brilliancy of its head, thorax, and thighs, exceeded that of the most polished metals. The name and place of this insect in your system, I earnestly wish to know *.

No. 6. is an insect called here a *Smith*. (*Silpha carolina*. Linn. Mant. 2. 530 ?)

I must now say something of an unknown animal, which you will find in a glass bottle, and which, I have no doubt, will afford you much satisfaction. The specimen here sent is the only one I ever saw, and I shall think myself fortunate if it reaches you in safety. When I first received it, the length was 37 inches, though the animal was then become somewhat contracted. At first sight I suspected it to be another species of *Siren*, but upon nearer examination, I found so many differences, that there proves to be no relationship whatever between them. Can this animal form a link between the *Lacertæ* and *Serpentes?* Is it at all allied to *Anguis quadrupes*, *Syst. Nat. v.* 1. 390 ?

It differs in many particulars from the *Siren;* most evidently in the following.—This animal has four feet, with two toes to each, without claws. The *Siren* has only two feet. It wants the gills and their wing-like coverings. It has no scales, nor, which seems to me very singular, any tongue! all which are found in the *Siren*. I have opened the throat, and satisfied myself respecting the presence or absence of gills. The following are the characters I have drawn up, of this ugly animal.

* This insect does not appear.

Head, rather long, depressed, tapering, serpent-like.

Mouth, extending half the length of the head.

Lower jaw, furnished with a single row of sharp, distinct teeth.

Upper, with four rows of similar curved teeth.

Upper lip, covering the under one.

Tongue, none!

Nostrils, two openings at the very extremity of the upper lip.

Eyes, dull, at the upper part of the head, on each side, covered with a thick tunic.

A thin retractile membrane covers each cartilaginous lateral spiracle, or orifice, by which the animal breathes.

Body, thick, nearly cylindrical, tapering, and keeled at each side, beyond the vent. *Tail* recurved. There is no *lateral line.* *Vent* a large opening, immediately behind the hinder legs.

Feet, four: two of them before, close to the spiracles, each with two toes, destitute of claws; two behind, at the bottom of the belly, with similar toes.

Inhabits deep ditches and lakes of fresh water.

I have now clearly ascertained that the *Siren* is oviparous, and that it never undergoes any transformation.

A few days ago I procured a fish that I had never seen before, the characters of which are as follows.

It came to me with the name of *Fat-back,* and

belongs to the *Acanthopterygii* of Artedi, the *Thoracici* of Linnæus. (Linnæus has here noted with his pen, Catesby 2. *t.* 1.)

Body, compressed, rather broad, covered with smooth scales; the *lateral line* running straight from the upper and posterior part of the cover of the gills to the tail.

Head, tapering. *Lower lip* longest. A single row of sharp, lanceolate, distinct teeth in each jaw. *Covers* of the *gills* clothed with smooth scales and marked with two red spots. *Eyes* near the lower corners of the mouth. *Nostrils* a pair of very singular perpendicular orifices, just before the eyes.

Membrane of the gills with seven rays.

Fins, eight. The *first dorsal one* of seven acute pungent rays, connected by a thin membrane, and received into a small furrow about the middle of the back. *Second* thicker, with 26 soft rays, three-cleft at the extremity. *Pectoral ones* of 15 rays each. *Ventral ones* of 5, placed under the pectoral. *Anal one* of 28 rays, the first of which are longest. *Tail* forked, of 22 rays.

Of all the fish caught in our seas, this is the most delicious, and the most agreeable to the delicate palate of an epicure.

I have lately received, from a friend at Surinam, a fish which turns to a frog, or rather the tadpole of a frog, resembling a fish. This transformation appears to me so curious and unusual that I send you the specimen in a bottle. You will find it in various progressive states, which, I flatter myself, may afford you entertainment.

I have described, and have lately sent to our friends Ellis and Pennant, a new and very rare species of river *Testudo*, known here by the name of the Soft-shelled Turtle, because the covering of its back, especially towards the sides, is of a softish, leathery, very flexible substance. This animal is found in the larger fresh-water rivers of East Florida, Georgia, and South Carolina.

This *Testudo*, the four-footed Snake, and the Surinam Frog, will prove, I hope, so many additions to the order of *Amphibia*.

What you saw in our newspaper, about Tea growing wild in this country, is without the least foundation, being one of the gratuitous lies of the printer of that paper.

You still express, in your last letter, an unwillingness to decide respecting the new genus like *Hamamelis;* for which reason I send you a few specimens of this plant, carefully collected, and preserved in spirits of wine. I request your opinion on this subject and all the above-mentioned, which will give me great pleasure.

I am truly rejoiced to hear that you have found a key to the class *Cryptogamia*, with which, hitherto unknown and much-wished-for elucidation, I trust you will speedily favour the world.

In the same parcel with the fishes, I have sent, for your examination, a Florida plant, unknown to me, which should seem to belong to *Gynandria*. Its appearance is handsome enough, especially in autumn, when the woods are much ornamented with its beautiful fruit and seeds. This specimen shows the flowers in a head, as they at first appear; but they are soon succeeded by a cone-like pericarp, nearly of the same figure as that of a *Magnolia*. At length the capsules burst into two parts, displaying the large shapeless seeds, which turn red as they ripen, and attract notice from a considerable distance. The height of the plant is 18 inches, rarely more. I am extremely desirous of knowing what it is, but dare not dissect this only specimen that has been brought me, which will be better in your hands. Formerly I have been in the way of seeing plenty of both fruit and seeds *.

Numerous insects, serpents, and lizards, are in the same box with the unknown animal, and other things above-mentioned, but in separate bottles. All these are consigned to the care of our excellent friend Ellis, who will, doubtless, forward them to you, with all care and expedition. May you live long, to benefit the literary world, and to the glory of science!

FROM THE SAME. [Latin.]

SIR, Charlestown, May 15, 1773.

Your last letter afforded me the sincerest pleasure, as bringing a good account of your health; news

* *Zamia pumila. Linn. Sp. Pl.* 1659. *Z. integrifolia. Linn. fil. apud Ait. Hort. Kew. ed.* 1. *v.* 3. 478. *Willden. Sp. Pl. v.* 4. 847.

not less welcome to me than to all good men. I am happy to hear that the natural productions I sent you proved acceptable; and especially that you are still so good as to preserve your wonted friendship for me.

I am always glad to be employed in your service, of which I trust you are by this time well assured. That I might give a proof of this, on the receipt of your letter dated Upsal, the 16th of January, 1772, I took measures to procure from East Florida, where it grows wild, some specimens of your *Zamia* *. I wrote to the Governor of that province, a great promoter of Natural History, and my particular friend, requesting him to send me some, carefully gathered, and well preserved, by the first opportunity. According to his usual kindness, he sent me what I wanted, as soon as the season of the year would permit; and from these specimens I have made out the following particulars and characters.

You mention in your letter, that the pollen of this plant is naked, on the under surface of each scale of the cone (or catkin). But the most careful examinations, under a microscope, have satisfied me that this is by no means the case. I was much afraid of committing a mistake, and leading others into error, and therefore submitted several scales to repeated investigation, always with the same success, for I saw no difference between them. The pollen is evidently contained in bivalve elastic capsules (anthers). It can scarcely therefore be re-

* See the last letter.

ferred, as you judge, to the Fern tribe, nor dare I assert it to be a *Zamia*. The construction of the female *spadix*, and of the pericarp, is very singular. The peltate heads of the proper perianths are externally so closely united, that they can hardly be pulled asunder without tearing. It is therefore scarcely to be understood how the pollen of the anthers can insinuate itself so as to fertilize the germens. For the perianths never begin to separate before the germens become swelled, exhibiting manifest signs of impregnation being already accomplished. The vacant internal space indeed, between the partial stalks of the perianths, affords the germens and styles full liberty to grow; but the very close union of the shields prevents any access of external bodies, or even of the most subtile vapour.

Being anxious to know more of this plant and its history, I have put the following questions to my friend the governor, by letter. Is there any *Spatha* or not? Are the male and female catkins always on the same plant, or from separate roots, and is there no variation in this respect? I have some suspicion that fecundation may take place sometimes in one way, sometimes in another. Do any birds, and of what kinds, feed on the seeds? Does the plant afford nourishment to any other animals? I am in daily expectation of an answer.

It remains with you to determine the genus of this plant, and that you may not want materials to form your judgment upon, I now send you dried specimens of the male and female catkins, as well as of the seed-vessel or cone, containing ripe seeds.

I am very glad that the most elegant shrub, called by me *Anamelis* *, has at length obtained its proper place; for I was much afraid it must have submitted to range under the banners of another, (*Hamamelis*).

Throughout the whole of last summer I was intent upon procuring you some of the Siren-like animals, as I still am, but hitherto I have laboured in vain. They are extremely rare; but under your auspices, nothing is to be despaired of. Perhaps a specimen or two may one day come into my hands, and enable me to make out enough to give the animal a place in your *Systema Naturæ*.

In the parcel now sent are two seeds of a tall tree, usually growing in watery situations, on the banks of the river Savannah, which divides Carolina from Georgia. The tree is rare, and hardly to be met with any where further northward.

The skin and eggs of a serpent are also here inclosed.

Farewell, dear Sir, you are the delight and ornament of our science.

Characters of the dwarf palm from East Florida. (*Zamia pumila, Linn. Sp. Pl.* 1657. *Suppl.* 443; now *Z. integrifolia.*)

Diœcia Polyandria.

Barren flowers.

Calyx. A *Spatha*? I am doubtful on this point.

* *Fothergilla* of Linn. Suppl. See the foregoing letters.

Spadix, or Catkin, simple, oblong, three inches in length, many-flowered, scarcely imbricated, consisting of about twenty rows of scales, more or less, which separate when ripe.

Corolla. None.

Stam. Filaments none. Anthers attached to the under side of each scale, numerous, globose, sessile, of one cell, and two valves, bursting elastically, and scattering a quantity of extremely minute white pollen *.

Fertile flowers in distinct catkins, and, if I guess right, on separate plants.

Calyx. *Spatha* none, as far as I have seen.

Catkin shorter and thicker than the barren one, cylindrical, usually composed of eight rows of scales.

Perianths. Scales peltate, club-shaped, with a square head, downy, permanent, being both perianths and receptacles.

Pistil. Germens two, angular, irregular, one at each side, attached to the perianth immediately beneath its square head, and turned inwards, perpendicular to the main stalk of the catkin. Style short. Stigma smooth, obtuse, slightly cloven at the side.

Pericarp. Cone oval, composed of 64 club-shaped perianths, or rather receptacles of the seeds, in regular rows.

Seeds. Berries about 128 in each cone, large,

* This description is more correct than that in the *Supplementum* of Linnæus.

red, irregular in shape, ovate, gibbous, or angular, single-seeded, two to each perianth or receptacle, slightly shaven off, or naked, at the base, and turned outwards. Kernel in a thin, smooth, bony, somewhat pointed, shell.

DR. GARDEN TO MR. ELLIS.

South Carolina, Charlestown,
March 25th 1755.

SIR,

Some few days after my arrival again in Carolina, your very kind favour was delivered me by Capt. Curling; since that time, what with settling afresh, and what with absence on some country jaunts, my time has been so taken up, that I have been prevented till now from having the honour of writing to you in return for so obliging an epistle. It is with the greatest pleasure that I find you approve of my pursuing the study of Natural History; as to any degree of accuracy, it scarce could be expected from so young a beginner, and one at such a distance from the proper helps and assistances of either men or books; as indeed there is scarce one here that knows a cabbage-stock from a common dock, but, when dressed in his plate, by his palate. But let me hope that the continuance of your correspondence may greatly improve me both in knowledge and accuracy. Let me assure you that both yours, and the approbation of other learned men, should I ever merit any, will greatly induce and encourage me to proceed in the study of nature. This year I have got the favour of several correspondents from Holland and Sweden, which does me the greatest honour. Let me, in the heartiest manner, present my thanks to you for your method of preserving plants in dried specimens; this is what I wanted much, and will carefully practise it this summer. Mr. Shipley wrote me it at great length, though I think the substance of it was in your own epistle. This last year I have made but few dried specimens. As my health obliged me to leave the province, I had not the proper opportunities, and if a short account of my peregrinations since I wrote Mr. Shipley last year would not be tedious, I should beg leave to subjoin it here as an excuse for my not having some things to send the Society as I thought then. About ten days after writing Mr. Shipley, I was taken violently ill of an acute inflammatory distemper, and, obliged, on a little recovery, to go to New York, in search of cool air. I had not been there many days before I was informed of that great philosopher and learned botanist, the Hon. Cadwallader Colden, Esq. so well known in the learned world both by his botanical and philosophical works. His botanical performances are published in the *Acta Upsaliensia* by Linnæus. Not only the doctor himself is a great botanist, but his lovely daughter is greatly master of the Linnæn method, and cultivates it with great assiduity. It was here that I first saw Linnæus's *Genera Plantarum* and his *Critica Botanica*. But I have never yet seen the *Philosophia Botanica* nor the *Amœnitates Aca-*

demicæ, nor the lately published work the *Species Plantarum*, which I have only yet heard of. As I was resolved to make the most of my northern journey, I set out for the Apalachian mountains, being resolved, as far as strength would permit, to lose no hour in searching out, nor leave any place unvisited, where plants were to be seen or met with. I would fain hope I succeeded pretty well, and I am sure brought many plants down with me, which we have not to the southward, though we have many others here in their room different from the northern or mountain plants. I met with some very curious minerals, too, such as *Lapis Asbestus* or *Cotton stone*, *Lapis Pyrites cubicus*, *Magnes*, the best that ever I saw, *Bastard Ruby*, *Talcum lamellatum*, *Crystallum hexagonum*, growing from their proper matrices, interspersed with some of the Emerald's matrices, and some true points just emerging. Some *Lapis Belemnites* or *Dentes petrificatæ*. Some mineral salts, and different boles. Some sulphur ore, and some cobalt with other ores. I have some small bits of these by me just now.

Of late there have many thoughts occurred to me relating to the reason of mountains abounding more with minerals than low countries, and I think I have hit upon some that seem probable to me, but it would be too tedious to recount them in a letter. I returned to New York greatly pleased both with my perfect recovery and my collection. I set out for the Jerseys with some Carolina gentlemen, and from thence we went to Pennsylvania. When we came to Philadelphia I met with John Bartram, a plain quaker, but a most accurate observer of nature. I met with Benjamin Franklin here too, a very ingenious man, especially in electricity. But Mr. Colden, Mr. Bartram, and Mr. Clayton are the only botanists whom I know of on the continent. After many jaunts, and some agreeable botanizing journeys, I arrived again in South Carolina in the end of December last, and immediately settled in a partnership with Dr. Linning's partner, as my worthy friend Dr. Linning gave up in my favour; so that what with settling, what with absence, and one thing or another, I have been hitherto prevented from acknowledging your favour; but let me assure you I will carefully cultivate a correspondence I value so much.

Let me here beg leave to present my heartiest thanks to you for your kind offer of having any new plants, that I might send you, drawn by Mr. Ehret, and presented with my descriptions to the Society. Sorry I am that I cannot just now send you some, for though I have the descriptions by me, yet I have not the dried specimens, else I would send them. It will be conferring the greatest honour on me, and I readily accept of your offer. This year I have been obliged to send some to some other Societies, who were kind enough to offer me the honour of being one of their number.

I should be greatly obliged to you if you would inform me on what terms, and after what manner, you admit the new members of the Royal Society. I think you justly observe the increase of botany in every kingdom of Europe but in Great Britain; and if I mistake not, we have a lively example of this in the King of Sweden, and his great favourite, Dr. Linnæus, sending out the ingenious Mr. Kalm to our provinces, where, after three years researches, he went home loaded with British treasures, and is just now publishing them in the Swedish language at the particular desire of the King. I suppose he means that not only the advantages of such discoveries, but the entire honour should be the invaluable possession of Sweden, and that we should by no means reap the advantages of their labours and public-spiritedness, so that we shall never be able to say of them *Sic vos non vobis*. Would to God there were some of our grandees of such animated inclinations for the improvements in natural knowledge. But this by the bye. I take it extremely kind that you should have mentioned these seeds, which will be agreeable to you. I think I can procure all of them next year. This year it was about the end of January that I received your letter, so by far too late for procuring any seeds but those that were then in my possession. This will convince you of the necessity of writing soon in the fall of the year; but indeed I should be extremely sorry that our correspondence should be confined to one letter in a year; I for my part would write you much oftener if it will be agreeable to you. You say you know several gentlemen that would exchange books for seeds. If they be seeds to be found here, they may depend on my procuring them if they agree to that condition. I shall beg leave to mention the books which my botanical collection consists of, and then you may judge how much I have been pinched for want. I have the catalogue of the Edinburgh garden, where I was first initiated in botany. I have Linnæus's *Fundamenta Botanica*, and his *Classes Plantarum* of my own, and Eulalio Francesco Savastano's four books of botany in Italian; but this is more valuable on the account of the language than sentiment. It was recommended to me by a Jesuit at Lisbon, where I bought it, but found it a trifling performance as to science, but musical enough in poetry. I have in loan the *Flora Virginica*, Tournefort, and Ray, which is all my collection. I have now commissioned for Linnæus's *Genera Plantarum* and the *Philosophia Botanica*, but have not yet had them sent me.

It was lately that I heard of Linnæus having published his *Species Plantarum*. Haller's system seems very good, and I think has several natural classes: he seems to have carried the *Monocotyledones* to their *ne plus ultra* of perfection, and by adopting Linnæus's principle, I think he has greatly improved the *Dicotyledones* by his divisions into the *Polystemones*, *Meiostemones*, *Isostemones*, &c. In a word, I think he has improved chiefly on Ray, Hermann, Boerhaave, Tournefort, and Linnæus; he has likewise taken the assistance of Ludwigius' principles, who is himself a most valuable and distinct writer. Systems, however, are endless things, and I believe they labour in vain, till once the essential characters of plants be carefully described. Sometimes in musing on this I have been led to think that the *Pericarpium* and *Receptaculum Seminis* is that part which will be the founda tion of Nature's system. At other times I admire the *Monocotyledones* and *Dycotelydones*, which seem to me to comprehend both Cæsalpinus' doctrine of the *Corculum Seminis*, and Linnæus's late thoughts of the new foliation of plants; indeed I think that they differ essentially so little from one another, that they seem to me to be different modes of the same operation of nature. I have lately drawn out some thoughts on this subject, but they are far too crude to be communicated. You mention your acquaintance with Dr. Hales, and his kind reception of the liberty I took with his name. His acceptance of my gratitude for his learned labours is very agreeable to me, and does me great honour. Please to inform him that if there be any thing here which I can serve him in, his simply intimating it to me will be a sufficient inducement for me to obey his commands. I shall this year preserve some plants of the *Halesia*, and remit them to you with a description after looking over all the characters again, so that I may omit nothing in it.

The enquiry you mention, as to the nature of *Corallines* and *Ceratophytons*, is admirably curious, and the present you design will do me the greatest honour and service, as I have never seen any thing worth notice on that subject, nor indeed on any of the marine productions, though we have many here curious enough, especially of the *Madreporæ* and *Milleporæ* and *hoc genus alia.* The method you mention of preserving them is very curious, and was really new to me. Since that time I spoke to several of the fishermen, but as yet to no purpose. Most or indeed all of them are negroes, whom I find it impossible to make understand me rightly what I want; add to this their gross ignorance and obstinacy to the greatest degree; so that though I have hired several of them, I could not procure any thing. I design in a few weeks to go down on the coast myself, and then I shall endeavour to procure some, and prepare them in the manner you desire. This last night I had sent me a piece of the *Madrepora*, which I really believe would weigh 40 pounds weight, all divided into kingdoms and commonwealths by high ridges running in various directions above the common surface about half an inch. To-day I have got another piece, but they are both old, and were found on the shore. The freight of these would be much more than their value, else I would send both. I have sent you one Hop-like matrix of the *Buccinum ampullatum*, which I think is very complete of the kind. I see by one of Gronovius's private letters, that he disagrees with Swammerdam in the use of this shell. Swammerdam thinks that it is the constant habitation of the Eremite-crab, and that God almighty created it entirely on purpose for a *receptaculum* to that creature. Gronovius says that this crab only uses it by chance or *pro re nata*, and not as its proper *domicilium*; for he says that he often found the crab in other shells as well as in the *Buccinum ampullatum.* I remember last summer I found one of them with a crab in it, but not knowing it, I put the shell into my pocket, and when I went home, put it on my table. To my surprise, when every thing was quiet, I saw the shell moving about, but taking it up hastily, could see nothing in it. Upon laying it down, and keeping every thing quiet, it began to move again; upon which I snatched it hastily up, but could yet observe nothing, as far as I could look with my eye. I then applied a hot iron to its axis, which immediately forced out one of the ugliest creatures that ever I beheld, and in appearance bigger than the shell in which it had contracted itself into so little a space. I was sorry that I had not opened it, as I might then have seen if it contained this Hop-like spawn, which I think would have fully explained the difference between Gronovius and Swammerdam; though it might not indeed have had it at that time, as there is no doubt one particular time when they have their spawn, as well as any other sort of fish. I have sent you some of these *Buccinums*, grown to their middle size; our coast abounds with these, and with most of the vasculous kinds, as the *echinate*, the *balanate*, the *conchoidate*, the *conchate*, and *lepate* shells. We have some of the *Monothalamias* and *Polythalamias*, but very few of them. I have just now by me a skeleton of a system of shells, which I have not yet been able to complete, both for want of a sufficient collection, proper books, and time. I have sent you a few of the different kinds of most that I have just now by me. A few days before I received your letter I had packed up a valuable collection for Sweden, and one for a gentleman in Maryland, to the northward here. I have likewise sent some to Mr. Henry Baker, who wrote to me by the same vessel that you did. I have sent you a pretty curious collection of seeds, some of which I hope may be agreeable. There are several rare and curious shrubs among them, and our beautiful Carolina Jessamine, of which yesterday I made out a very full, and, if I mistake not, a pretty accurate description. As it is very different from the *Jasminum*, I design to send it to you with a new name next season.

There are some seeds of all the antidotes yet known or used by the Indians or whites (unless of three), on the continent, for a rattle-snake bite; they are Nos. 9, 24, 41, 71, 75, 76. I mean these are the only valuable ones, unless three or four more, among which is the Seneka rattle-snake root. I made very particular enquiry about these things among the different nations of Indians, and in the different provinces through which I passed, as it was my particular desire to learn all their indigenous physic, being persuaded that from the meanest things, useful hints may be gathered; *inter stercora Enniana aliquando aurum reperisse dicitur princeps Poetarum.* In the whole I have sent one hundred different seeds, all which I collected (unless some melon seed) with my own hands.

As there are several of these whose medicinal virtues I have learned, I was once minded to collect my observations on some of the genuses and send you, but my letter has lengthened out too much. Before I go farther, I must beg leave to offer you my hearty acknowledgments for your present of seeds; any curious plants or flowers that you can spare will be extremely acceptable, and what I earnestly beg is, that you will be kind enough to let me know if these things that I send be agreeable to you or proper; and if properly preserved or packed up. Any directions, advices, or corrections on these things will be doing me service, and I will gratefully acknowledge my obligations to you. I have sent you some butterflies, with some camphor strewed round them, according to Mr. Shipley's directions. If these will be agreeable, I can send you any number of them, and I will preserve them in any manner you please. This summer I design to make a collection of our insects in general, with great quantities of which we abound, and next year I shall give you some account of my success. I have marked some of the shells, whether they be our coast shells, or whether I had them from the islands. Though I once designed to have classed and sorted them all, and named them, yet upon some reflexion I imagined you would easily know them much better than myself. I have sent you some of our common *Ceratophytons*, though not properly preserved as you desired, but I hope next year you will have some things after your own mind. I have sent one *Ceratophyton* spread out like a fan, which is here called Sea-fans; the crustaceous upper part is all gone, and now only the horny substance remains. I have sent some small pieces of the *Milleporæ* or *Madreporæ*, which indeed I take to be both the same, though I readily own I am too little acquainted with them absolutely to distinguish them. I have sent you one of the rostrums of a fish found on the Florida coast, which I take to be a species of the *Ziphias, rostr. apice ensiforme, pinnis ventralibus nullis*. I have been told that they are frequently found on the Carolina coast, though I have never seen any of them, and I have been all along the coast to the Florida shore. I have sent you the tail of a king crab, which are very plentifully found here. Last year, when I wrote to Mr. Shipley, I sent him some seeds, which were not all marked with those names that are commonly given them by authors; they were No. 14. which is now by Linnæus called *Ceanothus;* No. 18. was a species of Laurel called by the northern colonies All-spice shrub; No. 34. was the *Cornus palustris baccifera foliis lauri;* No. 57. was the *Collinsonia;* No. 77 was the *Melothria;* No. 86. *Ludwigia;* No. 93. is the *Hamamælis Gron.;* No. 95. is the *Syringa pumila baccifera;* No. 87. is a new tree; No. 98. was the *Euonymus latifolia;* No. 100. was the *Dioscorea.* These are a few of the things I have to mention to you at present, and till I again have the pleasure of hearing from you, shall carefully remember you, and long to be favoured with more of your directions in the prosecution of my studies of botany, &c. &c. In the mean time I have the honour to be, with great esteem, Sir, your most obedient and very humble servant,

ALEXANDER GARDEN.

P. S. About two years ago I wrote to Mr. Miller, of Chelsea garden, but never yet received an answer.

SIR, March 25, 1755.

After writing to you, and packing up some seeds for you, I thought it might be agreeable to you to have some of our Indigo seed, and I accordingly have put up some of the Guatimalee indigo, and some of the French indigo seed, and some seed of a plant which I take (from an imperfect description which I have had from some planters) to be a species of the same genus with the indigo. It produces a beautiful yellow dye after undergoing the same process as the common indigo.

I cannot help here begging that you will be kind enough to take a draught of the seed and *Pericarpium* of our Jessamine. I have sent both inclosed to you with the indigo seed. It is a most elegant and beautiful vine, and I think a new plant, unless what Mr. Catesby has done, which besides his print is just nothing. I have made out a description, which I shall send you of it next fall, with a dried specimen of itself; but as I cannot probably then get the seed, I cannot help begging that you would now take a print of the *Pericarpium* and seed, and keep them by you till I send the rest.

Please to direct to Dr. Alexander Garden, Physician, in Charlestown, South Carolina.

This morning I have got in from London a copy of the *Genera Plantarum*, *Amœnitates academicæ*, *Critica Botanica*, and the *Philosophia Botanica*; so that I may hope to make some greater progress this summer, though I have not been able to get Plukenet's *Mantissa* or *Almagestum*, nor any of Dr. Sherard's works, nor Dillenius's.

Besides the seeds mentioned in the letter, I have sent some *Pavia* seeds, *Cassine* falsely so called here; it is a new genus; some Red Bay; some Wahoo, by the New York people, Linden tree; some Swamp *Palmetta*; some Live Oak acorns; some Putchimon; *Zanthoxylum* or Tooth-ache tree.

South Carolina, Charlestown,
April 20, 1755.

Sir,

This informs you that I put one box for you on board of the Prince of Wales, Capt. George Curling for London, directed to Mr. John Ellis in Lawrence-lane, London. I should have been glad you had given me your particular direction, as I should be very sorry they should miscarry through any fault in that. There are upwards of 120 curious seeds, some of this province seeds; most of them I myself collected on the Apalachian mountains. I hope they will be agreeable. There is the *Coralla-dendron*, with all the antidotes yet known or used for the bite of a rattle snake. There are some curious shells, and such *Ceralophyta*, *Madreporæ*, &c. as I could procure here just now, with one of the stringy matrices of the *Buccin. ampull.* Since that time I have examined one of the fish that inhabits this shell; it is a most curious creature. There was one of the seeds marked *Sophora*, which I have since found to be the *Cytisus foliis fere sessilibus*, &c. Hort. Cliff. and Gron. 82. The rest are all right signatured. I put a box on board for you unmarked, and for which I have no receipt from the captain, with a good many sets of the sweet scented shrub which you desired, some of the *Zanthoxylum* of Catesby, some *Jonsonia* sets, and some Nutmeg myrtle, which shrub I never yet examined. As I have wrote you fully per Curling, I beg leave to refer you to that, and I am hopeful you will continue your ingenious correspondence. The enquiry you mention which you are engaged in, must be very curious, and very acceptable to the world. I have never seen anything on that subject, nor on the sea productions at all. I hope your next letter will come soon, and be full of queries, which may serve as so many stimuli to me for enquiry. I am just now preparing work for Mr. Ehret, which I shall send early in the fall. I have sent the receipt for your things to Mr. Shipley, and have wrote to Dr. Hales. I have the honour to be with esteem, Sir, your most obliged and very humble servant,

ALEXANDER GARDEN.

Please to direct to Dr. Alexander Garden, Physician, in Charlestown, South Carolina.

MY DEAR FRIEND, Charlestown, Dec. 24, 1755.

I received all your letters by Capt. Chisman and by Capt Ball, with the seeds, and by Capt. Cowie's ship, with the most curious cuts of the method of propagation of the Polypes. Each of your presents were extremely obliging, and singularly useful to me, and I do most sincerely return you my thanks, and I am fain to hope that I shall in time, in some measure, evince the sense which I have of the many favours which I stand indebted to you for. The books I value more than gold; particularly the *Species Plantarum,* which has enabled me to ascertain several species, and to discover some of Linnæus's mistakes, among which is that of his classing our yellow Jessamy amongst the *Bignoniæ,* which is absolutely a new genus, and as such I have carefully taken down its characters, and taken the liberty of calling it the *Ellisiana,* which freedom I hope you will excuse, as I can assure you it proceeds from no sense of any of your favours, but in gratitude to your freeing my mind from the error it laboured under in believing the Corallines to be vegetables. Believe me, Sir, that however much I think myself in your debt for other favours, that of having my mind relieved from error, and my judgment informed, is above all others in my sight; and in this point your kind endeavours have not been wanting. I shall soon send you the characters of the *Ellisiana,* with several others, which you may peruse, correct, and, if you judge proper, may present them to the Society, but only in latin, and if they will not accept of them in that language, I will send them elsewhere.

I have wrote to Dr. Schlosser, whose correspondence, from the character that you give of him, must be extremely advantageous to me. I have sent his letter inclosed to you, as I did not know his particular address. If a correspondence can be got with professor Allemand, and Van Royen, or Jussieu at Paris, all my desires would be fulfilled. You will be surprised at finding so short a letter, but when I tell you that this night I expect to be matrimonized, which affair has of late employed most of my thoughts, and you may easily believe I have no great goût for writing in the manner in which I propose to answer your kind letters, which answer you may expect in a few weeks by some of our first ships. Just now I have sent you a large box of curious trees; they are all fine thriving young plants, and will flower very soon. I am in great hopes that they will answer your expectations, and my intentions; they are very carefully packed up, and are in number 90 in all. I have likewise sent you a box of well cured seeds, which I hope will likewise arrive in good order. You have inclosed a list of them, and of the young plants. I am afraid that Capt. Ball will not take the young plants, as he is so full, though they be now on board of him; but to-morrow I will order them on board of the Prince George, Capt. Bostock, if they be not well provided in a good birth with Capt. Ball; so that I must beg that you will be so good as to enquire after them, as I cannot write you more particularly just now. The box of seeds comes with Capt. Ball, which will arrive in time. I must beg that you will be kind enough to present my compliments and thanks to Mr. Christopher Gray for his very obliging present, and assure him that I will do my best to supply him with some things, and I hope he may think these a specimen. But absence for nigh two months in the summer time, and then an affair of love commencing on my return, prevented me from attending much to my studies or collections this fall. You may expect some very fine plants of the *Beureria,* and the flesh-coloured flowered *Acacia,* which I have had in boxes for several months, but I do not chuse to risk so many on one bottom. You may likewise expect a letter pretty often, and some new genera. Frequent letters from you is the greatest pleasure that I enjoy. I have wrote to Mr. Baker by this vessel, and will soon write again; and to Dr. Hales and Mr. Shipley. I am now busied in copying over my journal to Saluda this last summer, which I shall send soon.

The French war quite puts me in the hip, as I can easily see it will prevent my hearing from you as often as I could wish, and vessels going home heavy loaded are much apter to fall into their hands than those outward bound. I hurry away to meet the parson and my dear girl, and must bid you adieu, and remain with the sincerest esteem, dear Sir, your most obliged and very humble servant,

ALEXANDER GARDEN.

I could have sent you more seeds, but Mr. Saxby told me that he was this day putting a parcel on board for you, which I assisted him in collecting.

South Carolina, Charlestown,
Jan. 3, 1756.

SIR,

A few days ago Capt. Ball sailed from this port, by whom I sent you a box of seeds, all in good case; and the same day I helped to assort and pack a box from Mr. George Saxby, my worthy acquaintance, most of which I believe are designed for you. You have there a very good parcel of the *Magnolia* seeds, which prevented me from sending any, as I had the care of preserving and sending those which I hope are for your use and service. By Capt. Ball I sent you a list of shrubs and young trees, which I put a board for you, but as he could not give them a birth below deck, I took them ashore, and now send them by Capt. Bostock, of the Prince George, who promises to take great care of them. Inclosed you have a list of them, and of the seeds sent by Capt. Ball. The seeds were all packed in good dry soapy mould. There are 90 young trees, all well packed in moss according to your directions. I hope they will all arrive safe, and give you satisfaction, than which I can have no greater pleasure. I am extremely sensible of the value of your present to me, and heartily render both you and Mr. Gray thanks, and I hope shall soon convince you how much I esteem your kind and obliging favours. I have been much retarded in my enquiries this summer by a love affair (as I wrote you in my last, to which I beg leave to refer you), but now as I attained all my wishes on Christmas-eve, I hope to be more in a settled way for the future, and will carefully cultivate your valuable correspondence. I wrote to your ingenious friend Dr. Schlosser by my last to you. I have still some fine plants of the *Beureria* (I made out a description of this last summer, and if Mr. Ehret pleases will send it to him) and the *Acacia caule hispido floribus dilutè sanguineis racemosis*. This grows wild about 80 miles back. I shall likewise by Capt. Chisman send some more seeds for you, in case that Capt. Ball should fall in with the French, for fear of which I have directed them to Jussieu (whose correspondence could I procure it would be very acceptable) at Paris. I shall write to Mr. Baker and Mr. Shipley (from whom I heard lately) and the Rev. Dr. Hales. Pray offer my compliments to any of those gentlemen, and Mr. Collinson; and I am, Sir, your most obliged and very humble servant,

ALEXANDER GARDEN.

P. S. By Capt. Chisman (who sails in a week)

you may expect a box of seeds, and some fine shrubs in another box.

My young wife promises some shells for Mrs. Ellis.

S. Carolina, Charlestown,
Jan. 13, 1756.

MY DEAR FRIEND,

I wrote you on the 23d of December by Capt. Ball, and sent you some seeds, with a list of them, and of some shrubs and young trees, which I afterwards sent by Capt. Bostock of the Prince George, by whom I likewise wrote to you the 3d of January, 1756. In my first to you, by Capt. Ball, I inclosed a letter in answer to Dr. Schlosser's very polite and most friendly letter to me. I am afraid that through an excess of good nature and politeness, you have given him too high an opinion of my poor abilities; but at the same time I beg leave to offer my heartiest thanks to you for procuring me so valuable a correspondent, whose epistolary acquaintance will always be most agreeable to me, first on your account, next on his own, and lastly for the advantages that I must reap by a continuance of his letters. The correspondence of these two other gentlemen you mentioned, *viz.* Van Royen and Allemand, will be extremely obliging, and I think will fully gratify my desire of correspondents, especially if I could add to them that of Bernard de Jussieu. You will no doubt readily think that it is odd in me, who live so far from the learned world, to have such an avaricious desire after new correspondents. I own it is really odd ; but I cannot help it, and I think that nothing is a greater spur to enquiries and further improvement, than some demands from literary correspondents. I know that every letter which I receive not only revives the little botanic spark in my breast, but even increases its quantity snd flaming force. Some such thing is absolutely necessary to one, living under our broiling sun ; else *ce feu, cette divine flame,* as Perrault calls, it would be evaporated in a few years, and we should rest satisfied before we had half discharged our duty to our fellow creatures, which obliges us, as members of the great society, to contribute our mite towards proper knowledge of the works of our Common Father. This past summer was mostly trifled away with me as to botany ; for what with business, and what with hours spent in attending the fair, I had few supernumerary hours to devote to botany ; however, I shall beg leave to refer you to my letter by Capt. Ball for an account of myself, and of my journey to our back settlements with the governor; to attend him on a journey to meet the Cherokee Indians. I have finished a copy of my journal thither, and as soon as I can get it transcribed, I shall send it to you and my other acquaintances at London, though I have wrote it in a letter to Dr. Huxham, the ornament of the healing art. It contains about a dozen and half sheets, and I hope will give you some entertainment, though it is not so full as I once intended; but the practice of physic now takes up more than two thirds of my time, and my beginning business obliges me to have great regard and constant application to it; our people often preferring a slavish attendance to any other qualification. There is nothing more common than to see a diligent and officious apothecary more employed than a physician; but I believe that is the case in more places than in this. I thank you kindly for your advice as to Linnæus, and I think I shall take it. I have (since I had his *Species Plantarum*) discovered many of his peculiar assumptions, as in the *Callicarpa*, which Mr. Miller had before called the *Jonsonia*. I have likewise discovered several errors (though these cannot properly be laid to his charge), as that of his putting our yellow Jessamy among the *Bignonias*, when it is so entirely different and distinct. This Summer I took its characters, and finding it a new genus, I took the liberty to affix your name by calling it *Ellisiana*, and shall send you the characters either by this or some other vessel going soon. I very gratefully own my obligations to you for your treatise on the animal life of the Corallines, and acknowledge with pleasure my being freed from the error of taking them for vegetables. This one great truth being so clearly elucidated, lays all true lovers of Natural History under indispensable obligations to the author, and I in humble gratitude judged it my bounden duty to ornament botany with a name that has contributed so much to advancing her purity. The main stalk is enough for her to nourish and support, and friendly indeed is the hand that drives away the caterpillar and vermin from pestering her native simple productions. Let her then wrap the pliant Jessamy round your temples, and allow her young sister to adjust the fragrant blossoms to your brow.

I likewise beg leave to present my thanks for the books which you sent. I heartily wish that the seeds which I sent, and am now to send, may be useful and agreeable to you and Mr. Gray, to whom I return my thanks and compliments, with assurances to him that if I live he shall not repent his generous gift. Books to me of all things must be the most valuable, as they more immediately tend to my improvement, which I value above all the riches in the world besides. This Summer Mr. George Saxby shewed me a list from you, which I helped him to procure, by giving him our country names for these you mentioned, and assisting him to some seeds, particularly the *Magnolia*, which I gave chiefly as I imagined he could send his much sooner than I could send mine, though after all they only went by the same vessel. He likewise sent you some of the *Schlosseria* or a new genus of the Palm tree. I shall send you its characters, and shall leave it to you to call it either the *Schlosseria* or *Halea*, in honour of my much esteemed correspondent Dr. Hales. Some time ago I called our wild Horehound *Halea*, but I am afraid that there is too nigh a relation between that and the *Eupatorium* to separate them, so that I would denominate this either by his name or by Dr. Huxham's, both which gentlemen I think greatly merit every mark of esteem which not only I, but every lover of science can confer. Poor indeed is this, but it is at least a testimony of a grateful mind. Some time since I called the *Solanum triphyllum* of Mr. Catesby, *Huxhamia;* but since I received the *Species Plantarum,* I find that Linnæus has called it the *Trillium,* for which reason I would willingly join either of these names to so beautiful a plant as this, which from proper and strict examination I am certain is a new genus. You shall have its characters, and then you may name it by either of them with my approbation. I am sorry that I cannot send you a specimen of this, as I can of some of the others.

I have observed your directions in sending seeds to none but yourself this year. I have sent you a good many fine plants of the *Acacia caule hispido,* &c. and of the *Beureria,* or *Frutex Corni foliis,* &c. I took the characters this Summer, and if Mr. Ehret pleases, I shall send them to him, to publish with his print of it.

There are many fine young trees in that box by Capt. Bostock, in number 90 in all, and they are put up as you desired with moss, &c. and have got a good birth below deck; I hope they will come safe, and answer your expectations and my intentions.

By a letter from Dr. Whytt, I learn that they have published the characters of a plant which I had sent me by Miss Colden, which she had called *Gardenia.*

I had met with the same plant some time before I received her letter, and finding it differ from the Hypericums (of which at first it seemed to be a species), I took its characters, and not many days after I received her letter, when I was at Philadelphia. I sent both her characters and my own, which I had made before, to Dr. Whytt, with some other things, and the Doctor thought them worth a place in your essays. He writes that he had never read your treatise on the Corallines, but daily expected it. As he is a very ingenious man and great lover of truth, the perusal of it must give him, or any such man, infinite pleasure and satisfaction.

I ought now to make some apology to you for not procuring you some specimens of our Corallines, which are numerous; but believe me, dear Sir, that I have never yet been able to spend one day at the sea side, though I am fully determined, please God, to be down some days this spring, and I hope shall give you some good account of my collections.

As to my being admitted a member of the Royal Society, I find it is a difficult thing, and as I but little merit it, and am not at all in a humour to pay, especially as I never shall have an opportunity to vote in the society; so I rest satisfied, and shall be pleased if I can contribute any thing to amuse my private friends, and send some few other things to the Edinburgh Society, who very generously offer me their favour. I own it would have given me satisfaction to have been admitted, and whatever my poor endeavours could have done, they should have commanded it.

Some few days ago I had a bunch of tubular membranous pipes, all joined at the bottom, brought me by a fisherman who picked them up on the beach. I was sorry that I had not seen them alive, but though I had, probably I should not have made much of it without the help of a glass. I must here take the liberty to put you in mind of your promise as to the water microscope for viewing these sea productions: it will be a most acceptable present, as were some glasses which Mr. Baker sent me this last year. Ludwigius's works and Rivinus, with Hermann's or Van Royen's, are the books which I want to look into most, and if Mr. Gray thinks that a traffic of this kind would be worth his while, I will very gladly enter into such a trade, and be bound to any conditions. I have spoke to two or three gardeners, and given them directions about some rare shrubs, and I hope shall be able to supply him to his liking. Any of Mr. Brown's curious cuts, or any curious cuts, will be very acceptable, or any good treatise on minerals, animals, shells, &c. especially if it be done in the systematical order, which method pleases me above all others. I think, if a system be any thing clear and ingenious, it is by far a greater help to the memory and judgment, than all other helps of descriptions, prints, cuts, drawings, &c. I have often discovered a vegetable from its class, when I am sure no description could have made me so certain. An ingenious system has always some essential character, but a draught has only the artificial appearance of the habit, extremely fallacious indeed! The man who gives the natural system, must be a *second Adam,* seeing intuitively the essential differences of things. Where is this man to be found? I have no one treatise on fossils, minerals, or chiefly upon fishes or shells. I believe Artedius is the best on fishes, but I do not know who excels on insects, of which we have myriads. If Peter Kalm's Canada plants be published, I should be glad to see it. I could have wished to have seen this performance before I had sent you my journal to *Saluda;* but, however, I content myself with the hopes that you and my other good friends will excuse any faults the readier, as I am but a young naturalist, and greatly removed from books and conversation. In this I have just given you the reflexions which arose in my breast from the phænomena that appeared during our journey; and if I should be wrong, I shall at least be pleased even to have dared at reasoning on these appearances; for I think that the mere contemplation and observance of natural objects, is in common to us with the brutes, but an enquiry and investigation of their several uses, and the economy of nature, is the business of rational creatures, and it may be of some use to accustom the mind to such enquiries, even if we should often fail in the true discovery. Peter Kalm had great opportunities, and if he made no good use of the observations, his works will be of far less use, and he will be rather esteemed an observator than a philosophic botanist.

Linnæus certainly merits this character. I have often bewailed the want of his various *Itinera*, which I find are published in Swedish.

In the September magazine I observe a letter from Mr. Miller to Mr. Crokatt on the *Toxicodendron*, with a translation of the Abbé Mazea's letter on the same subject. I should be glad to know the species of the *Toxicodendron* which either Mr. Miller or the Abbé means, for we have several species of it here. And the Chinese Varnish-tree grows no where with us but in the Back Settlements. I never saw it within 80 miles of Charlestown, consequently it was never known to Catesby whether it grew here or not, as he never was 30 miles back from the coast during his stay here; so that I cannot help thinking that the comparison between our *Toxicodendron* and the Chinese Varnish, by the German there hinted at, must be very vague.

Dear Sir, it is now five days since I wrote you the above letter, since which time I have been twice out in the country 40 miles on a visit to the sick, so that I have not had 10 minutes time since, till this evening that I am just returned, and am told the ship sails early in the morning, so that my fatigue and short time both prevent me from saying more, though I had twenty things more, besides letters to my other friends, Mr. Collinson, Baker, Shiply, and the Rev. Dr. Hales, to all whom I beg leave to trouble you with my compliments. I proposed to send you some characters of five or six genera, but I have only time to transcribe the characters of a new species of the Palm-tree. You may expect some more by Captain White. Please to communicate this genus to Mr. Collinson: he wrote me concerning the Palm-trees. I have sent you the seeds and the young plants of the *Acacia floribus racematim congestis, colore dilutè sanguineo, caule hispido;* 23 young plants, and the *Beureria* or *Frutex Corni foliis,* &c. 19 young plants, all alive and in good order; they have been in a box since September last;—you will please to call for them.

Let me beg of you to send me any new thing which you discover in the Corallines, &c. or cut of any curious plant, or bird, or shells. These are very curious, and what we can by no means procure here.

The young plants of the rhubarb grow.

My young wife promises to exert her utmost among the ladies here for some shells to Miss Ellis.

If you favour me with a short line on the receipt of any of these letters, it will be very obliging, as it will make me easy with regard to the safety of mine. With my heartiest prayers for your health and welfare, I remain with great esteem, dear Sir, your most obliged and very humble servant,

ALEXANDER GARDEN.

DEAR SIR, Charlestown, South Carolina, March 22, 1756.

I have now the pleasure of sitting down to write to you, for the fourth time since I heard from you;

and let me assure you, that there is no hour of my life gives me more pleasure than this, when I can with freedom communicate my mind to a distant friend, and be again entertained with his return. Sure, reverberated pleasures fire the breast; and make life, *life* indeed. I shall just mention the vessels by which I wrote, with the things I then sent. The first was with Capt. Ball, on board the Friendship, from this place; by him I sent you a box of seeds, with a list; and a letter enclosed for Dr. Schlosser, whose promised friendship I value much, and beg that you will procure me a continuance. If there be any thing that I can serve that gentleman in, I will most heartily promise and perform my utmost. I am sorry that I know little of his favourite study, *Minerals*, but I hope my good friend Mr. Baker will assist me in that, by his advice and some specimens, for I chuse to have only hints of things, and then to let my own fancy and imagination work. My next to you was by Capt. Bostock, aboard of whom I shipped a very large box of shrubs for you, of the *Magnolia altissima*, *Pavia*, *Jasminum leuteum* or *Ellisiana*, *Gordonia* or *Loblolly Bay*, *Liquidambar*, *Laurus baccis purpureis*, *Cupressus foliis deciduis*, *Liriodendron*, Sweet flowering Bay, five young plants that will all flower this year. Also the early upright Honeysuckles, the *Chionanthus*, &c. &c. I hope these, as well as the seeds, came safe to your hands. My next to you was by Capt. Cheesman, of the Charming Martha, by whom I sent you another box of seeds, in case the others should fall into the French hands. By him I likewise sent a box of the *Beureria*, containing 19 young plants, all in very good order, and matted up with moss, &c. as you directed.

This letter was chiefly designed as an answer to yours and to Dr. Schlosser's, though I had enclosed one to him, in my first to you. I told you in that by Cheesman, that I now was greatly pleased with my correspondents, and most willingly would rest satisfied, if I could by your assistance only add those whom you mention, Professors Allemand, Van Royen, and Jussieu; could I be only happy enough to have these acquaintances, I could employ all my time almost in serving both you and them.

I sent you by Captain Cheesman, in the same box with the *Beurerias*, 23 plants of the flesh-coloured flowering *Acacia*. I then told you that I had never yet been able to procure you any Corallines or *Lithophytos* alive, as my business has never yet admitted of my absence, though I hope, God willing, to spend some day soon on the coast. You were kind enough to promise me a water microscope, which will be a most acceptable present, as would Kalm's History of the Canada Plants, or any of Browne's cuts of the Jamaica plants, or Ludwigius, Rivinus, or Van Royen's works.

I have now, by my brother-in-law Mr. Peronneau, sent you a small box of some things, among which there are the following seeds: 1. *Lilium Martagon florib. erectis*, grows in meadows. 2. *Lilium Martag. florib. pendul.* 3. *Mitella.* 4. *Po-*

lygala, or Seneka Rattle-snake-root. 5. Spruce Pine. 6. *Thuya.* 7. *Ægopodium.* 8. *Apocynum*, with sweet-scented flowers. 9. A new genus. 10. A new genus. 11. *Monotropa.* 12. A new genus, *Mitellæ affinis.* 13. *Mimulus.* 14. *Aralia caulescens.* 15. *Triostiospermum.* 16. *Bartsia.* 17. *Gordonia*, or Loblolly Bay. 18. A new genus, *fructu tetrapteride.* 20. A plant that produces a very fine scarlet dye. 21. *Amorpha.* 22. *Aloe Carolinensis*; this grows wild in our woods; the leaves spread on the ground; the flower is beautiful. 23. *Beureria*; I never saw this seed till lately, that I had this sent me in the pod, when I took its characters, and have now sent you the seed.

I have filled up the box with a few shells, which I should be glad might be acceptable to Mrs. Ellis, but they are of so small consequence, that I have not courage to make an offer of them unless you will assist me in gaining her acceptance of them. Probably I may be better able to send her some at some future time. Mr. Peronneau has promised to present a Nonpareil bird to her, providing he can carry home two alive. We reckon this our finest bird, and if I should have one, I hope it may be acceptable. I have likewise sent a small bundle of the *Vermiculi marini*, which I took up the other day on the beach.

I come now to mention my journal to Saluda, which I performed last summer, during the hot months of June and July. I told you before, that I had wrote this in a letter to Dr. John Huxham, at

Plymouth, to whom I have sent a copy of it. In the mean time, I have sent you a copy of the introduction, which I beg you will peruse and communicate to Mr. Collinson and Dr. Hales, and let me have your opinions of it, if it be worth presenting to the Royal Society; which, if it should, I hope you will correct any errors either in the language or sentiments that you discover. I have wrote to Dr. Huxham, that I have asked this favour of you, if either of you should think it worth communicating. I have not yet been able to get it all wrote out, much less to copy it over, neither indeed did I judge it proper till I should know whether such an account would be worth while; yet I think I may say, that if any of it be useful, it must be what is to follow, as it contains an account of the vegetables and minerals that merited any notice during our march; however, if you find this worthy perusal, you need not keep it till you get the rest, as I should chuse it to be published in a series of letters. You will be kind enough to shew it to Mr. Baker.

As to animals, I have not mentioned any at all, as I am quite at a loss in that part of natural history, having no helps on that subject; yet I have ventured to send the characters of an insect to Mr. Baker, and he will from that see how readily I should make such collections, had I it in my power.

Since writing Mr. Baker's letter, I have been again asked by his Excellency the Governor of this province to attend him on another journey over the mountains to the banks of the Mississippi, and to set off in three weeks. I have not yet accepted the invitation, but I think that I shall, and in that case I shall have great opportunities of collecting some herbs, &c. The place proposed for the fort is six hundred miles from Charlestown, which will be a dreadful journey in the hot months; but the hopes of returning richly laden with the spoils of Nature and our Apalachian mountains, is more to me than soft ease or hopes of sordid gain. If we set off about the 20th of April, we shall return about the 20th of July, by which time I shall expect to hear from you.

I have sent you the characters of six of our plants here, which is all that I have had leisure to transcribe, as there are many days in which I have not one hour to myself, from the hurry of business in the practice of physic. I proposed to have sent you several others, especially the characters of that tree which bears the four-cornered or winged seed, which I have sent you. I have all the characters, but I must examine them before I send them, and that I have not time to do just now. However, I have sent the branch with the fruit, and a sprig with the flowers; so you may get your friend to draw it now, and I shall send you the characters early in the fall next year. All the rest you may examine just now, and present them, if you please, to the Society, only I must beg that you will correct any errors in the language or descriptions that you observe. As to publishing them in english, I think that would be absurd, and what I will in no wise consent to, though Mr. Baker writes me that the Society only chuses things in english. It was for this reason that I wrote my journal in english, but if the Society do not think it worth while, it shall be the last that I shall give leave to communicate.

As my giving any money (as I wrote you before) to be admitted a member, it is what I would not do to any society under the sun, as I always think that these things should be a matter of choice in the society, not of any pecuniary reward. There is no body of learned men in the world that I have a greater regard for, nor any that I should be more willing to oblige or to serve, than the Royal Society, but if they do not think that I merit a place as a foreigner, when I certainly am one to all intents and purposes, I think that I have no reason to mind them so much as my private friends.

I shall now give you a short account of the various genera whose characters I send you by this opportunity. The first, No. 1, is an anonymous plant (as I judge) which I hope you will be kind enough to denominate by some name as shall seem good to you, for to you I entirely leave it. The plant itself makes a pretty appearance, though the flowers be trifling. It flowers in May, and generally grows two foot high. I have sent you a dried specimen, which I hope you will get engraved, to be given in with it to the Society. The second, or No. 2, is likewise (as I judge) a new genus; I have sent you the seeds by Ball and Cheesman. The Planters call it *Supple Jack.* It is a beautiful vine, and sometimes grows to a large size, as thick as a man's thigh. Call this what you please, either *Huxhamia* or *Halœa,* or what you please, but be sure to give it a name.

The third, No. 3, is the *Cornifoliis frutex,* &c. of Catesby, which you now tell me Mr. Ehret has called the *Beureria,* which name I have still retained, and have affixed a description, as I imagined that you would be puzzled to get the true characters in England. If Mr. Ehret would join his draught to these characters, it would be better, and make the account complete; but if not, you may just insert it with the rest, as I had made out the characters before I heard of his draught of it.

The fourth, No. 4, is the *Ellisiana,* or our Carolina Yellow Jessamy; this is certainly a new genus, though Linnæus, misled by Catesby and other blunderers, has classed it amongst the *Bignoniæ.* I have taken the liberty of affixing your name, and I hope you will permit me to beg leave to have it presented to the Society under that title. I have sent a small dry specimen, which it might be worth while to draw and place with it. This has both flower and pericarp, which you will observe split at the top, into four valves, instead of two, as some have observed. I have given some history of the plant along with it.

The fifth, No. 5, is the *Gordonia,* which I call so in honour of my old master, Doctor James Gordon, at Aberdeen, a very ingenious and skilful physician and botanist, who first initiated me into these studies, and tinctured my mind very early with a relish for them. It is the Loblolly Bay which Linnæus classes among the *Hyperica*, just with as much propriety as I might join the Oak and Ambrosia, nay, indeed, there is not nigh such an affinity, as you will readily see, from comparing the characters of this and those of the *Hypericum*. I think you may take Catesby's cut of this, if he has one, to join with the characters.

The sixth, No. 6, is a new genus of the Palm-tree, which I sent you some time ago by Cheesman, and I hope has come safe to hand. I told you that you might call it *Huxhamia* or *Halœa*, in honour of either of my friends; which commission I now again renew, as I find that Linnæus has called that plant *Trillium*, which I sent to Dr. Huxham under the name of *Huxhamia;* however, I am glad that I have it still in my power to shew the Doctor how much I regard him. If you have called this by one of these, then the vine must be the other, No. 2, or Supple Jack. This is all that I have to say of these just now, only let me know what your opinion of them is, and I may probably be encouraged to go on.

I have sown your Rhubarb in several different places, and it has all come up very well, but there are only two or three plants that are left from the preying of a vile grub-worm, which gnaws all the roots and leaves; yet, nevertheless, I hope to make it thrive well before another year. I have likewise planted some Logwood seed, but I am afraid it will not thrive.

As soon as Mr. Peronneau (who brings you this) returns from England, he designs to make a pretty garden, and it will be much in his way to supply you, if you speak to him now.

I have sent you several dry specimens of herbs, of which I shall give you a fuller account afterward, but the bearer will inform you what little time I have to do any thing. In the mean time, I remain with the greatest esteem, dear Sir, your most obliged and very humble servant,

ALEXANDER GARDEN.

Please to present my compliments to Dr. Hales, Messrs Collinson, Baker, Shiply, and Schlosser.

Pray allow me to ask you the favour to shew the bearer Mr. Peronneau, your collection of sea plants and shells, that he may give me some account on his return.

MR. ELLIS TO DR. GARDEN.

DEAR SIR, London, June 14, 1756.

I had the pleasure of receiving your favour of the 22d March, 1756, from Mr. Perroneau, who has been so obliging as to deliver the seeds you mention, and the specimens of plants. I find he has had a very long passage; I hope, however, that the seeds will grow. Mr. Ehret is to call on me to-morrow to examine the specimens, and to give you his opinion which are new genuses; he is critically exact in examining plants.

I am very glad to see how well you dry the specimens; I hope you will go on to send me more varieties. The excursion you mention will probably furnish you with an excellent opportunity of meeting with very rare plants.

The loss we sustained by the distress of Captain Bostock, is really great. The rarest plants here are the *Beureria* and red *Robinia*, the upright Honeysuckles or *Azaleas*, the Swamp Laurel or deciduous *Magnolia*, called also Sweet-flowering Bay, white underneath*, though I am informed this plant scarce looses its leaves in winter; it is rare and valuable here, if of any size fit to blossom; the *Gordonia*, or Loblolly Bay; the *Myrica*, with Spleenwort leaves; the great *Magnolia;* the Umbrella-tree; the Yellow Jassamine and four leaved *Bignonia*. The common plants here are the *Pavia*, *Liquid ambar*, *Liriodendron*, *Cupressus foliis deciduis*, though the seeds are of use. The *Pavia* seeds or nuts must, as indeed ought most of the other seeds, be put up in dry mould or sand, to keep up their vegetative quality. Had the seeds of the *Nelumbo* been put into bottles of water, and some into sand or earth, it is possible we might raise them here; but on opening some of them, I found the germ withered, I mean the bud between the lobes. The trumpet-leaved plant is the *Saracenia*, which you have sent the specimen of. We have the *Amorpha* in great perfection here, and raise it from layers. I should be glad of what varieties you can procure of the *Sorbus*, *Cratægus*, or *Mespilus*, that grow on

* *Magnolia glauca.*

your highest lands; all plants are more valuable with us as they stand the weather, so that the mountain plants have the best chance. The pine cones, and oak acorns do well with us.

DR. GARDEN TO CHARLES WHITWORTH, ESQ.

SIR, Charlestown, South Carolina, April 27th, 1757.

Your much esteemed favour by Mr. Peronneau, came to my hands some months ago, and I before did myself the honour to acknowledge the receipt of it, and to send you a parcel of some of our rarest seeds of trees and shrubs, per Capt. Coats, who sailed from this place sometime in January last. I hope he arrived safe, and delivered his charge to Mr. Shipley, to whose care I took the liberty of directing them, as I was ignorant in what manner to direct for you at London; and this likewise will be delivered to you by Mr. Shipley.

Before I proceed to a particular answer of yours, I must first acknowledge the obligations which I lie under both to the Society and yourself, for the honour you do me in putting any esteem on the few hints which I took the liberty to offer to you on the producing of various vegetable colours. It is a subject which I have often thought on, and it gives me the greatest satisfaction to meet with your approbation. It seems extremely probable indeed, that some useful discovery may be the result of many and various trials, and I think the plan which you propose seems to be well designed for promising success. It must be the work of many hands, of many heads, and of time, but how is it possible that any person could be a niggard of time, when there lies before him a prospect of becoming the author of such a discovery, as may at once confer on him riches and honour, beget labour and bread to thousands, and become a valuable, useful, and ornamental commodity to Great Britain. If a course of twelve years experiments could produce this effect, would not the labour of every experimentator be well rewarded, and would not every one be entitled to some part of the honour of such discovery, who had thrown in their *mite*, and who, by their unsuccessful experiments had taught the one happy man what rocks to shun, or what labour to save, in gaining the wished-for prize, and expected discovery? Let us suppose that the consequence of such trials, was the discovery of a die of a different colour from indigo, but as valuable in its nature and quality; would not our latest posterity and different nations bless the happy æra? What are the immense returns of indigo now to what they were five years ago? Five years ago there was not five thousand weight of indigo, good and bad, made in this province; and this year there has been upwards of seven hundred thousand weight shipped from the different ports of this province. Now reckon this at 5s. 6d. sterling per lb. on an average, and it (this is a moderate computation) will amount to one hundred and ninety-two thousand and five hundred pounds sterling money of Great Britain.

And yet the quantity of rice which we make now is scarce less than what it was five years ago; how much richer *per annum* are we then, and of how much more benefit are we to our mother country; for there is at least one half of this that pays for the additional quantity of British commodities which we now consume, to what we did then. Our increase of luxury is nearly in proportion to our increased crops of indigo, and the value of labour rises daily, and of consequence the value of slaves.

But to proceed to the various parts of yours: as to proper accounts of the seeds which I sent you last year and which I now send you, I beg leave to refer you to Mr. Ellis, who will more fully inform you of the proper method of cultivating them in England than I can. I shall only observe that the *Zanthoxylum*, or Pellitory-tree, is very different from the *Parietaria* of Miller's Dictionary. The *Zanthoxylum* is a very pretty tree. The *Magnolia* is the finest and most superb ever-green tree that the earth produces. The *Catalpa* is a very grand tree, with broad leaves and beautiful flowers. It produces its foliage and flowers annually.

The seeds of the *Colocassia* may be thrown into a pond of water, and they will come up without any further culture, however deep the pond may be. Here follows a list of the seeds which I sent you in January last, by Captain Coats, directed to Mr. Shipley, though the Captain might leave them at the South Carolina Coffee-house.

1. *Magnolia altisima,* great Tulip laurel.

2. *Ellisiana*, or Yellow Jessamy of South Carolina.
3. *Aquifolium foliis integerrimis.*
4. *Chionanthus*, Fringe-tree, or Snowdrop-tree.
5. *Chamærops*, or Swamp Palmetto.
6. *Yucca foliis filamentosis*, very beautiful.
7. Umbrella tree, a most rare and beautiful shrub.
8. *Corallodendron radice nodosa*, Coral tree.
9. *Tetrapteris*, a very beautiful flowering shrub.
10. Swamp pine; this is the loftiest of our pines.
11. *Liriodendron*, or Tulip-tree.
12. *Acer quinquefolium*, five-leaved Maple.
13. *Aralia caudice spinoso*, Angelica tree.

These, I hope, came all safe to your hands, and you will find an account of each in the last edition of Miller's Dictionary, unless the *Ellisiana* and *Tetrapteris*, both which love moist soil and much sun.

I come now to your second letter, which contains a proposed plan for a society or company, that might mutually agree to make a number of trials on different vegetables, and to communicate mutually their success, and lay the same before such a number of your society, or such gentlemen as should form themselves into an American corresponding Society in London, in order to have your opinion and advice in our proceedings, and your assistance in procuring such plants and seeds as might be requisite for the furthering the design of each American association. You justly observe that there should be such a company in most of the capitals of America, at least there ought to be one in Georgia, one in South Carolina, one in Philadelphia, and one in New York government. This would greatly multiply your business in London, and it would divide the labour of the various societies here. From their various and different productions it would give greater hopes of success in the end. The great use of such associations might be, the pointing out proper articles for your Premium Society to exercise their benevolence on, and excite the industry and genius of others; at the same time it is scarce to be imagined but that Natural History and Philosophy must likewise be enriched by many unthought-of phænomena coming to light about the nature of vegetables and colours. It is surprizing to see how exactly the various phænomena of the Indigo beating, quadrate with, and confirm Sir Isaac's theory of the composition of the colorific particles of bodies. But of this, more in another place.

Agreeably to your proposed plan, I set about instituting here such an association as you intended, and after speaking to some of our most public spirited gentlemen, we agreed to publish an advertisement of our design, and give an invitation to all the country gentlemen, who would undertake to make any trials on the vegetables about them, and communicate them to us in town, who will lay them before you in London. This is all that we, who constantly reside in town, can possibly do, to excite and spur on the planters to make the necessary and proper trials, and in this way there shall be nothing left undone on our part. We hope to be able to have some useful hints to lay before you, but such things will take time; for my own part, if a dozen of years produce the desired effect, I shall think our time and pains well spent. As we published our advertisement in the Gazette, we imagine it may have reached the most distant parts of this, and probably some of our neighbouring provinces. Inclosed I have sent you a copy, which I hope you will take the trouble to peruse, and let us know what other steps we should take in order to make the scheme useful. We were obliged to make use of the British Society's name in order to have greater weight with the people, but we hope for their pardon for this freedom, as our intentions are for the public good. Sometimes I am afraid that I have carried this matter too far without some further advice or direction from you, as we may soon give you some trouble, and plague you with our Essays, if we should meet with any that we think may merit your notice. But I trust entirely to your generous and benevolent good nature, in excusing the trouble which we meditate for you.

You are so very good naturedly polite as to tell me that the short sketch which I gave you of my journey with our Governor to Saluda, gave you some pleasure, and you further desire me to insert any such remarks for the future. In obedience to your commands, I shall take the liberty to give you a short account of another journey on which I set out this last summer, in order to go over the mountains to the Upper Cherokees.

Mr. Glen, during his administration, always gave great attention to Indian affairs, and now we reap the fruits of it; while war and all its calamities have been harassing our sister colonies, we have been, hitherto, living in profound peace, and by our indigo have been amassing great estates. I before mentioned to you, that when the Indians made that solemn surrender of themselves, their lands and country at Saluda, they then desired Governor Glen to build a fort in their overhill towns, in order to protect their wives and children from the French, while they themselves might go to war against them. This had been thought of before, and enjoined the Governor from home, but the Indians would never give their consent till that time; neither did they ever before that time make any league, covenant, or treaty with Great Britain, much less (except the usual treaties of friendship and commerce) any surrender and submission of themselves, notwithstanding the groundless and false piece of history which was inserted some time since in the Magazines, about Sir Alexander Cumming's treaties with them. This gentleman, without any authority, but merely from curiosity, made a visit to the Cherokee country, and there happened, by accident, to be present at the making of a headman, who with some others he invited to go along with him to Great Britain, which invitation they accepted of, and this was literally the whole affair, as I have since learned personally from many who were eye witnesses. But this by the bye. On his return the

Governor earnestly solicited the Assembly to grant such a fund as would enable him to perform his promise, comply with the Indian's desire, effectually secure them to our interest, and establish the British power on the western side of the Apalachian hills. They saw the reasonableness and great advantages of this scheme; but very unluckily for this province, and very unluckily for the British interest, there happened to arise such animosities between the House of Assembly and Council, that the tax bill was not passed, and of consequence the sums which the house had voted could not be raised. This happened at a time when the proper season for building or never building, securing, or losing the affection of the Cherokees, was at hand. This unlucky incident but too justly alarmed every body, but both houses remained fixed in their determinations; and at last, to obviate the difficulty, a few private men subscribed (trusting to the Assembly for being refunded) two thousand pounds sterling in about three hours. The King had given one thousand sterling, and Mr. Glen intended to have advanced two thousand more out of his private purse, and to trust to the King and people for being refunded.

Being provided with this money, he resolved on the execution of his scheme, and accordingly got together about three hundred men, and procured provision, ammunition, stores, and, in a word, every necessary, with a very ingenious engineer, and set out from Charlestown 19th May, 1756, to execute one of the most beneficial projects for the three southern provinces, and for Great Britain, that ever had been planned before that time in America. He again did me the honour to ask me to attend him, which you may be sure I readily agreed to, as I promised myself such an immense range of botanizing among the mountains; and the more so as he did every thing to make the journey agreeable and useful to me, easing me of every trouble but that of collecting plants, the most delightful of all employments. We had advanced about 260 miles of our journey, through the desart woods, when an express overtook us, informing us of Governor Lyttleton's arrival, and bringing orders for disbanding all the provincial troops. Thus my hopes were blasted, Mr. Glen returned, and the scheme for building the fort was retarded for some months. Since that time the same plan has been carried into execution, and a fort built; but notwithstanding all the care of our present Governor, the French have been able (making a handle of our disappointing them and only amusing them) to make a considerable party in the nation, and have raised great commotions.

No man can be more assiduous than Governor Lyttleton has been to guard against those practices of the French; but what the consequences may be we do not yet know. Captain Thompson, of His Majesty's sloop the Jamaica, lately took off our coast and brought in here, a ship from Mississippi, bound for Old France, on board of which there was found a copy of a treaty concluded at Albama and Mobile, between the French and the disaffected party of the Cherokees. This was done during the interval of our being recalled, and of Governor Lyttleton's sending a fresh body of men into the Cherokee nation. We just now stand connected with two of the most numerous Indian nations on the continent, and our restless neighbours find it their great interest to alienate their affections from us; indeed it will be surprizing if they should be so far short of their wonted policy, as not to create jealousies and discord at a time when a diversion of part of Lord Loudon's forces would be of service to them. They are daily receiving fresh supplies of men from Old France, both at Cape Francois, and at Mobile, and New Orleans. Is it possible that this can be done, and yet not suspected to have the least apparent dangerous consequences to us? Do but reflect on the situation of the Albama fort, and the daily encreasing strength of the works, and number of the garrison. I should be sorry to be a prophet at the expence of this province, but I am afraid that in less than a dozen of years, this fort (supposing no bad events should fall out at the present juncture) will be to Georgia and Carolina, what Crown point is now to Connecticut and New York. Does Crown point influence or interest the six nations? Then Albama will in time command the Creeks, *nolentibus volentibus*, and will overawe the Cherokees, and oblige both to be subservient to that power, who will have both power and inclination to distress us. Would to God that these were all chimæras and creatures of the brain, but some little time may open our eyes when it may be too late. My paper prevents me from giving you an account of what governor Lyttleton is assiduously labouring out for our protection, and we are in hopes of some stroke being given by Lord Loudon. I am with great esteem, Sir, your most obliged and very humble servant, ALEXANDER GARDEN.

DR. GARDEN TO MR. ELLIS.

DEAR SIR, S. Carolina, Charlestown, May 6th, 1757.

I come now to answer your very kind, friendly, and obliging letters; but were I to tell you how often I have already resolved on doing it, and how often I have again put it off, you would readily imagine that I was prevented by some evil omen, or other, that could so maliciously contrast my repeated resolutions of sitting down to acknowledge the many obligations I am already indebted to you for. Every ship almost brings me a letter, and every letter a fresh instance of your friendship and generosity. Your last was delivered to me by governor Ellis, along with Mr. Ehret's plates; the governor, with a peculiar, but too partial a politeness, communicated the good opinion which you were pleased to entertain of me. He staid but a few days in this place before he set out for Georgia, but during that short time, I took frequent opportunities of waiting on him, at such times as more interesting visitants did not engross him. His arrival here was a welcome sight to us, but it was more, much more, so to the Georgians, a deputation of whose council waited on him here, and carried him over land to Georgia, where the united acclamations of the people received him, whom they had looked for so long, and so wishfully expected.

The capture of Ball was the most inconsolable disappointment that ever I met with; believe me, my dear Friend, that though my remittances by him, which I designed for you might not have afforded you so much pleasure as you seem to imagine; yet they were most sincerely designed as the offering of a grateful heart, in return for your invaluable presents to me by him. In my letters to you at that time, I gave you an account of my intended journey, and in what manner the arrival of our new governor put a fatal stop to us. Good God! is it possible to imagine the shock I received when the unhappy express overtook us just two days march on this side of the mountains. My prospects of glutting my very soul with the view of the Southern parts of the Great Apalachees was instantaneously blasted. How often did I think of the many happy hours that I should have enjoyed in giving you a detail of their productions? How often did I think of that secret pleasure I should have in being instrumental, though in the least degree, to the advancement of our knowledge of the amazing works of the Supreme Architect? How happy should I have been to have thrown in my mite, by adding one new genus or species to the vegetable or mineral kingdom! With what pleasure did I bear the sun's scorching beams, the fatigue of travelling, the cold ground for my pillow, and the uncomfortable dreariness of rain, when I had in view the wished-for examination of the productions of the mountains! We had advanced about 260 miles of our journey through the woods, when our hour was come that all our promised Elysium vanished, and left nothing but a blank, a doleful blank to me, and I may say, to every one of the company, for we were happily collected, and unanimity reigned amongst us. What will you think when I tell you that one of our company was a very accurate drawer, and he had promised me to do every thing for me, and according to my directions, that I should desire; so that in this one circumstance my loss was irreparable. But why do I insist on the most disagreeable of all the incidents that ever providence mingled in my lot? When I wrote to you by Ball, we had not heard of the declaration of war, and I imagined my letter would have got safe; else I should have made a shift to have sent a duplicate, though at that time I had scarce half an hour to spare, as my partner was confined with a violent fever.

That I may be able then to answer yours, I shall again begin with those letters which I received from you by Capt. Ball this time last year, after returning you my most hearty thanks for the many valuable things which you sent me, especially Mr. Cuff's glass, though I am almost ashamed to tell you that

I never yet have used it in the proper way; in a word, I never yet have been able to spare eight days to go down to the coast side; but I hope I may find an opportunity to surprise you some time or other. Notwithstanding Dr. Browne's not fulfilling his promise with his subscribers, still I think he deserves well of the learned world. Has he not given a much more full and accurate account of the island than Sir Hans Sloane? Surely then he at least deserves more esteem than him. His has been a useful book to me, and given me great satisfaction. Mr. Pott I think is an admirable author, and has given me great instruction; I am singularly indebted to you for him. As to the last article, Mr. Ehret's plates, they are indeed inimitable, and far exceed any thing which I could have imagined within the power of human art. What difference is there between them and Catesby! I can never pretend to make any kind of return equivalent to them.

Mr. Peronneau did not bring me Wallerius, but I have since wrote for him, and I expect him soon.

To begin then with yours of the 10th of April, 1756.

As to my calling the Jessamy by your name, it was entirely owing to my persuasion of its being a new genus, and I still remain in that persuasion; neither am I convinced nor satisfied of any affinity between it and the *Bignonia*. It differs in the *Calyx*, *Corolla*, *Stamina*, *Pistilla*, *Pericarp*, &c. and very essentially in each and every one of them. But I now have a plant which is entirely new, and the most superb lofty plant that ever I met with in America, which I shall beg leave of you to accept as a name-sake. I have already sent you some seeds of it; it is of the *tetrandria monogynia*, and is an annual, but if you would chuse a shrub, I have that beautiful one, which I call the Hop-like shrub, of which I sent you the seeds this year. I have declined sending specimens for fear of their being taken, for if they should meet with the same fate as those by Ball, I should at once turn bankrupt. Please to let me know if I may venture them.

I come next to the *Huxhamia* or *Schlosseria*, or *Palmetto Royal*, which you say is the tree *Yucca*. At first I thought you were right, and in my letter by Ball I very frankly acknowledged my error, and my obligations to you for your corrections, but on a further view of each of the flowers and the fruit, I do still think they are different. I sent you some of the seeds of each in their proper *pericarpiums* that you may determine. I dare say you will find them very different. And if this were the only difference I should think it enough to induce one to constitute two genera, notwithstanding that famous aphorism of Linnæus's, 176: "*Si Flores conveniunt, Fructus autem differunt, ceteris paribus, conjungenda sunt genera.*

The reason of this aphorism is plainly to establish his own system more and more, and that he might always be consonant with his once assumed principles. But I would ask a botanical philosopher what he thinks a country peasant would say if he was shewn these two kinds of fruit, and his opinion was asked whether he thought that they were the offspring of two fathers who were in every respect generically similar to one another? I think his answer would be, that as they appear so different, it scarce could be imagined that two fathers, exactly similar to one another, could have produced them. Supposing this then to be the answer, should not we be sorry to see philosophy contradict common sense? I am sure a metaphysician would be surprised to see such different effects flowing from the same similar cause, acting similarly. They commonly teach us rather to judge of the difference of the causes from the difference of the effects, and *vice versa*. But Linnæus cannot distinguish between the different genera of the *Pyrus*, *Malus*, and *Cydonia*, and yet every gardener in England can as easily distinguish these genera, as he can those of the geese, ducks, teal, &c. and he would be no less surprised to see a drake produce a goose, than he would be to see a quince produce a pear, or *vice versa* *.

But what is the reason of Linnæus passing over the *Spadix* in this genus? Is it of no consequence? In a word, as these still appear to me to be very different and distinct genera, I have taken the liberty to send you another copy of the characters of the Palmetto Royal, which I think are more accurate and succinct; but I still refer them to you for the last polish. The inhabitants here call it wild *Bananas*, and some wild *Plantanes*. It sometimes grows to 20 or even 30 feet in height, and the trunk

* The writer confounds *species* with *genera*.

divides into two or three large divisions or branches. The plant flowers in June and July.

I never could get any seed of the red *Robinia*, although I have some that flowered this year; the flowers dropt soon, and produced no seed.

The Senega does not grow in our low settlements.

Mr. Saxby's behaviour to you gives me great concern, but it is nothing different from his brother's here. He shewed me your letter and list, and your desire of my assisting him. This was inducement enough to me, and I assisted in collecting every seed which he sent, otherwise he could not have sent one, as he scarce knows a cabbage-stock from a common dock, did not his stomach inform him of the difference. But I have often met with instances of this kind, and am resolved now to keep my few discoveries for my friends.

I have Miller's last edition of his dictionary, folio, without the cuts.

You propose my sending you some of the rarest seeds by the packets, but our great distance from, and little intercourse with, New York, which is the port that the packet comes to, will quite prevent me. Before the war, we had vessels going and coming from New York by every fair wind, but since the declaration of war, I believe we have not had three vessels from that port. I am well acquainted with Mr. Colden, jun. who has the care of the office there, but the risque of sending things would be as great as to send them home at once.

In my letters to you by Ball, I have mentioned

to you my uneasiness at my not knowing sooner that any account of the *Toxicodendrons* would have been acceptable to the Society. When I was on my journey I met with two species of the *Toxicodendron*, which I never saw before nor since in this province. I then took some notes about them, but never thought of bringing specimens with me; indeed I have hitherto been very negligent as to specimens, as they always spoil so soon here, let you take what care you please. These trees which I saw were larger than any other of our *Toxicodendrons*, and had their leaves in fasciculi, growing from the branches, and the rhomboidal fruit in large umbels. The branches and leaves were very full of the milky juice. They both grow about 120 miles from the sea, in moist springy soil, on a rising ground. I should have been very happy to have had specimens of these, but I did not receive your letter, informing me of the great use of specimens, till I returned from my journey.

Do you think the *Catalpa* is really a species of the *Bignonia*, or not? I own I think they are very needlessly and unnaturally joined, but I shall be glad to have your opinion.

Your Rhubarb flourished here all last Summer, though it was intolerably hot till the end of September, when the sun and rain rotted and killed all the leaves and stalks. I gave him over for lost, but took care not to disturb the ground, and to my great surprise it came up this spring again with fresh vigour and strength, and looks very lively now.

This is the second year. What was sown in the wettest and moistest ground totally died, as I believe the roots were quite rotted with the great heat of the moist earth in the end of September. It must be planted here in a middling moist ground, and shaded.

I have lately seen a character of Dr. Russell's Observations on Aleppo in the reviews; it seems to be a very entertaining book.

I sent you a small parcel of the knobby rooted Coral-tree seed, quite fresh gathered, and in good case, by Ball, and a parcel of the seed of that new plant from *Ninety-six,* that I mentioned to you before; and I sent you some seeds of our Indian pink, which is a beautiful flower, and I think almost equals the Cardinal-flower in its lustre. I think this is a new genus, and very different from either the *Spigelia* or *Lonicera,* but should be glad to have your opinion of it.

All these were lost, as I suppose, along with a box containing 22 young Loblolly bays, 6 sweet-flowering bays, and 10 knobby-rooted Coral-trees; besides, two very pretty Nonpareil birds for Mrs. Ellis, which had been long accustomed to a cage. This I esteemed a great loss, as we seldom meet with such. I had sent at the same time all the various kinds of curious earths that I had picked up, in a box for Mr. Arderon, directed to Mr. Baker. This loss to me is irreparable, as I have no duplicates of many.

I come now to your letters by Mr. Peronneau, in

the second of which you mention the *Coilotapalus* of Jamaica as being of use in the granulation of sugar. It seems to be a useful plant for that purpose by Dr. Browne's account, but we have none of it here that ever I saw or heard of. Your conjecture, as to the alkaline salt of vegetables precipitating and granulating Indigo, is quite right; and I have, after trying it, recommended it to our back settlers.

Urine precipitates it, but what an immense quantity of these precipitants will it require to serve for one day, where a planter beats a dozen vats, each 16 feet square, with two feet of the liquid infusion in each. What the anonymous author mentions of the oil of olives, is certainly false, for all our planters use oil to every vat of Indigo, which they beat, but it is only to make the froth subside, which it does, and gives them a clear view of the surface of the liquor, by which they are enabled to judge more accurately of the various transitions of the colours from green to that blue, or purplish, colour where they design to stop. I never observed it, nor have I ever heard any of the planters observe, that it had any other effect, besides that of settling the froth.

Mr. Peronneau brought me a letter from Mr. Whitworth, and one from him to you, in both which he mentions that plan which you refer to in yours. In consequence of these letters, I soon engaged some public-spirited gentlemen here to associate themselves for the carrying it into execution; and we judged it convenient to publish our intention, with an invitation to all the planters to correspond with us here in Charlestown, who would on our part take care to lay every useful hint or experiment before you in London, provided that such trials or specimens as they might send us, carried with them a probability of becoming a useful commodity if encouraged. As I have written Mr. Whitworth very fully on this head, and inclosed him a copy of our printed advertisement, which letter I now inclose to you, and leave it open for your perusal. I beg leave to refer you entirely to that, which I hope you will put a wafer into, and transmit to him, if he is not in London. We will spare no pains to spur on the planters to make a number of trials on vegetables, and will from time to time give them notice of every thing that comes to our knowledge about them. The profits of Indigo have been of late so extraordinary, that it is in vain to propose any thing else just now. Indeed I know of no commodity that would agree with this climate, which could give them returns equal to one half of what Indigo would yield just now; and many of them to whom I have talked have told me, that by the culture of Indigo alone, they were able to buy silk apparel 50 per cent. cheaper than they could raise it, were they to employ nobody but children. But I have endeavoured to convince them that they might cultivate both the one and the other without their interfering, and that they must soon expect to see these golden days of Indigo at an end, and the price 50 per cent lower than it is at present. At this time we certainly send £.150,000 sterling value of Indigo to Britain, and we take the manufactories of Britain for every farthing of it. But to give you still a greater idea of the advantages of cultivating Indigo above any thing else, I shall only mention the article of Hemp. The premium which is offered from home, joined to that which the assembly here has given, amounts to upwards of twenty shillings sterling for every hundred weight that is raised in Carolina; and this is a real certainty that every planter has for the smallest or greatest quantity that he could raise, and yet, notwithstanding this, and the great price which the commodity of itself would bear, there is not one person who will trouble his head about it; so very obstinate are the planters in general, and so difficult to be put out of their own way. Nay, some whom I urged to make trial of Hemp, sending myself to New York for several bushels of seed for them, let it lie and rot by them after it arrived, without ever chusing to give themselves the least further trouble about it. Now, in such cases what can one do whose business confines him to town, but only to lament their folly. Colours of one kind or other are the only articles that will gain the planters good liking, and colours collected in a solid form like Indigo, are the most promising articles to bear a price that could reward the labour of a slave. Would a fine yellow extract bear a price with you?

Provisions here are in general very cheap. A bushel of good Indian corn may be bought from a shilling to eighteen pence sterling. Peas at eight pence per bushel, and potatoes at eight-pence per bushel, and often at one half of that. A quart of corn is reckoned a negro man's allowance for one day; so that a bushel should serve him a month, and twelve bushels for a year. During the fall months he may have as many potatoes as he can dig, without any value being set on them by the planter; so that through the year his negro will not cost him eight pounds currency, or twenty-three shillings sterling for provisions; and yet this negro will gain him from ten to thirty pounds sterling every year, supposing him only a common field slave, providing the planter uses but middling ground. Our soil produces corn, wheat, oats, barley, rye, and rice equally well, if the proper ground and situation be chosen for them; though wheat, oats, barley, and rye, grow best in our back settlements.

Besides Mr. Whitworth, I wrote very fully on this head to Mr. Baker, in answer to a letter of his sent me by the Society, and to Mr. Shipley.

I come next to the compliment which you are pleased to pass on the introduction to my journal. You may be sure it gives me great pleasure to see it countenanced by your good opinion, though I be very sensible that good nature and kind partiality might have been your inducements. But I can assure you, that your censure has often given me satisfaction, and I have often esteemed it a mark of your real friendship. Your ingenuous opinion and judgment in any thing which I have the pleasure from time to time to transmit to you, will always afford me the most sensible pleasure, and I shall always esteem it my happiness and honour to have such an adviser. Your communicating it to Dr. Fothergill, or any of your friends that you imagined would take the trouble of perusing it, is very agreeable to me, since you think that it may bear a reading. I designed to have continued it in a series of letters to Dr. Huxham, who I told him had my leave to present it to the Society, if he thought it worthy of their notice, but as you have advised me to put it all together, I have forborne sending any part of it home this year; and the rather so likewise, as I am getting draughts of some of the plants done here by a very ingenious young gentleman, which will naturally fall in as I have occasion to mention them. I shall be careful to make it as complete as I can, in the account of that part through which I passed. As I am in no hurry at all now to be an author, I have made myself quite easy as to its being buried awhile in oblivion; it may have the greater time to ripen. In the course of it I have been fuller in my physical and natural remarks than any thing else, but I am afraid you will think some of them too whimsical.

You tell me that you must curtail my descriptions, &c. My dear friend, I esteem this the sincerest mark of your friendship; let me assure you that no encomium whatever could have raised in me half the idea of the esteem which, through an excess of benevolence, you are pleased to honour me with.

I am now fully sensible that your corrections are right, and the characters of any plant which I make out now are not one half so long, and yet I think they are as expressive and descriptive, if not more so, than formerly.

I have now sent you the characters of the *Tetrapteris*; of the *Anonymos Bacchari affinis*, and of the *Hexandria tryginia* with the gramineous leaves, which specimens I sent you last year. The characters of the *Tetrapteris* are made out with all the care and exactness which I possibly could; and I have, according to your directions, made out english descriptions, which I hope you have added to the others. I am afraid to send any specimens in these perilous times, as I have few or no duplicates of any of mine; but if you think I might venture, please just to mention it only. I may in some future year send you some observations on the *Toxicodendrons;* though I have never seen any account of them. I never saw Kæmpfer's *Amœnitates Exoticæ*, though I believe it is a most amusing book. You greatly surprise me by Dr. Schlosser's treatment of you; his letter to me seemed to have more of the french vivacité and open-heartedness, than of the dutch closeness and selfishness. In one of the last reviews, I see a piece of his on lime water, that has been published in the last volume of your Transactions. Pray is he one of the Socii? Your present of minerals, which you were so good as to send me, was an invaluable present, and they seem to be very good of their kinds. My close confinement to town now prevents my rambling after plants, so that I shall have many hours to spend in the amusing study of Mineralogy. I just begin to have a relish for it, and as I have ordered Wallerius on the *Regnum Lapideum*, I hope soon to be able, from your character of that book, to make some progress. Pott has given me great pleasure and instruction. I think his treatise on light and fire is very ingenious. I forgot to mention the small piece of electricity which you were so kind as to send me. I think it is by far the best of any that ever I saw, unless Baccari, the Professor at Turin, who I think has wrote most sensibly on it. If there be any late well done thing on that surprising branch of philosophy, you will greatly oblige me to inform me of it. Some time this summer I will give you the trouble of an inclosed letter to Dr. Hales on that subject. I have not wrote to the doctor, but I have taken down a few notes, which I shall put together soon, and will inclose them to you for your perusal. I have engaged Dr. Linning to make some careful experiments on some vegetables, and especially Indigo, to lay before your Society. The worst of it is that the doctor is soon taken with trifling observations, and does not pursue them, else he might be able to do many things in that way, as he has turned a planter now altogether, and has quite done with practice.

There is one thing which I entirely forgot to mention to you, and that is, that you need not call the Loblolly Bay, *Gordonia*. I shall leave the denomination to you. You know all those who I think merit the compliment among my own acquaintances besides yourself, Dr. Hales and Dr. Huxham. After these I shall give you the trouble of denominating the others as you please. You may, if you think proper, call the Loblolly Bay *Huxhamia*, or what you chuse.

I have just now by me a very fine box of the early Honey suckle, 40 young plants; *Tetrapteris* 4 fine young plants; *Corallodendron*, 4 young plants that will flower; *Pavia*, 6 flowering plants; and other things. I put these up exactly according to your directions, and they will be sent by Ball or Curling, both whom we now daily look for. Pray offer my respects to Mr. Collinson to whom I wrote by Mr. Peronneau, and to Dr. Hales, &c. I remain, with the sincerest esteem, dear Sir, your most obliged and very humble servant,

ALEXANDER GARDEN.

I have inclosed you a list of a box of seeds that was sent you by Cheesman, and another by Capt. Coats of the Friendship, both of whom I hope are safe arrived.

MY DEAR FRIEND, June 12th, 1757.

Two days ago I had your kind favour by Capt. Ball, as I some time since had that by governor Ellis, and a little before by Mr. Peronneau with every thing that you mentioned, as Minerals, &c. but above all Pott's Lithogeognosis, a most instructive book, and the admirable plates of Ehret by governor Ellis. I have commissioned for Wallerius to the Crokatts, as I have likewise for some Thermometers, which I beg them to take your advice in purchasing from the best hand, as they are for a particular use, and I do not mind the price. I suppose Messrs. Bird's or Soisson's are the best; but I have left that to you. Accept of my sincere and hearty thanks for all your favours; they are indeed many, valuable, and highly useful to me; believe me I have a sincere sense of them. I hope you have received the finest parcel of seeds that I ever did or ever can propose to send you, either by Coats or Cheesman, both bound to London. They were shipped last January, and were carefully packed in earth, entirely agreeable to your directions, in kind, quantity, and in the method. In the parcel by Cheesman you will find several new kinds. I have written to you at great length, but I have not yet sent your letters away for want of proper opportunity, for which I have hitherto kept a box of young shrubs, *viz.* 4 young flowering *Halesia fructu tetragono;* 4 *Corolladendron radice nodosa;* 5 *Pavia;* some *Lychnis viscosa carolinensis;* some Indian-Pink root; about a dozen of the *Trillium* or *Solanum triphyllum* of Catesby; 20 flowering early Honey-suckles; and Henry Peronneau adds as many with his compliments. These make in all 40 young Honey suckles; most of them have flowered in the box, and are now all alive and healthy.

These will be sent by Ball for London, as he goes home under convoy. I must not miss this opportunity even though the season be bad; and I thought proper to inform you, particularly, as you have mentioned your insuring of them, though I imagine there will be no great danger, as the same ship of war conducts these ships home, as came out with them, and took great care of them. You will greatly oblige me by calling the *Tetrapteris Halesia*, and I have inclosed the characters which you will have by the first ships, and I hope you will correct and add them to the figure, if you think proper to publish them. I am just now finishing some reflexions on the common Meteorological Journals, in order to send them to Dr. Hales, in answer to a most obliging letter which I had the honour to receive from him some time ago. I shall inclose them to you, but the extreme heat of the weather, and my own weakly state of health, will not let me have them ready; I am afraid to send by Ball, which gives me great concern.

As to the packets and books from Holland, I never received nor heard of them till you wrote to me. I have never had a line from Dr. Schlosser since that one which you inclosed. I saw his *Tentamen Chemicum* in Phil. Trans. on lime water. I shall be singularly obliged to you for a copy of the cuts of all the plants, especially the *Halesia*. I am getting some done by a tolerable hand here, which will be inserted in the *Iter Saludiense* 2d part. It is not yet finished quite, and as it may not be agreeable to the Society to insert it in a series of letters, I am in no hurry about publishing it; but as it may amuse you, I will send it to you as soon as it is finished, for your own private perusal. You will hear from me at great length soon. I remain, Sir, your most obedient and humble servant,

ALEXANDER GARDEN.

There is likewise a small box for Mr. Baker to come by Ball, with a bundle of seeds enclosed for you.

The excessive heat obliges me to use an amanuensis.

MY DEAREST FRIEND, July 6, 1757.

My grief at my own and your disappointment is inexpressible. A few days ago I heard that both Captain Coats and Cheesman were taken, and with them the two most valuable collections of seeds that ever I could promise or even hope to procure for you. There was every kind that you mentioned in your letters to me, and many new and curious shrubs besides, as you will see by the inclosed list. Of all of them there were large quantities, of many four, five, two or three quarts. They were all carefully packed in well dried, soapy earth, exactly as you directed. My patience is gone. I shall never be able to make you amends for one half of your favours. I am ashamed to have received so many things from you, and thus to be baffled in every attempt I make to present you my small but grateful returns. They were all counter directed to Monsieur Bernard de

Jussieu ; so that if you correspond, I beg you will enquire about their fate. I would have given 20 guineas if you had only got one box. I never shall have any thing like them again.

Please to offer my compliments to Mr. Gray, and assure him that I have not failed through inclination or neglect, but through my evil, malign, and wayward destiny. But I have done for this war. There was a small parcel to Mr. Whitworth, too, that is likewise gone.

Good God! what is the meaning, that out of 21 Carolina ships that sailed from hence in January last and beginning of February, there should be 19 taken, and with them no less than five hundred thousand weight of our best Carolina Indigo. When I think of this, my own loss is cancelled ; but when viewed apart, it distracts me on your account. I have written to you fully by this ship, Captain Ball ; and by the Arundel man of war, to which I refer you. As I was sending a small box to Mr. Baker, I have enclosed a few of my remaining seeds to you, which you will get from him.

I have sent you a box of shrubs, among which are four young flowering *Halesia fructu tetragono;* four *Corallodendron radice nodosa;* five *Pavia;* some of our wild Scarlet Lychnis ; some Indian Pink ; a dozen of the *Trillium* or *Solanum triphyllum* of Catesby ; twenty flowering upright Honeysuckles, and Mr. Peronneau adds as many with his compliments ; these make 40 young Honeysuckles, most of them have flowered in the box, and are now alive and healthy.

I have sent you the characters of the *Halesia*, which I hope you will join to the print. I have sent you the characters of two others, of which I sent you the specimens last year.

Of late my health has been so impaired, that I have not been able to apply to any thing, though I have found two new trees this spring.

Now my dear friend I must come to a final eclaircissement about my having given you so much uneasiness in procuring my admittance into your Society. You will certainly credit me, when I assure you, that my own disappointment does not give me half so much uneasiness, as the plague and trouble it has given you on my account. My views are not now the same as they then were, and for that reason, it will very sensibly affect me if you give yourself any farther trouble in that affair. I am already quite ashamed of the plague I have created you, and although I well know and have experienced the benevolence of your nature, and your kind intentions towards me, in giving me your kind assistance in procuring that honour, yet the plan of the Society being such as it is, and no doubt for very good reasons, I most cheerfully acquiesce, and do most sincerely return you my humble and hearty thanks. Your favouring me with the continuance of your correspondence is now the great desire of my soul, and I find so much goodness, and such a fund of kind observations in your letters to me, that I am happy enough without thinking of any other new engagements.

Indeed my success in testifying the sense I have of your favours, has been so bad for this year and half past, that it has greatly damped my spirits.

Though my inclination to have had the honour of being one of your members, some time since, was very keen, yet I am satisfied that it is better to be as I am, for many very weighty reasons; amongst these are: 1st as the plan of the Society is such as admits no native of Britain but for a pecuniary consideration to be of the number, and as I now consider myself in every respect a foreigner, having little hopes of ever seeing Britain, unless just for a visit, and to return here again; for this reason I think it would be very needless in me to advance 25 guineas for the name, as I should never have any access to that instructive conversation, which you who reside in London must mutually enjoy at your meetings, besides the privileges of voting, &c. Hence, though the sum be inconsiderable, yet I judge it would be mis-spent. But all this is upon a supposition of my being received, which you will readily judge is a pretty bold supposition, God knows, having no other recommendation to trust to but the obliging interest of my friends, whom I chuse to trouble no farther on that head. 2d, as I live at so great a distance from books and men, and in a climate so ill adapted to study, I should never be able either to support the good opinion my friends may have conceived of me themselves, or might have given to others; and if I shall often trifle or err, I will much more cheerfully trust your obliging and benevolent temper for your friendly corrections than the Society's.

As to collecting seeds to sell to the gardeners, it is what I should not chuse to do, neither would my business permit me. What I may be able to collect shall be solely to serve my friends, amongst whom you have an indubitable title to more than I shall be able to collect these many years. I remain, with my respects to your lady, my dear friend, yours, &c.

ALEXANDER GARDEN.

South Carolina, Charlestown,
January, 18, 1758.

MY DEAR FRIEND,

Some few weeks ago I wrote you two short letters by two vessels that were then ready to sail, in which I gave you an account of the parcel which I now ship by the Fanny, Captain Brooks, for Lisbon and London; which is indeed a round-about way, but which I hope will be the occasion of their safe arrival. It was the seeming best of some few opportunities that are now in our harbour. It is in vain almost to settle any correspondence by the post on this continent, for it is so uncertain, so tedious, so precarious, and so ill managed from the northern provinces to us here, that not only mine, but all our mercantile correspondences are obliged to be carried on by the opportunities by sea. I have never yet received one letter by post from any of my acquaintance in Philadelphia or New York, though in some letters, by vessels, they often tell me they have frequently wrote to me by post. And in like manner they tell me that they never have had one from me by post, though I have often written to them by that conveyance. This is the case with every body here, so that we have now almost dropt it. Opportunities from this place to New York by sea, are now much seldomer than before the war, so that upon the whole, I apprehend that method of correspondence will not answer. I received both yours by our vessels, and one plate of the *Halesia*, which I greatly admire, but lament the loss of those which you mention to have given to the Crokatts; we have never heard any thing of them. My pocket thermometers, through careless packings, were all lost and broke, both external and internal tubes. I have never yet had a letter from either Schlosser or Linnæus, neither can I conceive by what opportunity it has been that Linnæus wrote, whose letters I should have received with that grateful joy as a son his father's benedictions. I once wrote to him, but I am still uncertain whether ever he received the letter. I have twenty things to write to you on these subjects, which I must defer to a future opportunity, when I shall write at full length. In the mean time I must beg that you will continue to communicate your literary news, for you alone are my oracle in these matters. I will, though now closer confined than ever, be still endeavouring to procure what things I can for you year by year. Pray offer my kind respects to all my friends with you, and tell Mr. Gray that it will not be in my power to serve him or you effectually till the war be at an end, as the uncertainty of conveyance quite damps me. Here is a list of what is in the box.

No. 1. *Aralia arborescens,* Prickly ash.

2. Contains, 1. *Callicarpa Linn. vel Jonsonia Miller.* 2. *Aquifolium fruct. verticillat. foliis longioribus splendidis integerrimis.* This is a much prettier tree than Mr. Catesby's Dahoon holly.

3. *Loblolly Bay.*

4. *Halesia.* I sent some young plants by Ball.

5. *Bignonia sempervirens, foliis qùaternis, vel ad genicula utrinq. geminis,* our prettiest evergreen vine.

6. *Magnolia palustris,* Sweet flowering bay.

7. *Chionanthus,* Fringe tree.

8. *Apocynum.*

9. *Magnolia fol. latiss. flore magno fœtido.* Umbrella tree. *(M. tripetala Linn.)*

10.

11. *Pavia.*

12. *Ptelea Linn.* very ill characterized by him.

13. *Laurus baccis purpureis.*

14. *Decandria Monogynia.* I sent half a pint of this last year; it is a very beautiful and new tree.

15. A pretty shrub whose name I do not know.

16. *Beureria*, or *Butneria*, or what you please, for Mr. Ehret.
17. I design to call this *Ellisia* or *Ellisiana*. You will see some account of this under No. 35 of the list of the seeds which I mentioned in my last to you to have been shipped on board of Cheesman. I beg you will enquire after it if ever any of it grew in Paris. I probably may never again get any of the seed, as it lies about 250 miles back in the country, and is certainly an entirely new genus.
18. *Icosandria*, a wild tree, whose name I do not know. It bears its flowers like the cluster cherry, but has a very different fruit. It is a beautiful evergreen. I am persuaded that you would be pleased with some sets of it.
19. *Catalpa*.

I will soon write to you at great length. In the mean time Mrs. Garden joins with me in our compliments to Mrs. Ellis and yourself; and I am, dear Sir, your most obliged and very humble servant,

ALEXANDER GARDEN.

South Carolina, Charlestown,
August 11, 1758.

SIR,

It is now some time since I had your obliging favour, and would have done myself the honour of answering it long ago had our communication with England been open, but for these four months past, an embargo on our shipping has entirely put a stop to all commerce or correspondence whatever from our parts.

It gives me sensible pleasure that you approve of the plan of erecting a corresponding society here, and that you should further pursue the design so far as to promote such a scheme with you. I think your idea of it is extremely clear, and your reasons for hoping success seem to be just and well grounded. From the steady and persevering firmness of the people on your side of the globe, you have the fairest prospect of success; but all enquiries and trials of this kind require time, which the volatile Americans will scarce bestow. I am sorry to inform you, that though we advertised our association and plan early in the spring of 1757, and gave particular directions to many of the planters in what manner to proceed in making essays on some vegetables, yet that year passed away without any one trial being made, and this is likely to do the same. I imagined that I could have been in the country last summer a few days, but a necessary attendance on my business in town would not admit of my absence, and a bad state of health has prevented me this summer; but I will still push the thing, for its utility is visible. The present confusion of war, low price of our commodity, and heavy taxes, dishearten the planters.

You are pleased to tell me that the short account of our affairs I sent you was agreeable to you; this gives me particular pleasure, that any thing that I could communicate might afford you the least satisfaction; and as you are still further pleased to desire a continuance of information concerning our Indians, and their connection, I have taken the liberty of writing you a few remarks on what I think we are still too slow and remiss in. I mean in making ourselves sufficiently acquainted with the various Indian connections among themselves, and a just and true geographical account of the country they inhabit. Our hopes and fears are just now so much connected with the good or bad success of the northern expeditions, that all our conversation turns on these matters; though, as yet, we have no certain account of any thing being done. The loss of an army or fleet to any potentate in Europe, could not affect him in near so sensible and effectual a manner as the loss of one battle by land or sea would do the British colonies in America. In Europe the loser's resources are always at hand, generally within his own kingdom, and thence readily to be procured and forwarded; while the resources of the colonies within themselves in America are of no real value or strength, and those from Europe are at a great distance, and liable to be retarded by many adverse accidents of wind and weather. Hence, the loss of a battle, or the failing of any attempt here, would be attended with much greater hazard and danger to our interest, than the like could be to any European kingdom. Upon a careful consideration of the history of our miscarriages and disappointments on this continent, we shall find that many or most of them have been owing to our ignorance of the geography of the country, and of the situation and connection of the various tribes of Indians within or near the territories which we claim. We had been at so little real pains, for example, to know the humours and connections of our old allies the Iroquois, that when this war commenced, I remember well, that the most judicious people about New York could not tell which side any of them would take, unless the Mohawks, whom (by chance only) Mr. Johnston was thought to have such influence over as to retain for us. While this was our case, it is certain that the French knew particularly the strength of every tribe, and the exact number of friends or enemies they had in each tribe; nay, not only their number, but their names and the share of interest that each had in his own tribe. This is a matter of great consequence, and what we have not yet attended to.

The Indians are a very shrewd and observing people, and take great notice of the method of distributing favours. They observe that the English do not know how to distinguish the good from the bad, warrior from the poltroon, and that the French are much sharper sighted, and more discerning; that they bestow their favours justly on those who merit them, and treat cowards like women. It gives the warriors the greatest concern to see any one, even of their own people, put upon a footing with them, and nothing can affront one of them more sensibly, than to see the same favour shewn to one of their common people as to them. Our extreme ignorance of the history of those nations that we are best acquainted with, often and daily exposes us to commit these blunders, and I myself have seen the bad effects of it.

Our ignorance of these things is not our only fault; we err greatly likewise in not getting a better and more accurate geographical knowledge of the continent. When the war began, we were greatly at a loss about the real situation and distances of many of the chief places on our northern frontiers, and more so about the situation of the back parts of Pennsylvania, Maryland, and Virginia, where I remember very well they were busied in getting maps and draughts of the country, after a fatal miscarriage had shewn them their error in not knowing these things before that time.

Further to the southward our ignorance of these matters increases; as if the increase of heat evaporated our desire of information. We know little, particularly about that vast tract of country which lies west of the Cherokee and Creek nations, between us and the French settlements; nor are we well acquainted with the various proper and important passes in the mountains, on this side the western Cherokees. Our whole knowledge of these parts is vague, gathered from such a rude and ignorant set of men as Indian traders, with some further knowledge which we have lately had from those employed in building and garrisoning Fort Loudon. But, alas! what can we expect from men of no knowledge in geography, nor any desire to attain further knowledge than the plain path backwards and forwards to Charlestown? As to our ignorance of the general geography of the country, I need only refer you to our late maps, where the southern provinces are laid down either from mere guess and conjecture, or from a still worse guide; I mean the French maps of D'Anville, De Lisle, and other French authors, whose business it is to deceive and keep us in our ignorance. Cast your eyes only on the projection of the Cherokee country, which is so laid down that every town of that nation falls either in North Carolina or in Georgia, while the kind geographers have, through a special courtesy, left a sort of path, of about 15 miles in breadth, for the inhabitants of South Carolina to follow Ferdinand to New Mexico, in quest of the remainder of their province. Yet, though the geographers may tell you that they discover these limits by particular good knowledge of our charter, &c. we rightly judge here, that most, if not all, the Cherokee country lies within our limits, and on this foundation, have contributed cheerfully for building Fort Loudon among them. If your general knowledge of these countries be lame, our particular knowledge is still worse, and very far from being sufficient for a general to be master of, who was to lead an army into those parts; or for a man who was to choose such situations for forts or settlements as would command the necessary passes and rivers that are scattered through this vast country, lying between the Cherokee hills and the Missisippi.

How soon we may have the seat of war transferred to these parts, is uncertain, and will that be a proper time to inform ourselves? Will it be safe for us, or will it be possible for us, to make a survey of their rivers, examine their courses and their falls, or search out the passes through the mountains, when a bloody minded set of savages are spread over these parts? How exact is the knowledge of the French concerning every thing of this kind that their plan requires them to know, in order to qualify them in the best manner for an effectual execution of it! How well situated are their forts for annoying us, protecting their own intercourse with, and overawing or commanding the various Indian nations that inhabit near them! Experience has taught us how far the French have exceeded us in this necessary piece of knowledge, and how severely we have paid for our ignorance; and yet we do not seem to learn from this experience. There is no person that ever I heard of, appointed by the government to set these matters to rights, and if I could judge of the case of other provinces from that of ours, we are in no good way ever to do it here, for I am pretty sure that there is not a surveyor in this province, from the chief to the lowest deputy, that ever has taken, or could take an observation of the latitude or longitude of any one place in the province; and yet, without this, the real distances or bearings of far separate places cannot be ascertained. Added to this, our knowledge likewise of the passes in the mountains, the courses of the rivers, and their chief branches, their falls, and carrying-places ought to be more exact and determined.

Fort Loudon lies between the main branch of the Tennessee and a branch that turns south-westerly towards Tilliqhua; and yet we never have thought of examining that river down towards its running into the Ohio, near the Mississippi. We know not the distance; we know not what falls or impediments there may be in it, to prevent our enemies from coming up it to surprise us. We know however, by the relations of some parties of the Cherokees, who have been lately down a part of it, that they are advancing towards us; but how easily, or how soon they may come, we know not.

It is a pity, indeed, that they should know the path so well to us, and we not know it to them, or how properly to prevent them from visiting us. Now is the time for encreasing our knowledge in these matters, and not when they have gained and cajoled our Indians into their interest. I am afraid, however, that these things will not be attended to till the French have built some forts higher up the north-east branches of the Albama river, and some on the most southerly and westerly branches of the great Cherokee river, by which means they will form a chain (if it be not already done) along the back of the Creek and Cherokee nations, that will effectually overawe these two nations, cut us off from any communication with any more westerly ones, command and secure their own innumerable small tribes of Indians, and be too strong for us ever to break. Strange indeed that none of the small vessels of the enormous fleet of Great Britain, should never yet have found the way up Albama and Mississippi! the reason may probably be, because in doing that, they would serve their country with little hopes of enriching themselves. To return; our next concern (if we would copy after the French, whose success has shewn the goodness of their measures) should be to learn the various connections of the Indian nations, not only with us, but among themselves; and thus we might be ready to forward and encourage the forming of such alliances as would promote our own interest with them, or early to prevent any connections or alliances that would weaken their dependance on us, wean their affections from us, or connect them closer with the French or their allies. These are matters that the French are very careful and earnest about, and that we have hitherto been quite neglectful of, insomuch, that we know not now what alliances or leagues are subsisting between the Cherokees or Creeks, and any other nation whatever; nor do we know what influence any other nation would have with them, in determining them what side to take, if the seat of war was amongst us. I am hopeful, however, that Mr. Glen's influence is so great with them, as to determine those that are gone to general Forbes to act with vigour. Mr. Glen set out from this place in May last, as a volunteer, to use his influence and best endeavours with them, and as far as we have heard has been pretty successful. The French have spies and agents in every Indian nation, even amongst the Creeks and Cherokees; we have none, nay, not a person amongst them in a public capacity, nor any but whom the thirst of lucre, or the shunning of justice, may drive thither. His Majesty has graciously condescended to nominate and appoint an agent for the Southern Indians, but how far the design and import of that appointment will be answered, by a man whose sole business is to cook good dinners for himself in Charlestown, time, and probably the defection of some one or other of these nations, will shew.

You no doubt will think that I have represented these matters too warmly to you, but who can see these things transacted in such a manner on our side, while our enemies are daily gaining ground, more through our indolence and ignorance than through their superiority and force? This province is just now happy in one of the best governors, and while we are under his mild and strict administration, encreasing our trade and commerce, it is a thousand pities that the public attention should not be a little turned on these matters, which scarce can be done but from home. Did ever the French allow their Indian agents to reside in Quebec or New Orleans? no, the Indians are to be watched in their own countries, and the negociations of French agents, who live amongst them, are to be rendered abortive or prevented. I am, yours, &c.

ALEXANDER GARDEN.

Excuse my using an amanuensis.

S. Carolina, Charlestown,
MY DEAR FRIEND, Feb. 17, 1759.

Your two last letters of the 1st of May and 12th September 1758 both lie before me. I most sincerely condole with you on your loss, and feel, with a heart full of grief, part of that sorrow and anxiety which must afflict you, on the melancholy situation of your family.

Since any of my letters to you can have come to your hands, I have received many marks of your esteem, and last by Captain White, I had your invaluable and truly grand present of the *Hortus Cliffortianus*. I never saw this superb and inestimable work before; neither do I believe that there is another copy in America, unless one which Mr. Clayton has in Virginia. Accept of my grateful acknowledgments for it; indeed this is almost all I can give. I am ashamed to have it so little in my power to send you many things that would be acceptable to you, and even what I have from time to time promised you; but I can sincerely assure you, that it is not want of inclination that prevents me. The method in which we are obliged to carry on our business of the practice of physic here, requires a constant and hourly attendance.

I must go regularly through your agreeable letters, but must first tell you what I wrote you by his Majesty's ship Winchelsea, Captain Hale, who had the misfortune to be taken, and thus the thread of my correspondence was broken.

First then, I enclosed a letter under cover to you,

to Mr. Whitworth, which I left open for your perusal, as you expressed some satisfaction at a former letter to him on nearly the same subject. This contained some observations on our public affairs, and although the face of things in America be much altered since, yet I have now enclosed another copy of it in this, and I hope you will excuse the trouble, and put a wafer in, after perusing it, and send it to him.

I sent you by captain Ball some specimens of our Cochineal plant, with some of the insect preserved in spirits, and some dried and put into a bottle. These I hope came to hand, and gave you satisfaction. I examined the insect, while alive, by your microscope, and found it answer pretty much to Dr. Browne's description, only what the Doctor takes to be the proboscis, is double in every one that I examined, and in some of them there is a long spiral filiform hair proceeds from the point of each proboscis. I did not observe this hair in all, but whether it be peculiar to some, or be the distinguishing mark of male or female, or whether I broke it off in taking away the cobweb-like substance, in order to see them, I cannot pretend to determine. In every thing else I think the Doctor's description seems just.

This is certainly the same species that we use in the shops. I imagine that the chief difficulty in collecting it is in cleansing away the cobweb. It is a prodigiously prolific creature, each of the females producing a vast number of eggs. The number I judge is some hundreds, and yet each of them seems large in proportion to the bulk of the mother.

The eggs are elliptical, quite smooth, shining, transparent, and of the same colour as the blood of the parent insect. You will observe that in powdering Cochineal there are always a number of small gritty particles, in which however the richest of the colour seems to consist; these are the eggs or young, and however big or large one of these insects may be, yet if it be not full of these eggs, the juice is of a thinner and more dilute colour, and will not give such a degree of tinge to paper.

As to the species of the *Opuntia*, on which they bred with us, I think it is different from that described by Dr. Linnæus or Browne, for the kind that bears this Bug or Cochineal; and I am sure it is very different from Plukenet's draught, t. 281. f. 3d and 2d, which Linnæus refers to, and which Plukenet affirms to be the *Opuntia coccinellifer*.

Linnæus says *Op. flore sanguineo* in one of his synonyms.—Ours is *flore flavo*. Linnæus says *Articulis ovato-oblongis*—Ours, *Articulis obversè-ovatis*. Linnæus says *Articulis subinermibus*—Ours, *Articulis spinis longis rigidis acutissimis munitis et penicillis spinorum urentium hinc inde in totam superficiem obsitis.*

In your last letter you desire me to give you an account of this insect, which I shall do if I live till the time, and God bless me with health; but the season in which I could best make myself acquainted with them, is the time that I am generally sick every year, that is, from the middle of May till about the middle of July. Indeed this situation was so irksome last summer, that it almost determined me to leave the province, and return to Europe; but, *forsàn et his quoque Deus dabit finem.* The above is a copy of what I wrote you on the Cochineal by the Winchelsea.

I shall be careful in sending Linnæus some specimens next Summer, in the manner he desires.

I likewise sent you by captain Ball some specimens of a *Protea* or *Leucadendron* which I found here. Pray let me have your opinion of it.

The Pistachia nuts which you sent me never came up; but the Scammony, *Coloquintida,* and Styrax promise well.

In a former letter you advise me to send birds to Mr. Edwards; but I am unacquainted with this part of Natural History, and know very few of the birds which we have here, though there be an immense variety; however, if he will mention such as he wants that are in the first (for I have no more than the first) volume of Catesby, I will endeavour, and probably may easily be able, to procure them.

The scheme of the Provincial garden is truly noble, and has a prospect of answering a good and great end. It will no doubt be highly beneficial to the province, and may in time be useful to Great Britain. I will most cheerfully lend my little assistance, but after talking of this matter with several gentlemen, we all were of opinion that if it could be recommended by your Society to the governor, council, and assembly, it would take at once, and be carried into execution with great spirit and life: as we judged that such a thing being proposed by any person here would not have weight enough to determine them to think well of the scheme. For this reason I have mentioned it but to a few, because I thought, if you should approve of this method, it is better that it should come at once upon them, and I am morally certain, from what I know of the people, that it will be received with open arms and great applause, and the necessary steps, such as you recommend, will immediately be taken. There seems to be a kind of necessity to drive the dull part of mortals to their own happiness and welfare. The task is irksome, but the reflexion of having intended and promoted a general good is the superior reward.

There is one thing which I must beg of you if you can procure it, and that is, one of Mr. Ehret's draughts, such as he gives to the engraver. The gentleman who draws for me has begged me often to write to you for one, that he may regulate himself by the method he uses.

I come now to your two last letters. I never could get any seed of the red *Acacia*. It grows only about two or three feet high, bears quantities of fine flowers, all which drop soon.

I never saw any other than the tree Palmetto and swamp Palmetto, between which I know very little difference but in the size. The Palmetto Royal is undoubtedly different from the Yucca, as you will see by the Pericarp or fruit of the Palmetto Royal, which I have sent among the other seeds. This grows to about 12, 15, or 18 feet high, but the tree Palmetto on the sea side grows to 30 or 40 feet high without a branch, and bears a fruit just like the *Chamœrops*, or swamp Palmetto.

I hope Mr. Clayton's book will soon be published; it must be very useful. He will have an account of our Pink root in it. I have sent an account and description, with a draught of this, to Dr. Whytt. Though I added a botanical description, the design of the paper is chiefly medical. I cannot think that this is Browne's *Anthelminthia*, nor any way related to it.

What you write concerning Linnæus desiring me to write to him, gives me great satisfaction. I gladly embrace the opportunity, and have written under cover to you. I beg you will forward these papers, after perusing and inclosing them. I have sent two copies of my letter to him, as I have done of this to you. I have sent him a copy of my characters of the *Ellisia*, which I have begged him to confirm in his genera, and give it the bishop's touch. I hope it will not reach him too late for the edition which he is now about. Please to offer my sincere respects to him, and hearty wishes for the continuance of his health and life, which are useful to so many.

I have inclosed a draught of the *Ellisia* to you, which I am persuaded you will be pleased with, especially when you consider that it was done in America. If it give you any pleasure, and if you allow me to call it *Ellisia*, I shall esteem the honour you do me in giving me an opportunity to testify, in some measure, the grateful sense I have of the obligations I lie under to you. I have not sent you a copy of the characters, because the letters to Linnæus are left open, and you will find the characters there, which I think you should transcribe; and let me beg you will have the draught engraved as soon as you possibly can, for I am afraid of what you mentioned concerning the french having some of the plants raised in Paris, and as it is an annual, they may publish it first. Will it be necessary to send the draught to Linnæus afterwards?

I must beg of you to send me half a dozen copies of this draught when it is done, for I confess I shall be proud of it, if it please you. I must send one to Mr. Colden, one to Bartram, and one to Clayton. Please to roll them round such a stick as silks are rolled on, for when they are folded they spoil greatly.

The seeds which I sent last year without names, were such as I did not know, and whose fruit only I had seen. There are some this year which I do not know, marked *anonymos* or *ignotum*.

What you observe concerning the Madder is certainly true. It would do well here, but it will be in vain to think of driving the planters to any thing till they see it in a Provincial garden. Their seeing the things grow which may be proposed to them, will have more influence in determining them to plant them, than all the advice that could possibly be mustered up otherwise.

I have not seen a plant of the pinnated *Toxicodendron* these two years. I have not been in the country for that time, but when I was sick, till yesterday, when I brought home some branches of a tree that I do not remember to have seen; it is an *Octandria* or *Decandria Digynia, calyce colorato octofido.* I have inclosed a small sprig to you, and one whose fruit is better formed, and more forward; pray let me have your opinion of this, what it is; it has some affinity to the *Chrysosplenium*, and some to the *Ulmus*, but differs greatly from both at the same time.

I have written frequently to governor Ellis as you desired. I sent him specimens of the *Halesia, Beureria,* and red *Robinia,* and a long catalogue of the common and Indian names of what I know you want.

There are many things in your last which I cannot possibly answer just now, but will write to you soon again. The *Salsola, Anabasis, Mesembryanthemum, Atriplex, Salicornia,* grow in the greatest plenty on our sandy coasts, so that any planter living nigh the coast might, if he would, make the experiment of the *Barilla.* I must beg one of Ehret's cuts of the fine double white flower, which you mention from the Cape of Good Hope. It appears grand in description, and truly deserves the name of *Augusta.*

I have sent you a small parcel of the flower with which the Indians dye red. It makes a surprisingly bright scarlet colour, which I myself have seen done without any other apparatus than just pouring boiling water on the herb for about half an hour, and then dipping the feather or wool amongst it. It never will wash out again. A lady procured this for me, but she unluckily mentioned her design of giving it to me to be sent over the great water, as they say, and as soon as they knew this they formed many excuses for not gathering it at all, and could not at last be persuaded to gather any, till the frost came, which destroyed its bright dyeing quality. This they knew well it seems before, but they think that when they communicate any of their knowledge to the white people, the plant or herb immediately loses its wonted virtue, and for this reason it is difficult to procure any thing from them. I send you this only, that you may examine the flower, and let me have your opinion of it. Try its dyeing quality. What is in the small paper I have had by me two years. I am yours, &c.

ALEXANDER GARDEN.

I have sent you a small dry specimen of *Ellisia* inclosed in the *Ellisia* draught; and I have sent you the only specimen which I have of the Loblolly Bay: this must not be called the *Gordonia*. Name it yourself, or let Linnæus name it.

Charlestown, Feb. 19, 1759.

Inclosed I send you a list of seeds shipped on board the Fanny, Christopher Brooks, Capt. and a bill of lading for the same. I have sent you copies of the same by his Majesty's ship, the Surprize, Capt. Antrobus, and the Penguin, Capt. Man, and by Capt. Mitchell of the Squerries, all which I hope may come to hand. By his Majesty's ship the Penguin I have sent you a most elegant draught of the *Ellisia*, and a specimen of it. I have likewise inclosed a letter for Linnæus, which you will please to forward after copying out the characters of the *Ellisia*, which you will find in my letter to him. I have likewise inclosed a letter to Mr. Whitworth, which you will please to peruse, and send to him if he has not already got a copy; but I am afraid both copies which I sent by our fleet in August last were lost. I have likewise sent you a specimen of the Loblolly Bay, which is the only one I have by me; this must not be called *Gordonia*. Name it yourself, or send it to Linnæus to baptise. I beg you will write to me soon, for nothing gives me more pleasure than your letters. Remember me to Dr. Hales, Mr. Baker, and Mr. Shipley. I am just going to write to Mr. Baker. What is your Society doing?

I refer you to my letter by Capt. Man and Antrobus.

By Capt. Antrobus, of the Surprize, I send you a duplicate of my letter by Capt. Man, and a duplicate of my letter to Linnæus, as likewise a copy of the list of seeds, also a copy of the bill of lading for the box, and some specimens of Mr. Rivers's tree.

I have likewise inclosed you a specimen of a tree which I found yesterday or the day before in the woods. It is neither *Ulmus* nor *Chrysosplenium*, though it has a connection with both. It is *Octand.* or *Decand. Digynia.*

MR. ELLIS TO DR. GARDEN.

London, March 25, 1759.

I wrote to you by the Prince of Orange, captain White, last September, which ship sailed towards the end of November. By it I sent you Linnæus's *Hortus Cliffortianus*, which I hope you received safe. I wrote to you under care of Mr. Clifton, the Attorney-general of Georgia, who sailed about Christmas last, desiring your examination of the Cochineal insect. I received the specimens by Capt. Ball, which pleased every body; but I found it was impossible for the Premium Society to encourage the collecting it, as labour is so dear in our colonies. And in yours it is doubtful whether the *Opuntia inermis*, or *Cactus coccinellifer*, of the Spanish, West Indies, will grow. I fear your winters are too severe. However, your attention to such an useful subject was generally applauded, both in the Royal and Premium Societies. I have picked out some male insects of this *Coccus*, or Cochineal, which I find are winged; but the females are mere bugs. I have drawn the male and female magnified.

I am now to inform you of some experiments I have made and am making on seeds, to preserve them sound and in a growing state; and the two sorts I have singled out are the oak acorns and chesnuts. The acorns that I covered, some with clay, some with gum arabick, and some with senega, after lying three months in a temperate room, were found dry, hard, and unfit for vegetation. Those that I preserved in wax, in wax and rosin, or in wax, rosin, and a little pitch, were quite fresh, as if just gathered, after lying three months in the same drawer. The chesnuts I have just now covered in the following substances:—1. some in tallow, from the kidney-fat of sheep; 2. some in common bees'-wax; 3. others in wax and tallow; 4. others in rosin and wax. These chesnuts were put into large gallipots holding a pint, and the substances, just melted so as to be fluid, were poured on them, so as to flow round and inclose them on all sides. These I placed inverted on a shelf in a cellar, but intend, when the weather grows warm, to put some salts on them, agreeable to a plan that Linnæus has sent me to preserve seeds cool in hot climates *. The hint of the tallow I received from Andrew Fletcher, Esq. the Duke of Argyle's secretary, who assures me he never yet received any of your large *Magnolia,* or Laurel-leaved Tulip-tree, seeds in perfection but once, and that was when they were sent in an earthen jar, with suet or tallow melted about them, so that they were entirely involved in the tallow. These, he says, came up, and grew amazingly in a fortnight's time after sowing. Whereas we know for

* See Linnæus's letter to Ellis, dated Dec. 8 1758.

certain that not one in a hundred, nay a thousand, ever comes up, that is not preserved in some such manner, and only put up in papers; which was the case with all those in the box of seeds you sent me last year. So that if you will be at this trouble, I will pay the expense. The *Æsculus Pavia,* or Scarlet Horse Chesnut, did not grow, though the seeds were fair to look at. This then is the only method that fact has yet convinced us will answer; so that I beg you will try it against next year. And, to save expense, I wish you would please to get some large gallipots, and put into each some of the following seeds (which are perishable, few ever coming up, out of the great quantities you send), pouring in fresh melted tallow on and about each sort. *Magnolias* of several sorts, particularly the Sweet-flowering Bay *(M. glauca). Pavia.* Loblolly Bay, or *Gordonia,* in the husks. Live-oak Acorns. Yappon, or *Cassine* berries. Sassafras, *Chionanthus,* and Palmetto berries. *Halesia, Beureria**, and the ripe seed-vessels of the Azaleas and Andromedas.

These earthen vessels I would have quite filled with the tallow, and then covered up close, and put into any part of the hold. The other experiments you may try in small parcels, if you think proper.

There are many sorts, those that are resinous, which constantly come up when sown, as the Pines, Cedars, Junipers, and Liquid Amber. But all such as are liable to dry and wither, such as stone-fruits,

* *Calycanthus floridus. Linn. Sp. Pl.* 718.

both small and great, acorns, and most small seeds, unless closely confined in their capsules, soon perish. Several of these may, with their capsules, be put into the tallow, to try its effects.

FROM THE SAME TO THE SAME.

[Postscript to his letter of March 25, 1759.]

I have your favour of the 21st of January last, as also a copy of the same. I am glad to find you have got the *Hortus Cliffortianus.* I must likewise thank you for the care you have taken in providing the seeds to send me by the Fanny, Capt. Brooks, which I hope will arrive safe.

As the methods of preserving seeds which I send you will probably make them grow when they arrive here, I must now mention the kinds I would have you collect and save for me; for, to be free with you, there are many of those you are so kind as to send that are of no use to the gardeners. For instance, the *Callicarpa*, Purple-berried Bay, and *Ptelea.* The two first are too tender for the open air, and the latter is very common with us.

The Large *Magnolia.* Umbrella ditto. Swamp ditto, white underneath. *Halesia;* this does not come up till the second year. Knob-rooted *Corallodendron.* Dahoon Holly. Yappon, or *Cassine vera Floridana.* Palmetto Berries. *Beureria,* or Allspice tree of Catesby. *Stuartia* of ditto, or *Malachodendron* of Mitchell. Loblolly Bay. Carolina

Kidney-bean tree. *Liriodendron*, some in tallow. Red Cedar Berries, or Great Juniper. *Azalea* seeds in their capsules. *Rhododendron*, this grows to the northward. *Chionanthus*. *Sassafras*. Evergreen, or Live Oak. The *Ellisia*, you were pleased to call so. Scarlet Horse Chesnut. Sundry sorts of Pine and Spruce Fir.

The gardeners and seedsmen can dispense with any quantity of the Red Cedar, Pines, and *Liriodendron*, as they are hardy, and need not be preserved in any thing but paper and a little soapy dry earth among them, not sand. I find sand dries them too much, unless in a cask.

The Loblolly Bay is a very rare plant with us, and has been sent lately to Mr. Collinson, in young plants, by Mr. Lamborn of your town. These are likely to do well. Not any of the seeds that you sent have ever yet appeared, for they are so thin that they perish immediately; but I am in hopes the tallow will preserve them in their capsules. Or perhaps some of them in that state, put into large pill-boxes, with a thick mucilage, or solution of that or any aqueous gum, poured on them when full ripe, and then dried well in the sun and air, so as to inclose them in a hard cake, may keep them in a vegetative state.

April 1. Mr. Ellis wrote to Dr. Garden for Chinquapin nuts in tallow; Coccoons of the wild silkworm *, and seeds of the *Opuntia*.

* *Phalæna Cecropia*

DR. GARDEN TO MR. ELLIS.

DEAR SIR, Charlestown, May 11, 1759.

Again the hour comes that I relish the most of any; I mean that spent in writing to you. If I could make this as agreeable to you as it is to me, my pleasure would still be redoubled. I have received all your letters, and by Capt. Rains part of the prints you kindly intended for me; some of them were lost in the Thames as they were putting on board. Let me again beg you to accept of my thanks for your noble present. I have had that elegant book bound in the neatest manner.

By this opportunity I have written to Mr. Baker, and sent him a box containing specimens of the seven different strata of Savannah Bluff. This vein runs across this and the Province of Georgia; but its depth and beauty are only seen here, where the river runs through it. Pray beg him to shew some of it to those people who make your English China. There is one stratum that is as white as alabaster, and not the least grittiness in it. In his box I put one of the Pericarpiums of the Palmetto Royal for you.

I have often wrote you my thoughts on a Provincial Garden here. I never mentioned it to the Governor, because I thought it would do much better if proposed by the Society. Such proposal (without any the least expence to the Society) would be well received, and have such weight with the people as I am certain they would immediately grant any sum for carrying the thing into execution. If the Society would take the trouble, in a few sentences, to point out a few of the advantages flowing from it, they would at once, American-like, be all on fire, and stick at no expense to promote the plan. Surely the best method of giving public money is for public benefit.

I will preserve some of the *Opuntiæ* with the insect on them for you and for Linnæus, to whom pray offer my humble respects and service. I long much to see his new editions. I have never been able to get his *Iter Oelandicum* or *Scanicum*, which I should be glad to see before I finished my own *Iter Saludiense*. I formerly begged you to burn the few sheets I sent you of the Introduction. I hope you will consign them to the flames if they have not yet met with their desert.

I shall try to get you some of the Liquidambar and Juniper Gums this summer.

You wrote me word that they had found the Tea plant in Georgia. This was the first account I had of it; but I scarce think it is so. They make such confusion of plants here, that I never pay the least regard now to any information, till I see them. What is here called the *Cassine* is a species of the *Ilex*, and a very beautiful species it is. We use it for hedges, &c. in the gardens. Indians make what they call their Black Drink of it. It looks most beautiful in the gardens, as the berries remain of a beautiful red, and in clusters, all the winter round, and even till the new fruit of the succeeding year push them off.

You ask if Olive-trees will grow here from the slip. I have consulted several people on this who have raised them, among whom is Dr. Moultrie, and they all assure me that all they ever raised were from the slip, and that they seldom failed.

I come now to answer what you ask about the heat of last summer. If I remember right, the highest that ever the mercury stood in my Farenheit was 93^{0}, which was on the 7th of July 1758. It stood upwards of 90^{0} for 12 days together at that time, and was never under 80^{0} from the 15th of June till the 13th day of September, when it fell to 77^{0}. About the middle of July the cool night air scarce brought it down to 87^{0}, and it often was at 88^{0} 30′, and 89^{0} at midnight, with the greatest calmness, stillness, and closeness of the atmosphere — not a breath of air stirring. In the day-time we often had small *Turbines*, and what they call Typhones. There were but few thunder showers, but when they came they were very dreadful, both from their impetuosity and the horrible thunder and lightning that attended them.

At this time, when the thermometer was exposed to the direct rays of the sun, the mercury rose gradually from its station in the shade to 120^{0}, and sometimes more — nay, I once saw it rise to 130^{0}, when held in the sun-beams for only three minutes.

With regard to Governor Ellis's observations, you do not mention whether they were made in the shade or in the open air. If it was in the shade, I confess it is more than ever I saw; but if in the open air, exposed to the sun, it is nothing at all extraordinary. What I have written above relates only to the observations made in the shade, where the sun's rays never penetrate, and where the air is no way influenced by them. — During this last summer I had an opportunity of making what appeared to me many new observations, which in time you may see, on the air. You will soon have an Introduction to a Meteorological Journal, which I will finish with what speed I can; but at this season my old complaint of a weak and sick stomach commonly seizes me, and it is already beginning to usher in its annual visit.

I have written to Dr. Hales, and inclosed it in yours. I beg you will present it, with my compliments, to him. I begged Mr. Murray to procure me some acquaintance with a careful bookseller, that I might be sure of being supplied with some little things which I might from time to time want. The Crokatts are not careful in these little things. They have totally lost to me a copy of *Wallerius de Regno Lapideo;* and they have likewise lost the four cuts of the *Halesia,* which you mention to have given them to send out; they never came.

I hope my letters and seeds by our spring fleet have come safe to your hand. Pray let me have your opinion of the draught of the *Ellisia;* and let me again beg you will procure one of Mr. Ehret's draughts, such as he gives to the engraver. Mr. Roupel, the gentleman who draws for me, wants much to see one of them, and no doubt it will be of great service to him.—What is become of the few new genera which I sent home before? I hope you will denominate them yourself. The *Diandria* with milky juice is a grand specifick for the lues venerea; so reckoned by many people here. I have never yet tried it; but our Assembly has bought the secret lately of this, and the secret of preparing the common decoction of the woods of the dispensatories, at the price of upwards of 400 guineas. You see how generous we are of public money! and how easily dupes may be duped!

May God still bless and preserve you, for the good of mankind and the advancement of natural history!

I remain, dear Sir, yours, &c.

ALEX. GARDEN.

Pray tell Mr. Baker to send on board captain Rains' vessel for his box.

You will soon receive a small treatise, which you may use as the author, a friend of mine, will beg leave to mention to you, in a letter which he has asked me to write along with it. If this fleet does not sail for a few days it will come herewith.

DEAR SIR, May 17, 1759.

I have taken the liberty of inclosing a letter to my sister to your care, and must beg you will be kind enough to put it into the Post-office for Scotland by one of your servants. Let me beg that you would forgive this freedom which I use with you.

My situation with regard to the want of a mercantile correspondent at London, to whose care I might direct such commissions, is often inconvenient for me. I have begged Mr. Murray, by whom I wrote to you, to procure me a correspondence with some bookseller, who will be useful to me in many respects. — As I wrote to you fully by his Majesty's ship the Success, Capt. Ourry, I shall beg leave to refer you to that. I have by this opportunity wrote to Mr. Baker, and sent him a box containing specimens of the several strata of New Windsor or Savannah Bluff. There is one layer or stratum that is of the whiteness of alabaster, and, what is surprising, it is extremely cold to the touch; and I am told that some people who have plaistered their houses with a mixture of this and a little clay have found them remarkably cooler than any other.

In my last I mentioned a small essay that a friend of mine had engaged me to send to you, and beg you to give to any bookseller for him. It is, as he informs me, now copying over, and if it can be ready by the time this fleet sails, it will be sent by his Majesty's ship the Mermaid, Capt. Hackman. I think there are some things in it that may please the common taste of these times; but you will judge better; and he is willing to submit its fate to you, as I told him I could not well plague you with it on any other footing. I am, as always, dear Sir, yours, &c. ALEX. GARDEN.

Dear Sir, July 14, 1759.

Captain Ball delivered me your favour of the 25th of March, by which I find you received my letters of January 1st and 21st, and before this time I hope you have received the seeds themselves. It grieves me much that Mr. Gray has not found them answer better, but, believe me, I intended them to serve him equal to what I wished or he desired. Some future parcels may succeed better, as I shall conform exactly to your directions in packing them up, and probably they may succeed well. You need not question that I will do every thing I can to obey and execute your commands, as it always is the greatest pleasure to me, at least to show my inclination; and if my power was greater, it would be all employed to do any thing that might be agreeable to you.

I hope you have received my letters by Captains Man and Antrobus, with the draught and specimen of your name-sake, and my letters to Linnæus, which I begged you to transmit. Pray inform me whether you think it proper that I should write to him in Latin. I did it this first time, as I did not know whether he reads or writes in English; but some foreigners are fond of showing their knowledge in the English, which I think was the case with Schlosser. I wrote likewise by the Squerries, Captain Mitchell; and by Mr. Murray, a young gentleman who went from us about April last. A little after I wrote to you again by the man of war Captain Ourry, and then enclosed a letter to Dr. Hales, from whom I had just before received 25 copies of his Essay on the Ventilators, &c. to be distributed to such persons as had opportunity, power, or influence, to put his many good advices in execution—this I have endeavoured to comply with. Pray remember my compliments to him and to Mr. Baker, to whom I likewise wrote by Captain Rains. I beg to be kindly remembered to good Mr. Shipley. You may be sure I read with great pleasure your observations on the method of preserving seeds; I hope it will enable me to be more useful to you in the article of seeds. I am just now beginning my collection for this summer. There are no pods of the red *Robinia* that ever I saw. It propagates easily from the root, and I am coming into a good stock, so that after this year I shall be able annually to give you a box, packed as you direct, and sent at a proper time.

In my last letter to Mr. Whitworth, there was nothing that particularly related to the Society. He had desired me to give him some account of our country, and as I was willing to comply as far as I could, I gave him some account of our situation with respect to the Indians, and of our ignorance with regard to the geography of the country; these were the things he wanted chiefly to know. I have not heard from him this long while. If I remember right, there were several things mentioned in my letter to him that it concerns us much to know, which we are still very ignorant of. I enclosed a copy to you, which I hope you read over and sent to him.

There never has been a plant of Madder planted here as yet, neither have I ever seen any person who shews the least inclination to begin. Rice and Indigo will alone go down.

A public garden is the only thing that will ever bring these things to bear in this place, and that supported and countenanced by our Governor and Assembly.

I have already written to you fully on this subject. What do you think of the method which I proposed? I have delivered your commission to Mr. Peronneau, who returns his compliments and thanks for your kind offer.

I wrote to you before, that all my thermometers which I wrote home for, came out entirely broken, owing to their being very ill packed up; besides, they seemed to be but clumsily made, and I apprehend not graduated with any care: I mean those that had the scales enclosed in tubes. So that as I have none just now that is fit for measuring the heat of water or liquors, I am afraid I shall not be able to gratify your desire of knowing the heat of our well water at different seasons. I shall try to get one or two thermometers soon, but unless they be good, I shall not trouble you with any observations made by them. My standard thermometer in the shade just now, at 1 hour p. m. stands at 86° 30′. This you will say must be comfortably warm.

As I have not any particular botanical news to write you just now, I shall give you an account of what I have put into my meteorological journal concerning the comet which appeared here some time ago. The following I have copied from my journal, part of which I once gave to one of our printers, who put it into the Gazette, but as this seldom reaches you, I may at least satisfy you where it appeared, and the time with us; as to more particulars or philosophical accounts, you must expect them from better astronomers.

On the 3d, 4th, 5th, and 6th days of April, a comet was observed here by many about 4 o'clock in the morning. It was nearly S. E. with its tail directed obliquely upwards towards the S. W. apparently of three or four deg. in length. By the account which I had of it then, for I did not see it at this time myself, I imagine it must have been in *Aquarius*, near its descending node. From this time it was not seen by any till the evening of the 28th, at 9 o'clock p. m. when I observed it for the first time that I had seen it, nearly in a line with the two foremost stars of *Corvus*, and the star *Omicron* in *Hydra*, about 3 deg. more southerly than the last mentioned, and a little to the west of it. It was in 6° 20′ *Libra*. Lat. S. 36° 30′. Declinat. S. 36° 20′. Its tail now appeared very long, seemingly to equal 8 or 10 degrees. But it was thin, rare, and diverging much. It was luminous. The direction was nearly horizontal towards the east. The head of the comet seemed to be almost surrounded by it, and seemed to be moving away in an oblique direction, whence I judged that it would soon, but gradually, become invisible to us.

On the 30th it was near the star *Chi* in *Hydra.* Long. 25° 30′ *Virgo.* Lat. S. 32° 30′. Declinat. S. 28° 50′.

On the 2d of May I saw it again, about 8 p. m. near *Alkes* in *Crater,* Long. 18 deg. *Virgo.* Lat. 27° 25′ South. Declinat. 21° 20′ South.

On the 3d of this month, the moon entered her 2d quarter, and obscured it so much every night, that I had no distinct sight of it again till the 18th, when I found that it had crossed the body of *Hydra,* and got into the lower limb of *Sextans Uraniæ.*

May 19th. It was a little above the two stars in the limb of *Sextans Uran.* Long. 3° 30′ in *Virgo.* Lat. 16° S. Declinat. 5° 55′. R. A. 149° 40′. This evening it was very faint and dim; and viewed with a telescope, was not well defined, but appeared like the stars called nebulous. It had been growing less and less visible and distinct to the naked eye, but did not appear to lessen its declination or latitude near so quick as it did when I observed it in the end of the last or beginning of this month. I saw it, however, tolerably well, till the 23d, when it appeared very faint and obscure, pretty high in *Sextans Uraniæ,* nearly in a line with *Cor Hydræ,* and the large star in the Lion's tail, called *Deneb Alased,* about two deg. lower than the equator. This was the last view I had of it, but by its course I imagine it might have been in its north node, about 15° *Leo.*

Thus much for the comet. I come next to tell you that I have now by me four specimens of the different species of the poisonous *Rhus* or *Toxico-*

dendron. I brought them from the country the other day, upon my being called out about 30 miles from town, which by the bye are all the botanical excursions that I can get now. I intend to have drawings of these plants, and to give a particular account of them along with the drawings, with such observations about their effects on animals, as I have partly seen, and partly had from good authority. I shall again look into the Transactions, and see what has been said of them; but as I have no connection with that Society, if you approve of it, I had rather send the draughts, specimens, and accounts to the Edinburgh Society. I shall send a copy to you of the specimens as soon as they are well dried, but I would only do it agreeably to your desire, for your own perusal, not for the Royal Society.

I have sent you a few plants and roots of the *Baccharis* or *Conyza,* which I sent to you before. You will remember that you desired some of it, to have it tried in your lying-in hospitals. This surely is the best method, and I should have been glad to have procured more, but this was all that I could get the other day, and I dug it up on the side of the path.

I must next apply to you in behalf of a friend of mine, who begged me to send you an essay, which you will receive from Captain Ball. It is entitled an Essay on the cause, origin, and progress of Love. He begs that you would peruse it yourself, and if you think it worth while as a piece of humour, or as a *jeu d'esprit,* that you would give it to any bookseller, who would print it on his own charge entirely, and make what he could of it, only giving the author 28 or 25 copies, ten of which to be neatly bound and gilt. The author thinks that Wilson or Dodsley would be proper hands, but he will willingly leave that to your judgment. All he wants is to be at no charge, and the above-mentioned number of copies. And if it will not bear these charges, he desires earnestly that after you peruse it, you would immediately burn it. The author thinks it had best be printed in an octavo pamphlet, a good type, and on a good paper. He thinks an emblematical frontispiece would be proper, but these are only proposals. He likewise told me to say something about the title page, but I protest I forget it. If you agree with any one to print it, you will desire him to write immediately to the author, directed to L. R. and give his letter to you, which I beg you would enclose under cover to me. You will be so good as not to mention the Essay's coming from America. I asked my friend to let his name be added, but he absolutely refused, and begged that you would not let the bookseller or printer know from what quarter it comes. I can tell you, however, that he is a Scotchman, which you would easily discover by his language, which he begs the printer would get corrected by some judicious and careful hand—have the Scotch idiom thrown out, and the punctuation rectified. He tells me that he has some other such frolicksome births, and if the world would only laugh, he would give them, but he is afraid of censure, thinking that people will take him to be too serious.

Whatever pains or care you take with regard to this, will confer on me a most sensible service, as I have engaged to endeavour to procure your friendship for the author. As he once read *Armstrong's* Œconomy of Love, and was highly pleased, he had some thoughts of dedicating it to him, but as he has no acquaintance, either personal or literary with the Doctor, he was afraid of giving him offence. He shewed me his Epistle Dedicatory. I pressed it from him, as I think it was such as the Doctor must have liked; but he told me he had thought more coolly, and must be excused.

Pray write me your sentiments of his performance. At his desire I read it over with all the care I could, and though I could be no good judge of the language, yet I think there are many scotticisms, where a small transposition of words would make it read smoother; but an Englishman alone can do this. I likewise met with singulars for plurals, and such like. These are errors common to all Scotchmen.

Since writing the above, I have again seen the author, and he again requests that in case of your not agreeing on these terms with a printer, you would take care that no copy be taken of it, but would have it immediately destroyed, especially as it affords a kind of key to some other of his papers.

I have sent you some specimens of the *Cratægus* or *Mespelus*, inclosed in the Essay, and one of the Loblolly Bay.

I have sent you a box of the Cochineal leaves for Linnæus, with the insect on them, which, if I remember right, was the manner in which you told me he wanted them. They are directed to you, examine them, and send what you judge proper. I am yours, &c. ALEXANDER GARDEN.

Having no more room on the other sheet, I am now to explain my meaning about the *Opuntia.* I have directed the box to you, that you might open it and look at the specimens, or, if you wanted, examine any more of the insects. After doing this, you will please to shut it up again, and transmit it to Linnæus, with my respectful compliments, and assure him that in whatever I can serve him, I will do it as far as I can. But, my dear friend, I must here again repeat an apology to you for neglecting many things which you recommend to me. I can assure you that my inclination is great to serve you in every particular, and the not being able is a greater disappointment to me than it possibly can be to you; for I am sure I must suffer more than you, for any neglect of study and application on my side. But let me assure you, that I am bound as if *manibus pedibusque,* by a servile kind of attendance on practice, which, from time immemorial, has been introduced and continued here, and every deviation is construed into neglect. The most pitiful slave must be as regularly seen and attended as the Governor. This custom I am determined to get the better of by degrees, but nothing of that kind can be done in a hurry. From seven in the morning till nine at night, I cannot call half an hour my own. Such is my laborious situation. I would not give you the trouble of perusing these bagatelles, if it were not to apologize for my neglecting some of your commands. Nevertheless, what I may not be able to do at one time, I will execute at another; and though I may not give a quick answer, yet I will keep a memorandum of all you want to be done.

As there are some things in my friend's Essay that relate to what he calls the characterizing mark of the animal and vegetable life, I have begged him to let me have one of his ten bound copies for Linnæus, and one for you, which I beg you will accept of, and transmit Linnæus's with my best respects.

Please to let me know how you think the *Ellisia* is executed by Mr. Roupel. I have had a number of things by me in specimens for these two or three years, but he is so plaguy lazy that I cannot get him to do more than one or two in a whole summer; so that any new plant which I get is often lost as to a drawing. I am, as always, yours, &c.

ALEXANDER GARDEN.

Some of the Governor's friends would, no doubt, be glad to know that he is just recovered from a peripneumonic fever. He is now quite out of danger.

MR. ELLIS TO DR. GARDEN.

DEAR SIR, London, Aug. 25, 1759.

Your several letters by the convoy, dated in May, came to hand, for which I am infinitely obliged. Your letters always give me pleasure, there is something so new and spirited in every one of them. I always read to the Premium Society those paragraphs that tend to the promoting of arts and sciences, and do assure you the members have a great opinion, not only of your judgment, but of your zeal in endeavouring to promote the good of mankind. But, in short, our Society has of late become a mere Society of drawing, painting, and sculpture, and attends to little else, as you may observe by a list of the premiums for this year, which I shall send you. But I am still determined to lay before them again the utility of recommending public Provincial gardens.

Mr. Baker has got his box, and intends writing to you. The earth you sent has been sent to Mr. Collinson before, but I shall desire him to show it to the china manufacturers.

The Georgia Tea plants sent me by De Brahame, are no more than the two sorts of *Cassine* *. You call them a species of *Ilex*. I wish you would dissect them carefully yourself, and also send me some specimens of the blossoms well dried, for they pass with us under another genus. De Brahame has likewise sent the Governor, who forwarded it to me, another species of *Halesia*, with two wings instead

* *Ilex Cassine*, α and β *Ait. Hort. Kew. ed.* 2. v. I. 278?

of four to the seed-vessel*. They grow about *Augusta*.

The observations made by Governor Ellis, were both in his house, and in the open air, but the thermometer was never exposed to the sun beams. The experiment made in the open air was under the shade of an umbrella, when he says it was up at 104, hanging before his face. Upon placing it at his breast, or under his arm, it fell to 97; so that he concludes the air he breathed was 7 degrees hotter than the heat of his body. This indeed is doubted here by many physical people.

I am obliged to you for the acquaintance of your friend Mr. Murray, who is a very ingenious sensible man. He now lodges at a bookseller's in the Strand, at the Plato's head, the names Wilson and Durham. They are honest and reasonable tradesmen. I am sorry the Crockatts are so negligent. I will send you some more plates of the *Halesia*, with the original drawing by Ehret, which I would not have parted with upon any consideration but to you, to whom I am under so many obligations. Indeed your compliment of the *Ellisia* (now a *Swertia*) I can never forget. You see Linnæus says I must be contented with one plant, which Dr. Browne has given me.

* *Halesia diptera, Linn. Sp. Pl.* 636. *Cavan. Diss. t.* 187. A plant very little known to botanists in England. Linnæus in *Sp. Pl.* has given a wrong account of the leaves, having been deceived by those of *Styrax grandifolium*, sent him with the true seed-vessel of *H. diptera*.

Mr. Roupel has shown an excellent genius in drawing. I am persuaded he has copied nature most exactly. When you get him to draw any more for you, the outlines will be sufficient, as in Plumier's American plants, and one blossom, leaf and fruit only, shaded. The rest can be finished here.

The genera you have sent are still by me. The Supple Jack is called a *Rhamnus* here. I have no specimen of the diandrous plant, with a milky juice, which is a specific for the *Lues Venerea;* and therefore beg you would send me some of it, with a drawing and dissection of the flower, and some few seeds. Mr. Baker has not sent me the *pericarpium* of the Palmetto Royal, but we all judge it to be a *Yucca* from the young plants. I cannot make any thing of your scarlet-dyeing plant. You can better examine it when fresh. I shall send Linnæus some of it. I have not seen the treatise of your friend, which you intend sending with a letter.

I here inclose you Linnæus's letter, which I received some time ago, so that now the correspondence is begun, you may go on. I am persuaded you will be happy in each other. His book of animals is come out, and I have it. He says his book on plants is just come out, but it has not yet arrived here. Mr. Murray tells me he has it in commission to purchase for you, or I should have sent it when it comes.

I wait for your observations on the male insect of the Cochineal fly; and also its economy, or the natural progress from the egg, till it becomes fit to be picked off for use.

I have a tolerable idea of it, but there are chasms in my account, which I am obliged to supply with conjecture. Pray send some of the ripe fruit dry.

If you send me any seeds, let them be few and choice, full ripe, and some preserved either in bees-wax or myrtle-wax, and some in suet. Also cones of your Swamp Deciduous Cypress. These keep without any preparation. Preserve likewise in wax a few large Magnolia, Umbrella tree, Small Magnolia, Evergreen oak, Chionanthus, Sassafras, and *Cassine* berries; and a few of the Loblolly bay seeds; for they are so thin that they perish before we receive them, and the others grow rancid, on account of their oil. In my letter to you last March, I mentioned preserving them in suet. I wish you would try both; for if we can hit on a proper method to bring them from Carolina, we may reasonably expect to bring Tea, and other valuable seeds, from the East Indies, by a still stronger covering. A friend of mine, Mr. Talbot, brother to the late Chancellor of that name, has been in China often, and assures me he has carried lemons, quite sound, from London to Batavia, and some even to China, by wrapping them round carefully with a cover of wax and mastick mixed together. Perhaps you have resins that, blended with wax, may do as well as mastick. It is worth trial for seeds at least. I find by Governor Ellis, that the acorns that were inclosed in wax, and afterwards in clay, moistened with a stiff solution of gum arabick, arrived in a vegetative state. If I live till the middle of February, I will send you the result of my experiment for preserving the Spanish Chesnuts. If these experiments succeed in preserving the seeds I now write for, it will encourage gardeners to import largely; for at present there is so much disappointment to them and us, that it disheartens people from trying. I think I shall be able to get you a very ingenious correspondent, one Mr. Lee of Hammersmith, an ingenious countryman of yours, who is now translating Linnæus's *Philosophia Botanica.* He will send you the books you want, for the seeds he wants. I sent your last cargo to Mr. Gray the gardener, as he had suffered so much by losses, in the former cargoes that were taken.

Your Pine seed, which you call the Highland Pine, is what we call the Great Swamp Pine, having the longest cones of any. Many of them came up, but perished soon after, which makes me think it would be better to send over the cones. The other Pine seeds are likely to do well. If you meet with the seed already taken out of the cones of this large kind, they may do better in wax or suet; but the cone is the most natural, and best covering for them. I have met with an ingenious sensible man of your parts, one Mr. Williamson, by whom I shall send the prints of the *Halesia,* and the drawing. He is curious, and has a true and just opinion of your merit. I got him a root or two of the Rhubarb, which loves a rich moist soil, and I am persuaded will do. I got him this plant, because it will soon produce seeds, and then you may propagate it as much as you please. I shall send him and you some seeds, as soon as they are ripe; with the Cork Acorns, and other things.

Your letter to your sister I sent in a frank, and assure you I never think of any trouble in obeying your commands. JOHN ELLIS.

DR. GARDEN TO MR. ELLIS.

DEAR SIR, Charlestown, Jan. 13, 1760.

I am now to write you a short answer to your most obliging, kind, and very ingenious letter of the 25th of August. It would be absurd to be constantly pleading excuses and making apologies for my not writing oftener; suffice it then to tell you, that I am very sensible the loss is entirely on my own side; for by writing to you seldom I must of consequence draw a letter from you but seldom; and yet I can assure you that when a London vessel arrives, and I find no letter from my friend, I have but little relish for any enjoyment this place can afford for some days after.

I rejoice with you at the progress of your Premium Society, and hope they will soon recommend strongly Provincial gardens. The use of them will be inconceivable, and it will annually grow more and more; pray do all you can to make them think well of this.

I shall soon send you the characters of the small Ilex, which you desire. Its being called *Cassine* is a great mistake. It makes a very good and most beautiful hedge, and may be kept as short and neat as the Box.

I never yet saw a two-winged *Halesia*, but I will look out.

Accept of my thanks for the plates, especially the original drawing of the *Halesia*. Mr. Roupel was struck with astonishment on seeing it, but he cannot yet well judge with what kind of pencil he (Mr. Ehret) does it, or can possibly do it, so fine.

Both Mr. Roupel and I are very glad that you liked the drawing of the *Ellisia*. I have carefully observed Linnæus's reasons for not changing the *Ellisia* of Browne, and for thinking that the plant which I sent was a *Swertia*, but I cannot join with him in the latter. I have given my reasons in my letter to him, which you will peruse, and be kind enough to send to him. I could, and I think with justice, have pressed him here, that in this, and indeed in all the *Swertias* and *Gentians*, he allows the fruit to be the best and most natural generical mark. This of consequence is an exception to the fundamental principle of his system, *i. e.* that the *Stamina* and *Pistillum* are not only the best classical but the best generical character, and very frequently, though fruits differ much, yet if the number, &c. of *Stamina* and *Pistillum* agree, he often ranks plants under the same genus.

Now I would beg to know why the Palmetto

Royal and *Yucca*, which have fruits so very unlike, should nevertheless be both *Yuccas*. I confess this puzzles me.

I am heartily sorry that I was too late with my *Ellisia*; it would have given me great pleasure to have had the honour of paying that compliment to you; but allow me to beg that you will mention any friend whom you may judge worthy of any botanical compliment, and I will send you a new genus, or you may take any one of those that you have, if you chuse either of them, to name them after your friend. The Loblolly bay I think is a very fine one, and would well suit a true botanist.

I have not examined the *Rhamnus* and Supple Jack, but if I remember the characters of the Supple Jack well, I am sure that Ehret is mistaken in thinking it a *Rhamnus*.

I have no specimen nor seeds just now of the *Diandria Monogynia* with the milky juice, but will endeavour to procure both this ensuing summer. As to any letter on its virtues to the Royal Society, which you advise, I think I shall scarce make it out. I am happy in corresponding with you, and shall always extend it to every thing that you desire to have from this place that I possibly can procure for you, but I have no desire to gain their applause or thanks, and yet I would willingly contribute my poor mite to the promotion or advancement of any thing that might be of public utility or benefit to mankind.

As in many other things, so I am particularly

obliged to you in your generous offer to take care of Linnæus's letters for me; this confers on me the highest obligation, and I will do any thing to repay you in whatever manner you desire.

I shall be much pleased to see these new performances which he mentions, especially those of his own, *viz.* the *Systema Naturæ,* in three volumes, the last of which he says will be published before this time. Likewise the last or fifth edition of the *Genera Plantarum*, the third and fourth volumes of the *Amœnitates Academicæ.* I must therefore beg that you will desire Mr. Wilson the bookseller to send me those books, and I shall get Mr. Biswick to pay him the value.

You shall have my account of the male Cochineal by the convoy in April.

Pray, if Mr. Ehret publishes any more draughts of plants besides those which you sent me, let me have a copy of them. You must excuse this demand, as I have no other possible method of coming by them.

Mr. Roupel is just now about draughts of five different species of the *Rhus,* which I intend to send to the Edinburgh Society, who have made some demands on me. I intend to make out as good an account of them as I possibly can. This I once intended for you, and if you will send them down without communicating any thing of them to your Society, you shall have them before any person else in the world. Indeed there is nothing that you could possibly command me to do which would not give me the greatest pleasure to execute; but as for your Society, I think that nothing in the botanical way ought to come before them. It is but the other day that I heard of their treatment of Linnæus, which I am sure must astonish every man who knows any thing of the merits of that surprising and superior genius. How in the name of God could they be so blind?

Is Dalibard publishing any thing? or are Jussieu or Du Hamel publishing any thing? or is Haller or any German botanist publishing any thing?

My friend's essay, which I mentioned to you in some former letters, was sent you by Capt. Ball, who I hope is arrived safe, and has delivered it to you, with several other packages for you. Pray let me know your sentiments on this essay, and inform me of its fate.

I come now to the seeds, of which I have sent you all that I possibly could procure, and I hope they will give you satisfaction. I packed them with my own hand in the manner you see in the inclosed list, where you will find that most of the valuable seeds are packed in different substances, agreeably to your directions. The *Magnolia palustris* is packed in bees' wax and tallow, in tallow, in myrtle-wax, and in a very strong solution of gum arabic. This, as it is a very difficult seed to preserve, will give you an opportunity of discovering which of these substances is the best for preserving it, and of consequence many others of the more tender plants.

I wrote to you before to ensure them, and I hope you will not repent it when they come to hand.

I would have sent Linnæus's letter, at least one copy of it, by this opportunity; but as I have not got some fishes sufficiently dry, I could not pack them without running the risque of losing them, so they must come by the convoy. I have a few snakes, &c. for him. I never before looked to our fishes attentively, but I am struck with astonishment at Catesby's blunders. Sometimes he forgets whole fins, &c. In a word, there is nothing can possibly recommend him but the specious beauty of the colouring of the plates.

If I had time, I believe Linnæus would soon make me have a relish for all the parts of Natural History, but my business in the practice scarce spares me half an hour, and I am obliged to steal the time I write this from the hours of sleep. We are just now on the eve of having the small pox. We have not had it for two and twenty years before, so that of whites and blacks there must be more than two thirds of the people to have it. Your box is on board the Union, Capt. Strachan. Inclosed you have a bill of lading. I am always yours,

ALEXANDER GARDEN.

I have sent you a number of very good specimens of the *Opuntia* with the fruit on it, in the top of of the box.

I will soon send you a few seeds by the next vessel.

Feb. 7, 1760.

That you may have some idea of what view our Assembly had of our late Governor, on his return from his ever memorable expedition against the Cherokees, I have sent you the following remonstrance, which was presented to him instead of a warm congratulatory address on his supposed success, which he expected. Though you will see that there was scarce time to write such a wished-for address, before he was obliged to call them together to think of protecting the province from the ravages and fury of that nation which he thought he had conquered; yet, notwithstanding the following remonstrance, the same house in about ten days, when they heard of his appointment to Jamaica, gave him an address. Such is the mutability of the human mind, and volatility of Americans!

The Remonstrance.

We his Majesty's most dutiful and loyal subjects, the Commons House of Assembly of this Province, sensible in the highest degree of his most excellent Majesty's great benignity and parental regard towards this province, and thoroughly convinced by long experience of his invariable endeavours to preserve to all his subjects their liberties and privileges, which are the most valuable part of their inheritance, cannot but lament, with uncommon concern, that they are constrained to make your Excellency acquainted with the deep affliction we labour under from the speech you delivered to us on the 13th of October last.

To be reproached with a want of love to our country, would render us private men, unworthy of those benefits which our birth-right, as british subjects, gives us a title to. To be but suspected of it, in our public station, must make us contemptible; but to be charged with it in the most open and extraordinary manner, and that by your Excellency, with whom this House has preserved a remarkable harmony, must (if true) prove us betrayers of our trust, and enemies of those who have confided in us as the guardians of their liberties and properties.

But a consciousness of the purity and uprightness of our intentions, makes us easy under this obloquy; and the duty we owe to our constituents obliges us to declare that we never shall implicitly, or against our judgment, comply with any demand made upon the public, even though we were sure of incurring your Excellency's censure. Yet it becomes us, whilst we have the honour to be Members of this House, to challenge and lay claim to that great and undoubted privilege which you have promised to protect us in; of our Members having free liberty of speech, to propose or debate any matter, according to order and parliamentary usage; and your Excellency's charging the majority of this House, in the manner contained in the said speech, is a violation of that most essential privilege.

It is with the utmost joy, even upon this present alarming occasion, that we declare our veneration for his most gracious Majesty's person, and our inviolable attachment to his interest and government; together with our steady resolution, firmly to support, with our best abilities, the welfare of this his Province; and that, as it is abhorrent to our nature to deviate from our loyalty, so it is repugnant to the sound principles of reason, virtue, and duty, to neglect the rights of those who sent us here, or our own fundamental privileges.

THE ANSWER.

MR. SPEAKER AND GENTLEMEN, February 14, 1760.

I have received with much pleasure your message of yesterday, wherein you acquaint me with your resolutions, to make provision for raising a regiment of a thousand men, to be employed for the relief of Fort Prince George, and to chastise the Cherokee Indians.

The message which you sent me on the 7th instant, contained expressions of that nature, as I must, in any times, but such as these, have taken an especial notice of. To be told that I had cast an obloquy upon you, are words, which it would ill have stood with my honour, as a private gentleman, not to have resented; and when addressed to me, in my public character, as his Majesty's Governor, I leave you to judge how I might endure them. But I have the satisfaction to see the King's service, and the welfare of the province, so much Promoted by your votes, for the very important services above-mentioned, that I shall suppress my own feelings upon the subject of the former proceeding, and give no interruption to the good work you have in hand.

W. H. L.

DR. GARDEN TO MR. ELLIS.

DEAR SIR, March 13, 1760.

The greatest fatigue, both of body and mind, has prevented me from writing to you by several of our last vessels, but be assured that no other cause could have hindered me.

Our Governor returned from the Cherokee country in January, as we then thought crowned with laurels; but, alas, bringing pestilence along with him, and having the war at his heels. The soldiers that came down with him brought a most fatal and malignant small pox from the Cherokees, which, in two or three weeks, began to spread in this little place so furiously, that the inhabitants were driven precipitately to have many more inoculated than could be attended by the practitioners of physic. Not less than 2400, and I believe not more than 2800, were inoculated in the space of 10 or 12 days. You may then easily judge of the hurry and confusion we were in, and the melancholy situation of the inhabitants of the town. Many at the same time were down with the small pox by the natural infection.

While this was our situation in town, the small

pox was spreading in many parts of the country, and the Indians who had played a game with the Governor, too well planned and concerted for him to see through, followed him close at the heels, and laid waste the whole back settlements with fire and sword, and continue their ravages with the greatest fury, notwithstanding the amusing compliments that are now passing between our Governor and his Assembly.

From one of the most flourishing provinces, we are, by a fatal piece of ambition, brought into a situation too terrible for us, who have a double enemy within ourselves to fear, viz. the small pox and the negroes. In this happy plight the Governor leaves us. Enough and too much of politics.

Inclosed comes a copy of a letter which I wrote to you in January; and if these unforeseen things above mentioned had not happened, I should have punctually complied with all my promises in it to you and Linnæus. I have not now a spare minute to pack Linnæus's fishes and snakes, but they shall come by the first sure opportunity. Of this you may assure him, and pray offer my sincere respects.

I have a thousand things to write to you, but they must be put off till a future opportunity, when I shall give you an exact account of the number inoculated here, and the success attending it. This I cannot be particular in just now, as many of our patients are not yet recovered.

Sorry I am that I cannot give you a more distinct account of our present melancholy situation; but for this last month I have not known what it was to eat a meal in peace, nor to enjoy one hour of undisturbed repose. At one time the lives of five hundred persons lay very heavy on my mind, and the necessary attendance that their cases required was more than my strength was equal to, but that scene begins to wear over, and I have just stolen a moment to give you this scrap.

If the fleet stay a week longer, I shall send Linnæus's fishes, and if Mr. Roupel can finish his draught, I will send him a new genus.

I am ashamed to give you the trouble of the two inclosed letters, but as I know your goodness in obliging me, I hope you will forgive me when I tell you, that they are from persons here, who have been some years from England, and though they have often written, yet they never could learn whether the persons to whom those letters are directed be dead or alive. Your being kind enough to make one of your own servants call on them and deliver the letters, will most singularly oblige me, and will be a greater favour than if done to myself, as I would do any thing in my power to serve the persons who wrote them. If they have any answer, please to inclose their answers in your first letter to me. Order me to repay this in any manner you please, and you shall find me ready. My sincerest good wishes attend you, and ever shall be yours, ALEXANDER GARDEN.

Please to let me know the fate of the two letters in your first after this.

I am told that our new Governor, Mr. Pownall,

will be at London before he sails for this place. Give me leave to beg you will be kind enough to mention me to him, as it may be much in his power to promote our Provincial garden, and some other botanical enquiries. You will forgive this liberty, or place it to my account to be regularly debited.

DEAR SIR, March 21, 1760.

My hurry and confusion with the small pox still continues, so that I have not a spare minute to put up any thing for Linnæus, and of consequence must wait till a future opportunity.

My worthy and kind friend John Bartram came from Philadelphia here to see me, about eight days ago. He has stayed with me ever since, and will continue about ten days more before he returns, when he proposes to go to cape Fear, and from thence home by land. He goes every day into the woods, and comes home at night loaded with their spoils. He has brought in a shrub that I never observed, nor do either of us know what it is; and as the flowers are not yet put out, I cannot have an opportunity of examining it. It is a most unlucky thing for me that my business prevents me from going up into the woods with him, as I could shew him all our herbs. Every one of them seems new to John, and he is often struck with surprise to find so many evergreens.

We cannot possibly spend any time together, except after dinner and supper about half an hour, but you may easily guess that this is an agreeable hour. Ever since I have been in Carolina I never have been able to set my eye upon one who had barely a regard for Botany. Indeed I have often wondered how there should be one place abounding with so many marks of the divine wisdom and power, and not one rational eye to contemplate them; or that there should be a country abounding with almost every sort of plant, and almost every species of the animal kind, and yet that it should not have pleased God to raise up one botanist. Strange indeed that this creature should be so rare!

I have at random sent you a few seeds of the *Chionanthus*, *Stewartia*, Dahoon Holly, &c.; probably they may keep till another year, and if they do not, there will not be much lost.

As we are soon to have Mr. Pownall for our Governor, I must beg that you will, when he is at London, take some opportunity of showing him the utility of a public Provincial garden here, and recommend it so strongly to him, as that he may push the thing with the Council and Assembly, who alone should be at the charge of a work, that will be attended with, and productive of, so many good consequences to the Province. I shall be obliged to you to mention me to Mr. Pownall, as it may be much in his power to promote some of my enquiries while he is with us, if he has any turn that way himself.

March 25th. I have been lately in the woods for two hours with John, and have shown him most of our new things, with which he seems almost ravished of his senses and lost in astonishment.

Though I proposed to write to you at length on our present situation, yet I find myself so pinched for time, I can but just observe, that our situation is such, I can see little prospect of much relief, till some brisk expedition be planned and executed against Mobile and New Orleans. It is true that some troops from the Northern Colonies may prevent many ravages and depredations on the settlements, but it will never be in their power to put an end to the war so long as the Indians are supported and supplied by the French. The axe must be laid to the Southern root, as well as the Northern root, of the French Polypus. Besides, it is much the interest of the Indians to prevent the French from being extirpated; and though they are cunning enough to join with the strongest, as the safest side, while there is nearly a balance between the French and English, yet I imagine they are sharp-sighted enough to see, that their own ruin must follow the entire ruin or destruction of either the one or other of these contending powers. If the French were totally routed out of America, what value would the friendship of the Indians be of to us, or what great bad consequences could flow from their enmity? It is surely very natural for them to ask this question of themselves. It is then their true and undoubted interest to support the contention and preserve the balance. While there is breath or strength in both parties, they will be courted by both, but if either side fall, they fall of course. These things are so obvious, that we should imagine them very blind not to see and attend to them.

Our present dependance is upon Governor Ellis, who is at present very active, and is leaving no stone unturned in endeavouring to kindle a flame between the Creeks and Cherokees. This plan, if it could take place, would save these Southern provinces; but I am afraid he will scarce be able to bring it to bear, and though I have the greatest opinion of his abilities and influence, yet I cannot think that ever the Creeks can be so blind to their own interest. They must see that their importance would be lessened if the Cherokees were too much crushed, or they may naturally see that their fall would soon follow.

I wish he may do more than I think at present that he will be able, but his measures are so well planned, and he is so pushing and indefatigable, that much may be expected. We have been for this month past in a state of suspense; our old governor embarrassed and taken up in settling his private affairs, and our Lieutenant-governor cannot act till the power regularly devolve upon him, so that our whole hope and dependance is on the Governor of Georgia.

This is our situation. You may now reflect on

the contents of my last letter to Mr. Whitworth, concerning our ignorance of the geography of this country, and the loss which that ignorance will be to us now; add to this our ignorance of the various connections which the Indians have, or their method of being supplied by the French.

It will appear no doubt strange to you, to be told that the French can distress Fort Loudon, or carry assistance to the Cherokees, as soon from Fort Detroit as we can relieve it from Fort Du Quesne, and much sooner than we can assist it from Charlestown. They cross lake Erie, and go down Sandusky river to the great Siolo river, where they have portage only of four miles; and from thence they go down stream all the way to the lower Shawane town, cross the Ohio, and pass through the gap of the Osiolo mountains, and get at once into the Cherokee country. This pass we have never yet seized, nor is its importance attended to.

You will laugh at me for spending so much paper in politics, and I confess it is idle; but these things make us look serious here just now, and I wish they could be seriously thought of at home. War is at our door, and when I consider the general ignorance of our people of the country, where the seat of the war must be, I tremble. I must beg that you will deliver the inclosed to Mr. David Wilson, and desire him to send me an answer by some of the first vessels.

My sincerest wishes for your health and prosperity always attend you, while I am, dear Sir, yours, &c. ALEXANDER GARDEN.

MY DEAR FRIEND, April 1, 1760.

If you knew how I am driven about just now, both day and night after the sick, you would laugh at my thinking of any thing of Natural History; but I was too anxious to send the fishes to Linnæus, to let this opportunity slip of putting them up, and drawing out their characters. Please to make an apology to him for my not writing sooner. I thought to have sent him several odd things, but I cannot possibly put them up just now. I wrote his letter in January last, but could never pack the fishes, as they were not dry.

I must still think that the *Ellisia* is a new genus, as you will see in my letter to him. But I am very willing to submit it to his discerning eye. I have sent him a new genus, which he desired I would, that he might name it after me, but that I have left to himself, as I am very sensible I am yet rather too much a Tyro to have that honour conferred on me. I have sent him a very good specimen, and will soon send you a very elegant draught of it, which I would have sent now, could Mr. Roupel have finished.

Please to forward the letters, and it will singularly oblige me, as it is you alone that I have to depend on to forward our correspondence. The bundle of the fishes is directed to you, which, after you have looked at, I hope you will put up again, with the two specimens, and forward to Sir Charles.

I should be greatly obliged to you to let me know his address, which I confess I cannot well apprehend; so I scarce know how to manage just now, neither do I know if my address to him may not give offence.

Please to let me know likewise how to direct his letters on the back. I am your most obliged friend,

ALEXANDER GARDEN.

The fishes are aboard of the Carolina, Captain White, by Mr. Smith.

DEAR SIR, April 1, 1760.

This comes along with the packet for Sir Charles Linnæus. I have put his letter aboard the man of war, inclosed in one to you, but as the bundle was bulky, I have begged my good acquaintance Mr. Smith to take charge of it to you. I have wrote to you fuller by the man of war, and remain, dear Sir, your obliged and humble servant,

ALEXANDER GARDEN.

DEAR SIR, April 12th, 1760.

This is my fourth or fifth letter to you since Christmas last, but every fresh opportunity of writing to you gives me new joy and pleasure, and makes me embrace it with true gladness of heart.

Inclosed, I send you a copy of my answer to Linnæus's letter to me. I sent you one copy by the Trent, Captain Lindsay, but for fear of miscarriage,

I have thought of sending you another copy, lest any accident should interrupt a correspondence which gives me so much pleasure, and which you have generously promised to countenance, favour, and establish. By Mr. Thomas Smith I sent you a packet of fishes for Linnæus, whose characters are in my letter to him. This is a new affair to me, for I never turned my thoughts to our fishes before he wrote to me, and I had but a short time to make a collection. But I intend him a valuable one next year, and as I trust he will correct me in characterising them, I am in hopes to improve yearly in my knowledge of that branch of Natural History. I must beg you will enclose and direct his letter, and I must entreat that if he favours me with an answer, you will forward it to me.

Ever since the beginning of February last, the small pox has raged here with great fury and with great mortality, especially in the natural way. The inoculated patients, of whom there has been a very great number, did not come so well off as might have been expected, at first, but of late I have discovered a method of treating them, partly owing to some hints given me by my learned and ingenious friend Doctor Adam Thompson, of New York, and partly by some bold trials on a negro of my own, which scarce ever fail of breaking the force of the disorder and its malignity, so far as uniformly to produce a mild pock, either by inoculation, or by the natural infection; but of this more afterwards. I am now much engaged in making further trials, so as to put this matter beyond all doubt; viz. that there is a real specific corrector of this disorder, by which it may be rendered so mild as that one life in a thousand shall not be endangered. This I hope to be able to make fully appear from both reason and practice. You may readily judge that I have good grounds to affirm this, or otherwise I should be far from writing so positively to you on a matter of this consequence.

Some days ago Colonel Montgomery, with 600 Highlanders and as many Royals, arrived here from New York, in order to go against the Cherokees. The latter followed Governor Lyttleton at the heels, with fire, sword, and devastation, and fully convinced him, and this province, that he laid a design to conquer them very ill, and executed it without judgment or discernment. Never was there a man more outwitted; never was there a province more abused. We have lost our money, our friends, and our character. All a sacrifice to ambition and undiscerning pride.

Colonel Montgomery marches from hence in a few days, and we expect, according to custom, great matters from a new commander; but whether it will be in his power to do much in carrying on a war in an unknown country, the very common geography of which is not known, much less the important passes through the mountains, &c. I leave you to judge. Governor Ellis is bestirring himself, with much judgment and spirit, to keep the Creeks neuter, if he should not be able to engage them to act for us offensively against the Cherokees.

The first will be a great service rendered to his country, the crown, and particularly to the southern provinces; but if he should be able to gain the last, and push them on to declare against the Cherokees, in that case the southern provinces will owe more to him than America owes to any man, besides Mr. Pitt and General Wolfe.

At present our situation is very critical; we are left by our Governor at a time when a powerful nation of savages is declaring war against us, which is solely brought on by his forward rashness, and which the province, and the judgment of every cool disinterested person, charges to his account; at a time when a sense of these things should have made him exert himself to the utmost for its safety and security, and made him think of every means to extricate us from the distress that his own ——— had plunged us into. This happens unluckily to be the period likewise, when the joint and united force of his Majesty's arms to the northward is required to reduce what remains of Canada, and of consequence the General there can spare little assistance to us; indeed, it is matter of wonder to me, how he could spare so considerable a detachment as what he has already sent, notwithstanding the importance of the service might evidently require more.

May 20th, 1760.

Some days after writing what goes before, I received your letter of the 10th of January, 1760. Good God, my dear friend, what did and do I feel on the subject of the first paragraph; yet scarcely can I say whether your misfortunes give me more concern, than your noble philosophic triumph over them gives me pleasure and joy. On the first I sincerely condole with you, on the last I heartily rejoice with you, and hope that before this time your affairs may again be in a proper situation, and yourself delivered from that embarrassment which must for some short space attend even the most philosophic mind upon such an event. Pray make me easy as soon as possible on this head. My earnest wishes, and heartiest and sincerest desires, for your prosperity shall always attend you, and till I again hear from you, I shall gladly hope to hear of your triumphing over the villany of mankind, and of your being easy in a handsome competency, for the cultivating of your keen philosophic spirit.

As there is a vessel to sail in a few days, I shall not at present answer any other part of your letter till then, but only take the liberty of mentioning a thing to you, in which I must beg and entreat your assistance. I ought, indeed, to make a double apology to you, both on account of the subject, and for the time at which it is proposed when you may no doubt be otherwise engaged. But for both, I hope your own good nature will readily find an excuse, and to that I trust.

The matter is this: a few days ago, the Honourable John Cleland, Esq. Comptroller of the Customs, or Surveyor General of the Customs, for South Carolina, Georgia, and the Bahama's, as he was stiled, died here, and since that time my friend Mr. George Roupel, has applied for that place, to the Governor here, and to Mr. Randolph, Surveyor General of the Customs in America. Lest, however, he should not succeed, of which he has very little or no reason to doubt, he has in the mean time written to some of his friends at home, for their assistance and interest, with the Honourable the Commissioners of the Customs.

It has several times occurred to me, that your mentioning the thing to Lord Halifax, with whom you have the honour of being acquainted, would be of the greatest service to him. Could you think well of this, it would confer on me a more sensible obligation than if done for myself; and I am certain that if you knew the man, you could not resist a strong impulse that you would naturally have to serve him. Honour, probity, uprightness, veracity, and noble generosity, are the distinguishing virtues that point him out a real gentleman, a worthy member of society, and a fit person for serving his King and country in such a trust.

He has served his Majesty as an Officer of the Customs, in this port, for I believe ten if not eleven years, but in a station far below his merit and capacity. In the office of Searcher he has borne an upright and umblemished character for that space of time.

This, then, my dear friend, is my request, that you would think so far well of assisting this worthy honest fellow, as to procure, by means of your interest with Lord Halifax, his Lordship's favour for him in this matter, for getting him appointed from home to the office of Comptroller or Surveyor of the Customs here. I need not tell you how quick and speedy an application of this kind should be, as there will, no doubt, be others, who will apply.

This matter is of such concern to me, that I must beg your answer by some of the first vessels.

Believe me to be, with the greatest sincerity, yours, &c. ALEXANDER GARDEN.

May 26, 1760.

Last night we had the terrible and dismal accounts of the Creek Indians espousing the quarrel of the Cherokees. These will no doubt pour down, like a torrent, on Georgia and our Southern settlements. This blow has been warded off for four months by the good and spirited management of Governor Ellis, but the Cherokee and French interest have prevailed. The detaining the Cherokee chiefs by Mr. Lyttleton, under the pretence of hostages, and afterwards butchering these people in cold blood, has laid the foundation of the bloodiest war that has been kindled as yet in America. The great unhappiness was, that Mr. Lyttleton pledged his word and honour that those should be delivered up to their own people, and set at full liberty as soon as he reached Kewohee. But when he came there, and found that he could not bring them to deliver up any voluntarily of these, he then detained twenty-four of these very men, to whom he had pledged his faith for their safety. These he called hostages!—they were, to a man, put to death in cold blood—*Hinc illæ lachrymæ.* Nothing but the speediest assistance, both to act against them by land and sea, to Mobile and New Orleans, can save Georgia and the two Carolinas from utter destruction.

MY DEAR FRIEND, Charlestown, June 1, 1760.

I wrote you, a few days ago, a few lines in the end of a letter, which I had begun to you some time before, but which the want of any vessels going for London had made me lay aside till then.

Never did a letter give any one more surprize, concern, or grief, than yours gave me. I need say nothing more than to tell you that I have felt very sincerely on the subject of the first lines. The noble, magnanimous spirit which you show, in regarding these misfortunes in their proper light, gives me the greatest joy, and the highest sense of your value and merit. It is a great addition to this joy, to find that a short time will again bring matters to a proper bearing, and restore you to the philosophic train. May God prosper your designs, and give success to whatever plan you have of making yourself and friends happy!

Ever since the beginning of February, I have been so engaged with the small-pox, that I have not had a moment's time to myself, from a constant hurry and confusion of business. That matter is nearly concluding, and one or two weeks more will put an end to it. Then I shall be able to give you some account of these things you desire; particularly you shall have some of the Cochineal insects on the *Opuntia*, packed in the manner you order, and all the observations on it which I have ever made; which are, indeed, but very few.

I am sorry that I was deprived of the pleasure of hearing from Mr. Baker, by the ill-founded report of my death. Pray offer my respectful compliments to Dr. Hales and him. You three are my only acquaintances and friends in your city; and in you I rejoice and am happy, without the least desire of extending the number.

I hope your public-spirited labour about the Provincial gardens will not be lost. It will be of the greatest utility, and if it be warmly recommended to the various Assemblies, they will no doubt contribute to the expense. I am very glad to hear that Mr. Pownal interested himself in it, and I hope he will be able to convince his brother, our Governor, of its high utility to this province, as it no doubt will be to every province; nay, the variety of climates and different latitudes, absolutely require the thing to be general. As I hope in a few years to be able to disengage myself more from the hurry of business, so I then hope to be able to give more attendance and assistance to that plan; and it may be greatly in Governor Pownal's power to promote my schemes.

I am greatly obliged to, and highly honoured by the Royal Society, for whatever favourable sentiment they may vouchsafe to entertain of me, but for whatever account of the *Toxicodendrons* I can make out, I am, in a manner, engaged to lay it before the Edinburgh Society, which I shall do in and through you.

My friend has been much pleased with what you say about his Essay, and begged heartily, when I saw him last, to thank you for the trouble you had taken in reading it. He continues still in a fixed resolution to conceal his name and place of abode, but declares his willingness to hear any objections against the doctrine. In answer to your query, he says, that if he had used the word *spiritus* in the Essay, it would have been as a synonimous term for *anima;* because, he observes, that we translate the Greek word νȣ̃ς, *animus,* which answers to our rational soul, and the Greek word πνευμα, *spiritus* or *anima,* which is common to us and the brute animals. He imagines that πνευμα is the breath, life, *spiritus, anima,* or animal life; but that νȣ̃ς or *animus* belongs to a higher class. This was his answer, as nearly as I can remember.

He hopes you will favour him with a further account of its fate, and if you approve of its being published, he proposes Mr. Wilson's doing it, if you think proper, and if he agree to the terms of doing it wholly on his own account and charges, only giving 25 copies to the author, ten of which are to be neatly bound and gilt. One of these, as I wrote you before, he intends for you, and one for Linnæus, which I begged of him. Whatever line you write to him, is to be directed to L. R. and inclosed to me.

I received your seeds, preserved in wax and suet, and have planted part of them, of the success of which you may expect a very particular detail and account.

Two days ago we had certain intelligence brought us that the Creeks were our enemies, and had begun by butchering the whole Creek traders that were in the nation. This nation consists of 2500 fighting men. The Choctaws, who have always been in the French interest, consist of 7000 fighting men; the Cherokees of about 3700 or 4000 men: in all, about 13,000 external enemies. About 70,000 negroes in our bowels! Our strength consists of 1200 men with Col. Montgomery, gone against the Cherokees; and we muster about 8 or 9000 men in the province. This is our happy situation! The ports of Pensacola, Mobille, and the Mississippi are open to a clandestine trade from New York, Rhode Island, and Carolina. From these places the French are supplied with blankets, powder, bullets, paint, &c. the most destructive traffic that ever was invented. Not a man of war to look into these harbours! The French pouring in their regulars on our backs from Fort Detroit, &c. They cross Lake Erie, come up Sandusky, down the Great Siolo, cross Ohio, and up Great Sandy Creek to the upper Cherokees. From thence they cross over to the heads of Albama River, or to the heads of Flint River; and may, and no doubt will, cover our backs before we be awakened, and may be and are ready to pour destruction on us before we open our eyes. We think of nothing, know nothing, trouble ourselves about nothing, and shall sit still till the Provinces of Georgia, South Carolina, and Virginia be overrun. You have forgot us at home—you think not of our situation—you think not of the incredible number of Indians that are behind us—you think not of the French assiduity in gaining them—you set your minds on Canada—and you forget that by gaining the Southern Indians they will gain more friends and we more enemies than would equal thrice the number of all the Northern Indians, French and English. We have but a handful of white men and 70,000 negroes. None of the Northern Provinces have negroes, but are almost all Whites. We know not whether our Indians or negroes be our greatest enemies. Think of these things, and believe me to be, yours, &c. ALEX. GARDEN.

Mr. Ellis wrote to Dr. Garden June 13, 1760, sending him Ehret's print of Warner's Jasmine, by the name of *Gardenia,* given by Mr. Ellis, who had written to the same effect to Linnæus. Mr. Ellis also informed Dr. Garden that the Premium Society recommend Provincial gardens; that he would recommend Dr. Garden to Governor Pownall; and would order Wilson to return his MSS.

Sent him a specimen of Dossie's Silk-grass; requesting his opinion. Also an account of the success of the seeds inclosed in different substances. Tallow grows rancid with heat, and spoils the seed. Myrtle-wax answers best. Loblolly Bay grows from seed. Mr. Ellis desires it may be called *Gordonia*, after Mr. James Gordon of Mile-end. Begged for some pods of *Yucca* with filamentous leaves; also resin of Red Cedar; and an account of the *Rhus* and *Toxicondendron*, whether any varnish really comes from it, and whether as good as Japan. An account of the Cinnamon in Guadeloupe, from Woulfe. Du Hamel's two books on Vegetation.

DR. GARDEN TO MR. ELLIS.

MY DEAR FRIEND, July 16, 1760.

It is no small comfort to a man in distress to have one to complain to, or to communicate our sufferings; and though these things that I am now to mention to you do not particularly affect me as an individual, neither does it fall to my hand particularly to take notice of them; yet what I feel for my fellow-subjects, and for the distressed situation of our province, and the confidence which I have in your friendship, make me gladly take an opportunity to lay open our distress to you, and hope for your pity at least.

It might be thought an invidious task to point out the source of this present war, which now makes our unhappy province a scene of blood, because it is generally and well known here to have all been owing to the blind ambition and misguided passion of one man, which produced a series of blunders in the management of Indian affairs, till at last our present war is the fatal consequence, at a period of time when that person could be no longer a witness of the bloodshed and murders which his, I am afraid, wilful mistakes have occasioned and produced.

But be these matters as they may. It is now an undoubted fact that a most bloody war, not only with the Cherokees (among whom we are certain there has not been a Frenchman nor French Indian for these 12 months past — this is a great proof that the war is owing to our own mismanagement), but with the Creeks; and, it is more than probable, already with the Choctaws. These three nations make between 8 and 10,000 fighting men. We have about 6 or at most 7000 musterable whites, of from 16 to 60 years; and we have about 70,000 negroes, not at all in a good temper. Col. Montgomery, who came with 1200 men to our relief, is again ordered by Gen. Amherst to the northward, and is now actually on his return from the Cherokee country to embark. Judge now, my friend, what we shall be to these savages; and judge with what heart 2000 or even 3000 of our Militia can march against them, when they leave their wives and children amidst 70,000 intestine enemies, guarded only by 3000 remaining Militia! But what can we do when we go? We never can propose to conquer them while they are supported by the French; nay indeed we shall be happy if we be able to prevent them from overrunning the province. It is not to be supposed that the French will lie by spectators. No, the game will be too good for them. Their numbers are now daily increasing in Louisiana, from the vast numbers of emigrants from Canada and about the lakes; and these Southern Indians are at least five times the number of all the French Northern Indians put together, which have distressed these numerous provinces and powerful armies for these five years past.

What a happy turn of affairs for the French! especially while the mouths of the Mobille and Mississippi are open to them. And this we are too certain of from the vast number of New York, Rhode Island, and Philadelphia vessels that have been seen going there, and to Pensacola, a small Spanish port in the Gulf of Mexico, just by the mouth of Mobille River. Not the least stop has there been to this trade till they have thrown in all the Indian trading goods that they could get; and now that the French magazines are full, their time comes for assisting and pushing on the Indians.

Col. Montgomery has given the Cherokees some smart drubbings, killed many of their men, taken many prisoners, and burnt all the towns in the lower part of the nation, but was not able to penetrate through it to the relief of Fort Loudon; so that unhappy garrison must fall into the hands of the merciless savages. Most of the difficulties arose from the Colonel's ignorance of the various passes, &c. through the nation; and indeed our gross ignorance of the geography, of the Creek and Cherokee countries, will be the fruitful source of many rebuffs to any army that attempts to reduce them. This I well foresaw several years since, and strongly represented it in my last letter, about two years ago, to Mr. Whitworth, and often used to mention it to Mr. Lyttleton, but not a word has ever been said of it; so that now we have these things to study and to learn by dear blood, which we ought to have known exactly long ago, or at least before we attempted any thing of this kind. It would have been, and still will be, much cheaper for the Government to get capable men to make proper observations upon the rivers, their courses, the passes, mountains, &c. of those countries that are as yet at peace with us, than to have these things to be done by the armies that may soon march for their reduction (for it is not now to be doubted but that all the Southern Nations will engage against us), whose knowledge will always be purchased by the lives of many poor wretches, who must fall in every defile, or difficult pass, through which they must go.

The American war is so far from drawing to a period, that it will just spring up afresh, and much more in favour of the French, than ever; for the number of these Southern Indians is so great that they can want nothing but French officers to ruin Georgia, both the Carolinas, and Virginia, and that in one year.

If the Ministry neglect Mobile and Mississippi much longer, it will for ever be out of their power to apply a remedy. We shall be lost, and no possibility of recovering the provinces remain.

On the 21st of August, 1759, I caught a male Cochineal Fly, and put it into the aquatic microscope.

It is very seldom that a male is met with. I imagine there may be 150 or 200 females for one male.

The male is a very active creature, and well made, but slender in comparison of the females, who are much larger, more shapeless, and seemingly lazy, torpid, and inactive. They appear generally so overgrown, that their eyes and mouth are quite sunk in their *rugæ*; nay their *antennæ* and legs are almost covered over, or so impeded in their motions from the swelling about their insertions, that they scarce can move them, much less move themselves.

The male's head is very distinct from the neck. The neck is much smaller than the head, and much more so than the body. The body is elliptical, and something longer than the head and neck together, and flattish underneath.

From the front there arise two long *antennæ* (much longer than the *antennæ* of the females), which the insect moves every way very briskly. These *antennæ* are all jointed, and from each joint there come out four short *setæ*, placed two on each side.

It has three jointed legs on each side, and moves very briskly and with great speed. From the upper part of the tail there arise two long antennæ-like productions, which are projected out from behind the body to four or five times the length of the insect. They diverge as they lengthen. They are jointed, very small, slender, and of a pure snowy white colour.

It has two wings, which take their rise from the back part of the shoulders, and lie down horizontal, like the wings of a common fly, when the insect is walking. They are small at their first origin, and seem round and strong; then they spread out wider, so that they lie over one another when the insect lays them along its back. They are much longer than the body, and have several strong nerves, particularly one that runs from the neck of the wing, along the external margin, and inarches with a slenderer one that runs along the under and inner edge. They are quite thin, slender, and transparent, and of a snowy whiteness. When put in spirits, the wings stood erect.

The body of the male is not of so deep a colour as the body of the female, nor near so large.

How can Linnæus class this among the *Coleoptera?* Vid. Syst. Nat. edit. 6ta.

How can Browne class it among the *Hemiptera?*

Does it not belong to the *Neuroptera* or *Diptera?*

The following is an account of the seeds which I received by Mr. Robinson: —

1st. The Pistachia Nuts in wax were just spoiling — they would have come safe on a short passage and earlier in the year. There are not any of them come up.

2dly. Cork Acorns in tallow, just spoiling None of them yet come up.

3dly. Ditto—ditto—first put into water and then into wax, quite spoiled and rotten.

4thly. Ditto—ditto—put into wax, quite rotten and spoiled.

From this, tallow seems to be the best; but they did not come early enough; for, from the time they were put into the ground till about ten days ago, we had continued drought.

Yours, with great sincerity, ALEX. GARDEN.

Mr. Ellis wrote to Dr. Garden Nov. 12, 1760, telling him that all his letters, to the 16th of July, were received, and lamenting the unhappy situation of the province. No interfering with Mr. Pitt—merchants have applied, but had no satisfactory answer. Must wait till the business is finished in Canada, and then possibly they may be attended to.

Mr. Ellis could not ask a favour of Lord Halifax. When the Governor comes over he may venture to address him.

The Premium Society have received Dr. Garden's letter, wherein he acknowledges the receipt of the cuttings. They were drawn up, but the Society approves of his method of treating them. Linnæus will answer his letter soon. Warner's Jasmine is to be given to the Royal Society by the name of *Gardenia*. Linnæus found out the true characters from a single blossom, met with accidentally amongst his specimens. Gordon the gardener made £.500 by it, first and last, being so elegant a plant.

Solander promises seeds of the true Rhubarb, that in the gardens being false.

Acknowledged the receipt of the description of the male Cochineal Fly, which agrees with Mr. Ellis's specimens — of seeds from China — Tea-seeds sown in tubs. Mr. Ellis ádvises that the difficult-growing seeds should be sown in a tub or box of earth; and small plants of *Magnolia* and Loblolly Bay planted. Enquired after other plants of Catesby; also about the *Opuntia*, for Linnæus. The recent *Asterias* from Dr. Bruce.

DR. GARDEN TO MR. ELLIS.

MY DEAR FRIEND, About January 1761.

I have not written to you since July last, which is indeed a great blank to me. But I will repair it soon, when I shall write at great leisure and length.

Since that time, I have received your agreeable favours of Nov. 12 and 20th. In answer to which, take the few following notes at present. Your compliment of the *Gardenia* was most acceptable to me, and you need not doubt I shall gratefully remember it. Has Linnæus adopted it? I am surprized that I have received no letter from him. By this opportunity I have sent him a very large, and I think well preserved parcel of fish. I hope they will be agreeable to him. O my good friend, how many blunders and gross misrepresentations have I seen in Catesby! — gross beyond conception!

I have made out characters of most of the fish I have sent, merely that Linnæus might correct me; but I am grieved that I cannot hear oftener from him. Pray be a mediator, and rectify all this. You will see the characters inclosed with your letter and my letter to him. I should be glad that you saw the fish, but I am afraid the unpacking them will be of no service to them. I have put colocynth and aloes about them, to keep off the insects, who were my great antagonists while I made my collection. I spared no pains to procure these for him; and this in a horrid country, where there is not a living soul who knows the least *iota* of Natural History. I confess I often envy you the sweet hours of converse on this subject with your friends in and about London. How must you enjoy Solander! O my God!——

I have likewise sent a box of snakes of many different kinds; I believe many more than ever went from America before at one time, and a much greater variety: they are all carefully preserved. The fish are in the long box which has the brown paper about it, and the snakes in the square box. I must beg you will take particular care of these, as I hope, if they arrive safe, they will convince our great Northern father and leader, that I have taken as much pains as I have had pleasure in collecting for him.

I have ordered Mr. Ogilvy, under whose care they come, to pay whatever charge there may be in unshipping and re-shipping them, which I beg you will permit him to do, only charging yourself with the direction and care of the proper method of conveyance. I beg they may be accompanied with a letter from yourself, making an apology for the incorrectness of the language of the characters, for the transcribing of which I was obliged to use an amanuensis, who has committed many blunders. Many of them I corrected; but as he never brought me the copy till this day, I have only had time to give them a hasty reading over, and my letter to Linnæus was finished.

I have got the first and second volumes of the late edition of his *Systema Naturæ.* Admirable books indeed!

Mr. Ogilvy is a gentleman whom I must take the liberty to recommend to you as my particular friend; and whose conversation must be agreeable to you. Whatever civilities you shew him will indeed be particularly done to me. He will take charge of any letter or package for me. As his stay in London may be for some time, he will be very useful in forwarding any thing, and to him I shall for the future direct your letters, or any little things, if you do not stay in London, and if it may be agreeable to you.

As this letter is bulky, being inclosed with my characters of the fish to Linnæus, and my own letter to Linnæus, I have on that account given it in charge to my acquaintance and friend Mr. John Gregg, of this town, the young gentleman of whom you wrote to me some time ago. He will deliver it to you himself, as he will, soon after his arrival, do himself the pleasure of waiting on you. As he told me that he would endeavour to procure you as many of our shrubs and plants as he could, I have endeavoured to tell him which of them he could procure; and he has been diligent in making up several boxes, which I hope all arrived safe to your hand. When he returns I will assist him as much as I can, that he may supply you further.

In my last letter to Linnæus, I sent him the characters of a shrub, which I took to be new, and which I imagined he was to call the *Gardenia*, if he found it so, but your politeness has prevented him. He desired me to do this himself. I now send you Mr. Roupel's draught of it, which I think is well done, and exactly copied from the life. If you think it worth while, pray have a print of it, or a copy, and send this to Linnæus.

If you have any real botanical friend to oblige, put Linnæus in mind of him at my desire.

Let the Loblolly be, as you desire, a *Gordonia;* if it be agreeable to you, it makes me happy.

I have never yet been able to finish my account of the *Toxicodendron*, partly for want of time, and partly from my not being able to procure specimens of the different species for Mr. Roupel to draw them. The total want of any assistant, or of any person in this place who has more real and proper *gout* for that study than a common horse, prevents me from getting many things which I otherwise might.

I hope the box of seeds came safe to you by Strachan; if it did, I am satisfied and happy, in the hopes of its proving agreeable to you.

There are three boxes directed for you now; Nos 1 and 2 are for Linnæus from me; No 3 is for yourself; it was sent to my care from Mr. William Clifton, from Georgia. It came too late to go to you by our last ships, so that I have been obliged to keep it till now, much contrary to my desire, as I am afraid there may be things in it which may be spoiled. But it was unavoidable.

I must beg that you would send me two or three more plates of the *Gardenia*, after it has received Dr. Linnæus's sanction; or any other of Ehret's curious productions.

I showed to Mr. Bartram all my new plants, and gave him specimens; at the same time I told him that I intended to publish them all by themselves; those he mentioned to you were some of them. Several of them I had brought into my own little garden. An account of all which you will have soon.

I am sorry that I cannot send you the *Toxicodendrons;* you may depend on a copy as soon as they are done: but I must likewise send a copy to Edinburgh, as the gentlemen there have done me the honour of electing me one of their members; so that I stand indebted to them for what further I can procure for them.

I am glad you did not shew my letters to any one. I write to you as my dear friend. But though these things were mostly consistent with my knowledge, yet I would by no means meddle in them further than with a friend. It lies out of my way, and I am no politician.

Our wretched Assembly will not think of Provincial gardens. Mr. Pownal must be instructed properly to manage that point with them. They are now busied in finding out excuses to the General, Mr. Pitt, and the Ministry, for their bad behaviour last year. There are a few pitiful snarlers here that are daily quarrelling yet with General Amherst, Colonel Montgomery, and every officer that has nobly exposed his life to preserve their wretched beings.

Yours, ALEX. GARDEN.

My respectful compliments wait on your new friend and son of Apollo, Solander.

Pray let me know what price seeds or plants of our shrubs would bear with you, as I have a man in view to apply himself wholly that way, if there was a prospect of its succeeding.

You can easily put the proper directions upon Nos 1 and 2, the boxes for Linnæus. I should be glad to know the manner in which you direct to him.

Pray send me one or two young plants of the *Gardenia.*

MR. ELLIS TO DR. GARDEN.

MY DEAR FRIEND, London, April 8, 1761.

I received your very kind letter of the 20th of November by Capt. Strachan, and am as much pleased with the very fine collection of seeds you sent me, so curiously preserved, as you can be with the *Gardenia.* Linnæus has actually adopted it among his new genera, which will be published in his *Addenda;* and the Royal Society, which still makes it more public here, has ordered my account of it to be printed. I gave in Linnæus's characters, with those of the *Halesia;* and it will surprize people when they know that a nurseryman, James Gordon, in less than three years, has made £.500 from four cuttings of a plant. Every body is in love with it, and you may depend on having a plant of it from me; for as it is a double flower, besides being the native of a warm climate, it produces no seeds here.

I have bought two more coloured prints of it, which shall be sent by the ships in June; for I fear these stragglers may be taken by the enemy. It has given great jealousy to our botanists here, that I have preferred you to them; but I laugh at them, and know I am right; for, without flattery, you have done more service, and I have obliged more people through your means, than they have in their power to do. Continue to look out for new genera; and send some dry specimens of *Stuartia,* Loblolly Bay, and your rarer plants.

Tell Mr. Gregg I am infinitely obliged to him for his politeness, and shall write to him fully soon, before he sets out. — Pray let me into the history of the *Rhus* and *Toxicodendron.* You know how much I shall be able to triumph over Miller's blunders by that means.

I never got the bottle with the male Cochineal Fly. I have one drawn that I accidentally discovered, together with the female, but wish to know more of their œconomy, to make the account perfect, as it is an interesting subject with mankind in general. I received a curious parcel of shrubs, through Mr. Gregg, from Governor Ellis, who is still in New York. I long to see him as much as I fear the people of Georgia do; for I find the Wolf king does not behave so friendly to them now he is gone.

I have told Solander, who continues here a year longer, that till Linnæus writes, you will send no more specimens; which I approve of, for these affairs should be reciprocal.

I must now tell you of a great misfortune you have had. The New York packet, with Tea-seeds in wax, sent by me to you and all the Governors on the continent of North America, was taken by the French Dec. 27th last. They were all packed in sundry letters, by Mr. Hampden our Post-master-general. I am in hopes some of what I have sown will come up. I have an Oak acorn now growing, that was preserved a year and a quarter in bees'-wax.

DR. GARDEN TO MR. ELLIS.

My dear Friend, April 26, 1761.

This will be delivered to you by my worthy friend Mr. Charles Ogilvie, who I must beg leave to introduce to you, not doubting but his good sense will soon recommend him to your civilities and esteem. As he is to be some time in London, he will take particular care of all my little commissions, and whatever letters or papers you may have for me you may give to him. From him you will receive two boxes for Linnæus, one containing his fishes, &c. the other a very large collection of snakes, lizards, alligators, &c. These I hope will come safe to hand; and I doubt not but they will be immediately forwarded to that most worthy Professor. I have written him a long letter, and sent him a very full copy of the characters of our fish. The collection, I think, is pretty considerable; at least I am sure I spared neither time nor pains in procuring them. I should be glad that you had an opportunity of seeing them; but I am afraid that packing and unpacking them will hurt them. My surprize is great that I have never heard from him, as it is now so long since you mentioned that he was then about to write. Pray beg of him that our correspondence may be more frequent; for though I am harassed both day and night with business, yet I would always steal away some hours for improvement. My letters to him, and my characters of the fish, with my letter to you, inclosing a very fine draught of a new plant, done by Mr. Roupel, all come to your hand by Mr. Gregg, who is a worthy, good young lad, and who seems to be happy in the thought of being with you for some time. As he has been employed in collecting for you, I have given what assistance my confined situation would admit of; that was, to tell him the names of all our plants that he brought to me: and he has indeed put up, with much industry, several good boxes for you, which I hope came safe to hand, and were agreeable. When he returns I shall assist him still more, in showing him our new genera, that he may collect for you.

If you could make out a print of this new plant I should be glad, and then send the draught to Linnæus. It is the draught of the new genus which I sent him last year, and which he told me he would call *Gardenia*, though your kind and obliging politeness to me has prevented him. You may at my desire put him in mind of any of your own true botanical friends. — I requested you before to let the Loblolly Bay be a *Gordonia*, as you intended. — Pray prevail on Mr. Gordon to let me have a plant or two of the *Gardenia*, carefully put up. He may depend on what return I can make him. I should be glad to have two or three more prints of it, after it has received Linnæus's finishing sanction.

The other day I received a letter from Dr. Whytt, informing me of my being elected a member of the Edinburgh Society; so that I shall be obliged to prepare some paper of thanks, to convince them of the sense I have of such a mark of respect and honour. I am sorry to hear my old Professor, Dr. Alston, is dead there. The old gentleman was a keen opponent of Linnæus, and yet, for all that, was a sensible botanist in general, though particularly a Rajian or Tournefortian. What has Hill wrought himself up to? I am told that it is something considerable.

Give me all your literary news.

Remember me in a particular manner to Dr. Solander. How happy should I be in having an hour or two's *tête-a-tête* with you both! If seas and mountains can keep us asunder here, yet surely the Father of Wisdom and Science will take away that veil and these obstacles, when this curtain of mortality drops; and probably I may find myself on the skirts of a meadow, where Linnæus is explaining the wonders of a new world, to legions of white candid spirits, glorifying their Maker for the amazing enlargement of their mental faculties. What think you of this time, my dear friend? Shall we have a hearty shake of the hand, if such practices be fashionable, or in the mode? Believe me, I long to see more of my God, and to know many of my friends that I am afraid I cannot meet elsewhere.

To return. You will find a box for yourself, marked N° 3, from Mr. Clifton of Georgia. It has, through unavoidable accidents, lain by me some weeks; but it came too late to hand to be sent to you sooner than by this fleet.

I am, with great truth, yours, &c.

ALEXANDER GARDEN.

MY DEAR FRIEND, July 25, 1761.

My last letter to you was of the 26th April, by Mr. Ogilvy. In this I gave you an account of all my former letters, boxes, papers, &c. for yourself and Dr. Linnæus, from whom I long much to hear; and I am surprized to find his silence is so great and long. I then begged of you some prints, and one or two plants, of the *Gardenia*. This I shall esteem a great favour, of which I have received many from you. One letter from you, after so long a silence, will be a great refreshment to my languid and exhausted spirits.

I have at last met with a man who is to commence nurseryman and gardener, and to collect seeds, plants, &c. for the London market. He is a sensible, careful man, and has a turn for that business. He shall receive all the advice and assistance that I can give him. I must beg your interest in his favour; that you would bespeak what custom and commissions you can procure for him from your gardeners or nurserymen, or from any gentlemen who may want what our province affords. He wants much to be acquainted with Mr. Gray and Mr. Gordon at Mile-end; and I must beg that you would procure some commission from them to him. He is to employ his whole time in procuring whatever may be ordered. His name is Young; and any letters for him inclosed to me, will be taken care of. I must beg that you would endeavour to inform me, on his account, what the prices of our several seeds are, or the value of young plants of Loblolly Bay, *Azalea*, *Umbrella-Magnolia*, *Beureria*, *Magnoliapalustris* *, *Halesia*, *Stuartia*, and such like. You will oblige me much to inform me of all this by first opportunity.

* *M. glauca Linn.*

I have been much importuned to beg you to inform me of the following questions; *viz.*

What ingredients, added to lime made of oyster-shells, will make a terrace-mortar, that will stand the sun and rain in Carolina, for flat roofs to houses?—Flat roofs would be of the greatest utility to us in Carolina.—What are the just proportions of each of the ingredients?

What is it that the Spaniards, Portuguese, and Italians terrace their roofs with?—and how make the floor to bear that terrace?

I must beg you will be good enough to make some enquiry among your friends conversant in these things. The gentleman who proposes the questions is my good friend, whom I should wish to oblige by an answer to them.

Believe me yours, ALEX. GARDEN.

MY DEAR FRIEND, Feb. 26, 1762.

I have just now read your three last letters, which I have often done since I received them, and every time with equal pleasure. O my God! if it were my good fortune to pass one summer with you in your retreat at Mr. Webb's, or any where else, how happy should I be! But my lot, for some years at least, will be this barren neck of sand on which we live. Our last summer was remarkably wet in the beginning, and as remarkably hot in the end of it. A violent epidemic putrid bilious fever distressed the town much, and confined me within its walls the whole summer, so that I had not one trip into the country. My close confinement to business has almost made me forget my practical botany; and nothing remains but an inward burning desire and love of that delightful study, which my necessary attention to business will not permit me to gratify, at least for a time.

We are now at peace with the Cherokees, and have been so for these seven months past; and we have great reason to think that it will continue. Col. Grant has effectually served us, and done very essential service to the British interest; and yet we have many of these surly, ill-tuned minds amongst us, whose wayward dispositions make them live in pretended doubts and uneasiness about our situation. How happy might we be if we knew in what our real happiness consisted!

I am much obliged to you for the prints of the *Gardenia*, with your letter on that and the *Halesia*. But I never got those copies, which you mention in your letter by Captains Curling and Cheesman, July 3d. I went on board several times, but the Captain had lost them. This was a great loss to me. I am wholly indebted to you for this high compliment; and I have a very just sense of your politeness on this head. My hearty and real thanks are likewise due to you for the two plants which Mr. Gordon sent me. One was quite alive, and I think will do admirably well. I must beg that you will offer my thanks for the same, and he may depend that I will not forget what he wrote to me about. If you can possibly spare any of the seeds which I have sent by Capt. Mitchell, I beg you will let him have some of them. The quantity is but small, though they are all that I possibly could procure, as I have been much confined this fall, having myself had a severe touch of our putrid bilious fever, but I happily got the better of it.

The seeds are packed in layers, in the following order, beginning at the bottom of the box:

1st. *Ptelea*, which I imagine is yet a scarce and rare plant with you.

2d. *Chionanthus*, or Fringe-tree.

3d. Dahoon Holly, and the *Magnolia palustris* *, or our Swamp sweet-flowering Bay.

4th. *Magnolia altissima*, Large Laurel.

5th, is the new genus of which I sent you the characters and draught for Linnæus last year †. It is a very pretty flowering shrub, approaching to a tree. Pray let Mr. Gordon have some of this seed.

6th, is the *Corallodendron* seed, very fresh and good.

Some *Nelumbos* and some *Azaleas*, with several other seeds, &c.

7th, is the *Ptelea* again in a large quantity.——

N. B. I have sent you a very uncommon ear of Indian Corn or Maize.

* *M. glauca, Linn.*

† See p. 302 of this volume.

In this layer, among the *Ptelea* seeds, is a phial with the male Cochineal Fly in spirits, where it has been for two years. All these are in a box on board of Capt. Mitchell, of the Squerries.

I have likewise sent you the Silk Grass Hammock, which I hope will give satisfaction. It will be at least a curiosity, and a proof to your infidels what America can produce.

I am extremely obliged to you for the trouble you take in conveying my packages and letters to Sir Charles Linnæus; but I must beg that you would permit my friend Mr. Charles Ogilvie to pay the expences and charges on them, which I have desired him to do. I am very happy in finding that they were agreeable to Linnæus; and I am sure the last package will be more so. I received his last letter, but daily expect another by your means. I have now by me a fresh collection, which I shall describe, pack up, and send as soon as ever I can get a few spare hours. There is nothing here that I can procure by money but what I would cheerfully get for him or you; but there are many things that would require me to look after them myself, and these are the things which I cannot yet procure, as my business of physick confines me to the walls of Charlestown; but that period will pass over.

I expected to have heard from Dr. Solander, by what you wrote me, but I suppose he might have been too late for the fleet. If I have half an hour to spare, I will inclose a line for him, which you will be kind enough to deliver.

I shall be happy to see the Loblolly Bay called *Gordonia*, in honour of our friend Mr. Gordon. It is certainly a new genus, as is likewise the plant which I sent you a draught of.

Please to remember me to my friend and acquaintance Mr. Gregg, when you see him. I have written to him by this fleet. I have now only to beg leave to offer my respects to Linnæus and to Dr. Solander, and I am, my dear friend, your most obliged and humble servant, ALEXANDER GARDEN.

N. B. The Hammock is in a package by itself, directed to you, on board of Captain Mitchell.

MY DEAR FRIEND, January 20, 1763.

Though I never had an inclination to write you a longer letter, yet I never had less time to do it. I am sensible that this is an old-fashioned apology. but, upon the word of a botanist, it is true at present, At the same time, I know I am indebted to you for two letters, both which are lying near me, but I cannot possibly look into them for fear of finding more matter for writing than at present I can undertake. With your last, I received Linnæus's long and most obliging letter, written after he received my last parcel. I have now a parcel for him, and if he knew one half of its contents, they would make his teeth water, but I will not write till I send them, and when I shall find time God knows.

Dr. Solander's favour lies yet unanswered. If I

can command a few minutes to-morrow night, I will at least convince him that I want not inclination to be one of his acquaintances. I have some few odd things for him. I had written to him before I received his letter. A choice letter it was. It contained, besides the most genteel and polite offer of that gentleman's correspondence, a full and curious account of London literary news.

Good God, you must have a droll set of large periwigged doctors in London! I scarce think that the present College will advance physic more than that plain, honest, but sagacious individual Sydenham did all alone. Enough of this, and may the Lord pardon this fit of blasphemy; but I thank God that I am not among them, though, to be sure, it is an honour that I shall never attain; but I would not forego the pleasure that I have in freedom, for all the gorgeous shackles of their jog-trot practice.

By this vessel you will receive a parcel of seeds, and I hope you will remember my friend John Gordon at Mile End. Especially let him have all No. 21, or the Dwarf oak. He wrote to me particularly for this plant when he sent me the *Gardenia*, whose sudden death I take to be no good omen for the continuance and duration of my botanical name and character; but if I do not outlive it, I shall be pleased, and if I do, I shall certainly make myself happy in some other acquisition, if it should only be like the former, imaginary! You shall soon have a plaguy long letter from Sir, yours, &c.

ALEXANDER GARDEN.

The small-pox has just broke out amongst us, and is a fresh plague to us, though milder than the last time.

November 19, 1764.

MY FIRST AND STILL DEAREST BOTANICAL FRIEND,

No less than four of your favours now lie open on the table before me, and appear like so many living witnesses rising in judgment against me for one of the foulest neglects I was ever guilty of—the want of punctuality in answering so dear a friend's favours. I will not venture to offer an apology, but expect forgiveness from your clemency. I have only to say, that the present opportunity which I embrace to answer them, is a time full of much mental pleasure, but of much bodily pain and distress. I have been confined for these five weeks to my house, with an attack of nephritic pains, attended with a troublesome nervous remittent fever. This period, when pain would permit, has been wholly appropriated to answer my correspondents and friends letters; and, indeed, it is the only time which I have to apply to that agreeable purpose, so that in reality it is a time of enjoyment to me, especially as I am at leisure to receive and enjoy visits from my friends here, whom at other times I am obliged to avoid, merely on account of constant avocations about business. In Charlestown we are a set of the busiest, most bustling, hurrying animals imaginable, and yet we really do not do much, but we must appear to be doing. And this kind of important hurry appears among all ranks, unless among the gentlemen planters, who are absolutely above every occupation but eating, drinking, lolling, smoking and sleeping, which five modes of action constitute the essence of their life and existence.

Now to your letters in order; in the first of which you give me a very entertaining account of the *Uraki*, and the use of its seeds. I wish to see Solander's paper on this subject, but I have never heard from this gentleman for these eighteen months, though I then wrote him a long letter, and sent him all the insects and little odd things which I had by me. I am really ashamed in neglecting your commission about the Cochineal Fly, and though I will not again make a promise, I believe you may depend on having it; but it cannot be got till next summer. I had a parcel of the plant with the animals on it, this last summer, and though they were in my balcony till they rotted, I never could command an hour to hunt for the male.

I intend to keep a note of what you desire from me, in my pocket book, that it may serve as a memento to me. It is often forgetfulness, in the hurry of business, that occasions my neglect.

I was amazed to learn, by your letter, that the Captain of the Boscawen should have denied his having a box of seeds for you. Amongst them there was a parcel of the Turkey-oak acorns, for Mr. Gordon at Mile End.

I come now to yours of October 28, 1763. In that you tell me that Solander saw one of the *Ligustrum's fructu violaceo* of Catesby, which was sent over by Mr. Bartram. I believe that I before wrote you that Bartram had two specimens of that tree from me, and it must have been one of these which he sent, for he was not here when the tree was in flower. It grows in my garden, and that was the only one which he ever saw. I have never yet seen the fruit of it, as it has not yet borne fruit since I brought it from the woods. It bears plenty of flowers every year, which soon fall off; and, indeed, it appears to be a *Ligustrum*, but I do not think that this tree is what Catesby meant by his *Ligustrum fructu violaceo.* You will see, by the inclosed, that I think he meant another plant, which I have sent you as the *Ellisia;* and I have added my reasons for thinking it a new genus, and his *Ligustrum.* Please to let me have your sentiments of this, and whether you choose to accept it for a name-sake, for I have a great desire, and shall have much pleasure, if you will allow me to have the honour of naming a plant after you. Only give me notice soon, whether this is or is not a new genus, which I take it to be, and whether you like it, and I shall soon furnish you with another, if this does not please. I am now entirely out of my little smattering of botany, having had no practice in that way for these four or five years past. I have almost forgot the meaning of the terms, though your most agreeable letters are so many spurs to me.

Agreeably to your desire, I have spoken to Mr. Young, and given him your directions and my best advice; so that I doubt not but his seeds and young plants will be good, and his prices much lower.

You desire me to recommend some person to you, that is curious. This is the hardest task that ever you gave me. I really know of none such nearer than Mr. De Braham to the southward, or Mr. John Bartram to the northward. If any other exist, I really know them not, nor have yet heard of them, but if I do, I shall certainly endeavour to set on foot a correspondence between them and you.

I have my best and sincerest congratulations to make you on your being appointed King's agent for West Florida. This must be just in your way, because, as you justly remark, it will enable you to glean every thing curious in the vegetable kingdom from these unexplored regions.

Can you let me have one of Solander's catalogues of the Musæum, and likewise Solander's draught and papers on the single *Gardenia?*

When you can learn the fate of the plant, of which I sent Mr. Roupel's draught and specimens, seeds and characters, to be forwarded to Linnæus, I shall be glad that you would communicate it to me. Shall I beg the favour of you to mention it in one of your letters to Linnæus? If you could do it, I should esteem it a favour, because I am anxious about it, as I took it to be a new genus. Linnæus never mentioned his having received it to me, nor the least word about it.

I now come to mention yours of June 9, 1764. In that you write me that you had received a letter from Linnæus, and that he enquired after me, and was afraid that I was ill, because I had not written to him. All this surprised me much, because you know that I not only wrote fully to him, but sent him every thing that I had been able to collect for him. These I sent by favour of Mr. Murray, directed to your care. When Mr. Murray arrived in London, you were not in town, and he wrote me that you had desired him to put them into Dr. Solander's hands, to whom I likewise wrote largely at the same time, and sent specimens of many curious insects, and every other thing that I could collect. Now, my dear friend, ever since that time, I have never heard from Dr. Solander in answer to mine to him, nor have I ever heard from Linnæus, informing me of his having received, or not having received, my packages and letters. Mr. Murray wrote me that he had absolutely delivered them to Solander. I have written to Mr. Murray to wait on you about this matter, because I am afraid that, by some mistake or misfortune, I have failed in getting them conveyed to Linnæus. There was a valuable collection of fishes, with a copy of my characters of them. These would have almost compleated my account of our Carolina fishes, and their miscarriage gives me more pain, grief, sorrow, uneasiness, &c. than I can describe to you. The specimens I can never make up; a copy of the characters I have sent to Linnæus again, inclosed to you, to be forwarded to him, to shew him that I was not to blame, and that I had done my best to be punctual in acknowledging the honour he did me in corresponding with me, or in sending me his commands to be executed by me. I have likewise inclosed to you a copy of my letter to him, which I wrote along with these things. I have not directed it, because I know not his new and noble address. Pray inform me of it, in your first letter to me, and give me all the information you can about that great botanical luminary. Put a direction on my letter to him, and let it be forwarded, because the characters of the fishes may be agreeable to him, even though I have had the misfortune to have the specimens miscarry; and it will convince him that I was not guilty of so gross a neglect as that of not answering the most polite and obliging letter of his, which you were so good as to forward to me. I likewise expected to have had some notification, or diploma, of my admission as a member of the Society of Upsal, under which title he is pleased to address me in his last letter.

Now I imagine that when one is elected a member of any society, he must have a regular notification of the same from the secretary, in the name of the society. As yet I have had no other notice of my being a member, but Linnæus addressing me in his letter by that appellation. Please to assist me in this, and set me right.

As I have been confined to my chamber ever since Mr. Lloyd arrived, I have not seen him. To-morrow I intend to call at his lodgings, and ask him to spend a day with me, when we shall have much chat, *tête-à-tête*.

In one of your former letters to me, you desired that I would name the Loblolly Bay *Gordonia*, after Mr. Gordon at Mile End. I answered you that I had no objection. However, if it is not done, you may, if you think proper, omit it, because I intend to put this, and some others, into a little *Fasciculus Stirpium Carolinensium rariorum*, or what the French call a Bouquet, which I may send into the world when the hurry of the day is a little over with me, which time I hope is approaching and almost at hand.

I am very sensible of your friendship in your warm expressions concerning my health, and would willingly think of going to settle in England, if I were not afraid that the charge and expence of maintaining a family, and educating children, were more than my small finances would admit of; and as to business, I fancy it would be the next thing to a miracle, that an American should get any business, in my way, in London, where there are already so many men of eminence and merit. What I labour for at present, is to get what I think would just be a modicum to maintain and educate my young family, which I think I could do, in a retired snug way, for 500 pounds sterling *per annum*.

Permit me, my dear friend, to ask your opinion on this matter. I am sensible that this is presuming on your friendship and good nature; but to whom can one apply in a case requiring advice, but to a friend? Forgive my impertinence, and let me beg your candid opinion in answer to these two questions.

1. What might a person, in a retired, private station, require to maintain a family of five or six people, all included, except servants, in or near London, *per annum*; including likewise the schooling and education of three or four young ones, of the above-mentioned number of five or six persons?

2. What chance might a person at my time of life, about 34 years, have to get a little business in the practice of physic in or near London, in order to assist in defraying the above expence? These are my doubts. I confess that I could wish to go to Great Britain to enjoy life and leisure, not to labour anew. But I find my health break fast, and my constitution to waste; and for that reason, if there was the least prospect of success, I would risque something. Shall I beg an answer when you have a little leisure? Write freely to me, and be assured that I shall esteem it, among the favours which I have received from you, one of the greatest which ever you shewed to, my dear Sir, yours, &c.

Alexander Garden.

P. S. I hope the length of this, and the inclosed, will make up in part for my former neglect. Sickness gives me this happy time of leisure.

My dear Friend, Dec. 10, 1764.

Since writing the inclosed, some weeks have elapsed without any vessel sailing for London, which gives me a fresh opportunity of addressing you, and at the same time of informing you of my daily recovering my strength and health. Since writing these letters, I have frequently had the pleasure of seeing Mr. Lloyd, who is very clever and sensible, and has, in many of my enquiries, given me much satisfaction. He proposes to set out in a day or two for East Florida, where I wish he may meet with things to his liking. By all the accounts which I can learn, the lands in that Government are in general excellent, unless for a few miles round Augustine to the land side. A constant spring reigns through the year. Some of our Carolina gentlemen, who went there to see the Governor, and take up some lands, are just returned; and they tell me that the whole face of the country, now at this time, has the appearance of spring. They give the greatest encomiums of the lands to the southward; and they tell me that on the Keys and towards the Cape, all the West Indian productions are met with, and amongst others, mahogany in the greatest plenty. They are positive that sugars will grow well, and that four cuttings of indigo can be obtained every year. We can only get two here. This is the chief of what I have been able to learn from them about Florida; and as I know them to be gentlemen of great veracity, candour, and sincerity, I am sure that I can depend on what they say; and the more so, as two or three of those who went, and with whom I have conversed, are the best judges of lands and planting that we have here. For about 20 miles near Augustine, the lands are sandy Pine-barrens, nor as yet produce any thing; but beyond that, all choice lands for every thing. Mr. Roll, I am informed, has not chosen such lands as Americans would have chosen. He is stationed on St. John's river.

Just as I was writing, Mr. Harry Lloyd called to take his leave of me; he sets out to-morrow for East Florida; he desires his best compliments to you, and says he will write to you from Augustine. I have put labels on some of the specimens to distinguish them; and I have put the seeds into a box, with paper between the layers, to preserve them. Some of the seeds are globular, and some oval, as you will see by those marked in the upper layer of seeds in the box. There is no other mark that I can find of two species. All this large packet goes under cover to Mr. Charles Ogilvie, who will deliver it with your seeds, in order to save expences and postage, as the vessel goes to him.

Yours, ALEXANDER GARDEN.

MY DEAREST FRIEND, South Carolina, May 18, 1765.

There is no time in which I find myself more universally easy, happy, free, and high spirited, than when I sit down to write to you. I really cannot account for a certain vivacity which I immediately perceive on such occasions, even in my dullest moments. But be this as it will, as soon as I take my pen to address you, I find new life, new strength, and new spirit to pervade, animate, and invigorate my whole frame. Thus you see my writing to you is really a favour and good office done to myself, and that you owe me nothing on that head.

Your last letter to me, by Captain Ball, now lies before me, to which accept of the following categorical answer. 1st. From Mr. Harry Lloyd, whom I found extremely lively and entertaining, I received your new prints of some *Serratulæ,* really beautiful and curious, and I only wish that I could add to them. Probably I may, but it will not be this year.

2dly. If I come to England, and settle there, I shall, till then, reserve what I have to say on the small-pox, agreeably to your advice; but to the point, what could an American doctor expect?

3dly. Linnæus's enquiries after me are most obliging, and by the inclosed you will see that I have not been insensible to them, for I have sent a pretty group of new genera; and I have, as you will see by my letter to him, only reserved the naming or baptizing of one for you, which I want to have called *Ellisia.* As I am well satisfied that every one is new, I want you to choose one which you would have for your name-sake. If I could take the liberty of advising you, it should be No. 2, or No. 4, No. 5, or No. 6. The two first are shrubs, and very beautiful indeed. The two last are pretty trees. No. 6, particularly, is an evergreen, which you will see in Catesby, and then you may choose; but let me beg that you would be pleased to fix on one of them, as nothing would make me happier than that you should be pleased to accept from me the compliment which you so generously bestowed on me. Permit me then, through your favour, to obtain this earnest request; and I beg that you will settle and determine the point with Dr. Linnæus.

I have left my letter to Linnæus open, that you may see what passes between him and me, and favour me with your judgment thereon. I have sent him several of my new genera, which, indeed, I confess I intended to have reserved till I should have an opportunity of publishing the whole together. But an accident, which you are before now, no doubt, acquainted with, made me send him these; *viz.* last winter Lord Adam Gordon spent two months with us, and he then engaged a gardener, one Wilson, who had resided here some time, and to whom I had shown many things, to go home and live with him. This man carried many things from hence, and among others, all these genera, whose characters I now send. He was directed by Lord Adam Gordon to carry them immediately to Mr. Robertson or Robison, the Queen's gardener, where, no doubt, they will be seen, will flourish and be examined, and as I had seen and examined them long before, I thought that I might therefore send them to Linnæus, to insert in his new genera, as discoveries of my own, which in reality they are. As my last packet, which was delivered to Solander's care, miscarried, I thought of begging the favour of you to forward these to Linnæus immediately, knowing well that I could depend on your punctuality and exactness in all matters of trust.

I have sent a copy of the above characters to Dr. Hope, Professor of Botany at Edinburgh, to whom I soon intend to send some more for their Society, who some time ago did me the honour to elect me an honorary member.

There are likewise sent the characters of what I take to be a new animal, to Linnæus, and two specimens of the same, in spirits in a phial, which I must beg that you will forward immediately to him. I have sent the dried specimen of this animal to you, from whence I took these characters. It was very large when alive, and now you will see the most essential characters of it still, in the dried specimen. You may either keep it for your own use, or present it to the Musæum, only giving Linnæus time to see and examine it first. Pray how does Dr. Solander go on in his catalogue? I am sorry that he never answered my last, as it breaks our correspondence—I hope without a cause.

Permit me to enquire what is Linnæus's address and title at present; you wrote me that he was ennobled, but you forgot to mention his title or address; let me beg you will mention both in your next.

4thly. You tell me you are surprized that I overlooked a new species of the live oak, which John Bartram found near Charlestown. Let me assure you that John Bartram received from me these very specimens, some of the *Phillyrea*, and

and many others, from my *Hortus Siccus*, of which he has, it seems, made a different use from what I apprehended. Yet, after all, he is an excellent man and I forgive him, because it is a matter of little moment who declares the glories of God, provided only they are not passed over in silence.

O my friend, how I rejoice at your present situation, in which I consider you as being in the way of daily receiving new intelligence of the works, the wonderful works of Jehovah, from all our new, great, and glorious acquisitions. Long may you live to enjoy the same! Pray continue to give me all your botanical and literary news.

Ever since October last I have had very ill and irregular health, being seldom well above two days at a time. I am now just about leaving this province, either to visit East Florida or the Cherokee mountains, in quest of health; whichever way I go, be assured that while I live and breathe, an ardent enquiry into the nature and properties of the sublunary manifestations of that God, of that tremendous and alwise Contriver of all things, whom I must soon see, will be my earnest desire and delight; and a communication of whatever I meet with, to you and Dr. Linnæus, will be the solace of my retirement after my return.

I believe I have specimens of most of these new genera by me; but as I think they will be unnecessary, and am, besides, in a very great hurry at present, in preparing to set out, I have not looked them out to send to Linnæus or you.

In one of Linnæus's former letters to me, he directs to me *Socio Societatis Upsaliensis;* but he never sent me my ticket or diploma of election. Let me beg you will put him in mind of it.

Pray how does your Royal Society prosper? have you many men at present who have the eager and ardent desire of investigation among them, or who are animated with the love of enquiry?

Believe me to be for ever, your assured friend,

ALEXANDER GARDEN.

[No date.]

MY DEAR SIR, Between May 18 and July 15, 1765.

An opportunity again offers of addressing you—a happy moment for me. I thought, by this time, to have been in the Cherokee hills, but the defection of two members of our party, out of four, has disappointed us for a time. My last to you was of the 18th of May, to which permit me to refer you for my letter and communications to Linnæus, and for all my enquiries and queries about him, the Upsal and Royal Societies; since that time, I have had the happiness of a letter from you, dated the 13th of Feb. 1765, a new mark of your regard and attention to me.

I observe all you say concerning the *Prunus Padus,* and I really expected it all before you sent it. But I hope I have made up for it in my last to you and Linnæus. You say nothing of my critique on Linnæus's *Dodecandria*, *Icosandria*, and *Polyandria* classes. It is what I long have thought, and I should have been glad of your sentiments on them.

I shall be glad to hear of the fate of my new genera; what is your choice, and what has become of the former genus which I sent him, and of which you have Mr. Roupel's drawing. I have another species of the same, a still more elegant and beautiful plant, though its leaves are not quite so broad, but the flowers excel in beauty*.

I tremble to think of Linnæus's late danger, but now raise my voice and warmest gratitude to the Great Author of Nature, for his recovery, and the preservation of so useful a life. Such a genius is truly a *Rara avis in terris*, very thinly sown.

It gives me the greatest pain to think that I never heard from him on my last collection of fishes, and I eagerly hope that you sent him the copy which I inclosed to you, in my last of the year 1764, containing the characters of these fishes, which would be some small reparation of his and my loss in having them miscarry.

You make me very happy in the flattering account you are pleased to give me, of a prospect which I might have in settling in London. Happy, happy, indeed, should I be, and I hope shall I be, in spending the latter part of my days near to, or with you! In you I shall always depend on a friend, and I think I shall shortly have the pleasure of addressing you as such, face to face.

* Probably *Styrax lævigata*.

I long greatly to see Solander's catalogue, which I doubt not you will furnish me with as soon as possible. Please to offer my respects to him in the kindest manner, wishing him all manner of success in every undertaking.

I shall be much pleased to see your account of the British *Fuci*; and though I will not venture to promise you a male Cochineal fly, yet I will this very afternoon go in search of one, and continue it till I find one.

I am much obliged to you for the hint you gave Wilson, but he is grown too lazy and careless, and does not mind my orders, so that I must apply to some other. I have never seen any of Linnæus's late performances.

I have now only to beg leave to recommend to your acquaintance, my friend Mr. George Hall, a gentleman of this place, who now goes to England with his family to reside. He has a great desire to visit the British Musæum, and begged that I would on that account give him a line to you, to procure him some opportunity. I am persuaded you will find him well worth your acquaintance, as he is very intelligent, and can give you much information concerning Carolina and America that you may depend upon. He is a person of extensive trade here, and well acquainted with the interest of the province.

Believe me to be for ever, your assured friend and most obedient humble servant,

ALEXANDER GARDEN.

P.S. I have begged Mr. Wilson to send the latest editions of Linnæus's *G. Plantarum*, of which I have already the fifth.

The latest edit. of the *Syst. Nat.* of which I have already the tenth.

And the latest edit. of the *Spec. Plant.*

Also the eighth, seventh, sixth, and fifth volumes of the *Amœnitates Academicæ.* I have the four first volumes already.

Let me intreat the favour of you to send him a line, informing him what are the latest editions, that he may not send me what I have already.

Yours always, A. G.

South Carolina, Charlestown,
July 15, 1765.

My dear Friend, et mihi magnus Apollo,

I protest that I am in much better health and spirits than when I wrote to you last; and I am now determined that you shall not complain any more of not receiving letters enough from me *per annum.* I was in your debt — I acknowledge it — but now that I am well, I will not only pay my debts, but am determined that, by great industry, I will try to bring you in my debtor.

First of all let me inform you that I have had Mr. Bartram for my guest for these nine days past. He went this day for Cape Fear, from whence he returns to me in about three weeks, and then he proposes to set out for East Florida. I have had many conversations with him, and have endeavoured to give him all the light and assistance I could, into the nature of the hot southern climates, and their productions. I have been several times into the country, and places adjacent to town, with him, and have told him the classes, genera, and species of all the plants that occurred, which I knew. I did this in order to facilitate his enquiries, as I find he knows nothing of the generic characters of plants, and can neither class them nor describe them; but I see that, from great natural strength of mind and long practice, he has much acquaintance with the specific characters; though this knowledge is rude, inaccurate, indistinct, and confused, seldom determining well between species and varieties. He is however alert, active, industrious, and indefatigable in his pursuits, and will collect many rare specimens, which, from their being sent home, will give you a good idea of the country productions. He is well acquainted with soils and timber, and will be able to give you much light on these heads. He appears to me not very credulous, which is one great matter. His collections and specimens all go to Mr. Collinson, where you will have an opportunity to see and examine them. I have given him many specimens here, and made him well acquainted with the appearance and common habit of most of our *plantes qui naissent aux environs de Charlestown.*

This I hope will render his enquiries into the

Florida plants more certain and accurate; and I shall rejoice if it is of the least service to him. He tells me that he is appointed King's Botanist in America. Is it really so? Surely John is a worthy man; but yet to give the title of King's Botanist to a man who can scarcely spell, much less make out the characters of any one genus of plants, appears rather hyperbolical. Pray how is this matter? Is he not rather appointed or sent, and paid, for searching out the plants of East and West Florida, and for that service only to have a reward and his expences? Surely our King is a great King! The very idea of ordering such a search is noble, grand, royal. It may be attended with much use to mankind, much honour to the Royal Patron; and it will be a further illustration of the power, wisdom, and goodness of our great Heavenly Father. My heart warms over the notion of his Majesty's attention to, and high discernment in, a matter that relates to the advancement of the glory of God, and the good of mankind. How happy must the people be who are governed by such a king!

But while we are on the subject of the Floridas, pray permit me to enquire what is doing, in relation to the various claims, that private persons make, to the lands of East and West Florida, under the purchases of these lands, which they made, agreeably to our Treaty of Peace with the subjects of Spain? I have often thought of mentioning this, but always forgot it; though now the matter becomes so serious in America, that I should think it was high time for those in power to attend to it; either publickly and openly to shew these claims to be illegal and void, or else to settle with the claimants, so as to give certainty to new settlers. I can assure you, that from what I see, or can learn, and I have been very attentive to what has passed or been agitated in conversation here, the lands of East and West Florida will never be settled, till these claims are either declared good and valid, or illegal and invalid. No new settler will ever choose to sit down and improve, where there is so much as a shadow of danger of his being molested and turned out. An uncertain claim to a landed estate or possession, or a title liable to litigation, is of all the most precarious, least desired, and least sought after; and this very matter, whatever the Ministry may think of it, is and will be an unconquerable bar and lett to the settlement of these two provinces. No man will be satisfied with the warmest assurances of a governor, when another person tells him that he may indeed sit down and cultivate there, but must remember the lands are not his own.

The expedient of shutting up the Courts of Justice against these claimants, cannot be the spirit of the British Government.

I wish so sincerely well to the British interest, I wish so sincerely well to the settlement and improvement of these Colonies particularly, that I have taken the liberty to mention these things to you, as I know you are concerned as an officer in one of them, and wish, as well as I or any man can do, to the extension of the King's dominions. Had it not been for the claims of two or three men, the Province of East Florida would, before this time, have had at least a thousand substantial settlers from this Province and North Carolina alone. And now those who were going are really so intimidated, that they rather choose to rest where they are, without lands, than to go there upon an uncertainty. Do you not think this requires the attention of the Ministry, and of every well-wisher to his country? It will give me much pleasure to have your sentiments on this; and if it is agreeable to you I will, in any future letter, give you a much fuller state of the matter, as I have daily opportunities of hearing the subject discussed in our clubs and companies.

But to the point. I mentioned before to you that I intended, when I could get a little leisure, to collect as many of these plants which appear to be new genera, as will make a *Fasciculus Rariorum*, which I will send to seek its fortune. I am now beginning to make out my list, and in looking over my notes, I find, that I have at different times sent several of them to you. I do not exactly remember all of them, but I will just mention those few that I recollect; *viz.* the Loblolly Bay; the Supple Jack (which you told me Solander once took for a *Rhamnus;* this I am vastly amazed at, as the Supple Jack has a very distinct calyx and corolla, one of which, not only the *Rhamnus*, but likewise all its relations, *viz.* the *Frangula, Paliurus, Alaternus, Ziziphus*, &c. are said to be destitute of); the Palmetto Royal; the

Yellow Jessamy; both which I am well satisfied are new genera; the *Catalpa;* a very beautiful herbaceous *Hexandria trigynia,* sent you in 1757, by Capt. Ball, with several others. Now, my dear Sir, what I would beg of you is, to know whether you have ever communicated any of them to Dr. Linnæus, and whether he has inserted them into his late works, or if you have communicated them to any Society? Because, whatever you have communicated, or choose to communicate, I would not put into the Catalogue which I intend to send in quest of adventures. I hope you will be so obliging as to set me right in this point.

By the bye, I must inform you, that I think there is a great affinity between the N° 2 of these genera which I sent to Linnæus, and the *Prinos* of Linnæus; and yet I think there is difference enough in the *Pericarp,* to constitute a distinct genus. I have now sent you some seeds of N° 7, of the genera which you had to forward to Linnæus. I really think this is a most uncommon and curious plant, and heartily wish you may be able to raise the seeds.

Please to let me know whether specimens would be necessary to be sent to Linnæus, along with the genera, when I send any.

I have several new genera by me, which I think will give pleasure to the botanist, though I believe none more than one I met with the other day, which I at first took for a *Peplis,* but, on examination, it turned out quite otherwise. I have found great plenty of the *Diodia* of Gronovius, near

Charlestown. *Apropòs:* mentioning Gronovius puts me in mind of enquiring after the new edition of his *Flora Virginica*, which I understand, from John Bartram, has been published some years, and that he has a copy, sent him, either by Gronovius the father, or the son, with many emendations and additions. Pray let me beg you would be kind enough to desire Mr. Wilson to send me a copy of it, and if at any time you see him, to beg him to send it by the first vessel. I wrote for it long ago, and for all the late editions of Linnæus; but he grows careless, though I always desire him to call on Mr. Ogilvie, merchant in London, immediately for his money. As I desire much to have this book, I must again beg that you will order him to send it soon.

My dear friend, I have now to tell you that my prospect of seeing the Cherokee hills this year is all over. After I had made up a party, and after I and some others had, each of us, put ourselves to near £.40 sterling expence, those who were absolutely necessary to complete our party and assist us onward, fell off, and discovered that they could not go, only two days before we were to set off. Thus, my dear friend, have I been twice baulked in the hope, view, and expectation of seeing those stupendous mountains. But there, or here, or any where, be persuaded that I shall always be happy in assuring you that I am, yours, &c. ALEXANDER GARDEN.

I have had at least 200 plants of the *Opuntia*, all covered with females, and have searched for male flies, but we have not been able to catch one; he is so nimble and active, that unless he is intercepted some how, in the down or cotton, he gets off too quickly to be caught.

MY DEAR FRIEND, Dec. 16, 1765.

Now am I sat down to write you the most extraordinary letter which you have ever received from me, as to its contents. Heavens! what will you say when I tell you that we are quitting you as fast as we can, and that this may be my last letter to you for many months?

The fatal Stamp Act is likely to put an end to our intercourse. You have imposed a taxation in America, which the Americans say they will not receive. Every colony upon the continent has risen in opposition to King, Lords, and Commons, on this occasion. We were among the last who fell into this ———. But what shall I say? — Another time may be fitter to write of these things. — The opponents call themselves the Sons of Liberty; and among their number here you might rank those of the *first note*, from the *very highest* to the lowest, making now and then some exceptions. Whether there is much policy in having Americans governors, the politicians on your side of the Atlantic must determine. From the smallest beginnings great things often arise. The die is thrown for the sovereignty of America! Will she then be on her own bottom, or will she be subject to the controul of Parliament? The most lenient, mild, and soothing methods, or the most vigorous measures, are the only cure for this strange ———. The courts of justice are shut up — navigation and commerce stopt — the produce unvendable, and credit going to decay — and all correspondence with Britain at an end! — All this brought about in the most surprizing manner. Nay, I am far from being certain that this letter can reach your hands without being looked into. The number, rank, situation, and employment of the *Sons of Liberty* are so extensive, that they fill almost all places; and their jealousy is extreme. Bad and oppressive as the Act is in itself, it is represented by a few designing men as being ten times worse than it is; and this alarms the people to distraction. Why were not many copies dispersed to satisfy the people? Why were ———

Alas! — enough of this! —— Now for Natural History.

Your agreeable letter of the 11th of September now lies before me; and first I most sincerely thank you for forwarding my letters, &c. to Linnæus, and then sending his letters to me. I hope my last may be agreeable to him; and I am greatly pleased with your observations on the *Mud Iguanas*, but I am still persuaded they are a perfectly-formed animal, and no *larva*, as you suspect: though as I pay the greatest deference to your opinion, I will suspend any positive answer, till I have some future opportunity of examining them more narrowly and carefully.

The next paragraph of yours relates to Doctors Hope and Skene. This last gentleman is my old schoolfellow; we studied philosophy in the same class at Aberdeen for three years. I applied to the study of physic and medicine; he pursued his philosophical studies for three years more, and then I persuaded him to repair to Edinburgh, and to study under the Professors of Medicine, where I left him; but we have never corresponded, notwithstanding our former intimacy and connexion. He was always extremely clever, and I doubt not is so now in a high degree. Pray let me beg that you will offer my compliments when you next write to him.

My dear friend, I will implicitly follow your very obliging and most friendly advice concerning my *Fasciculus* of Carolina plants. It shall be exactly as you advise; and I return my best thanks to you on this head. Indeed I owe you a thousand obligations for favours conferred on me; but, of all this, the last is not least, as it may save me pain, uneasiness, and dishonour. It will be some time before I can complete it; and when it is done you shall have it.

I have received two letters from Linnæus — the one you inclosed in yours, and one I had by the packet. I am much indebted to him for the favourable opinion he is pleased to entertain of me; and I shall be very happy in thinking that the few things I sent him have given him satisfaction. I shall long much to see the new edition of the *Systema Naturæ*. How miserably Monsieur Buffon falls foul of him every now and then ! many of his criticisms seem to be envious and trifling, though he is a valuable author. Let me ask, in what estimation is Buffon with you ? The grandeur and elegance of his book appear to me often to be the valuable part of the work. I mean, that in many things he is short, superficial, and trifling ; witness many of our American animals ; and his affected retention of the Iroquois and Brasil names, merely out of opposition to Linnæus, is ridiculous and puerile. I admire his general dissertations very much ; and on many animals he is full, clear, decisive, and satisfactory. He writes in an easy, agreeable, lively style, and is often truly a painter. Thus you see, Sir, I have with much forwardness given my own opinion, while I beg to know yours, which will more certainly determine and regulate my judgment of his merit. I am vexed at him for snarling so at Linnæus. Plague on it, why cannot they agree ?

I have just got the tenth, eleventh, twelfth, and thirteenth volumes of Buffon ; and Wilson has sent me Jacquin, but he omitted *Gron. Fl. Virginica.*

Poor Harry Lloyd is dead. — I shall be much obliged to you to know your sentiments, and what you can learn, of the Florida purchases. I have much reason for enquiry, as I really want to try a scheme in that province soon ; and I should be glad to know, if possible, how matters stand, as the uncertainty of that point at present not only retards me, but some or many hundreds of others.—We hear that it has been very sickly in West Florida, and that many of the late-arrived troops have died there. Query, whether it would not have been more for his Majesty's service and interest, if these two battalions were here in this town, rather than in the fens of Mobile? — How many thousand things have I to say to you! but I must say them on some other occasion. — If we send you any more ships, I shall send you some specimens for Linnæus of those things which I before sent him. — Pray let me have all your news, and what Solander is doing; to whom please to offer my compliments.—May I ever expect a diploma of my election into the Upsal Society? — Linnæus once directed to me as a member, but never said a word more. How do you address him now? and how do you direct to him? — Farewell!

Yours, ALEXANDER GARDEN.

MY DEAR SIR, Aug 6, 1766.

I think I have often told you that when I write to you I forget the ills and mishaps of life for a time; it gives a temporary suspension to all my cares. Your last to me gave me great satisfaction, both in what you communicated, and likewise as it contained a valuable letter from our friend Linnæus.

You can fancy how much I am pleased with his letter. I have left mine to him open for your perusal, where you will see what I have written to him on this *Animal bipes rarissimum, monstrum horrendum informe.* It is certainly a new genus; if it were a *larva*, we should surely see some of them in their perfect state; but we have positively no animal corresponding to it.

Besides this, I have sent him specimens of all the new genera of last year, and of two other genera, one of which I had indeed sent him before; it is the *Decandria monogynia* of Roupel; and there is a new species of it. The other is a new tree, or such a tree as I remember not. I hope these things will give him great pleasure. He has called one the *Hopea;* and, as I cannot have the pleasure of naming one after you, I hope you will command one for any of your friends. You formerly wanted the Loblolly Bay for Mr. Gordon at Mile-end. If you still think so, you should take it. I have sent you some good specimens of it, which you will find under your cover of Linnæus's specimens. As the specimens sent him are very tender, I had a board put behind them, to secure them against harm, or danger of being crushed or broken.

I must beg that you will be so obliging as to forward these things speedily to Linnæus, as I could wish they may have a place in the new *Systema Naturæ,* which is about being published. I hope therefore you will, agreeably to your usual politeness and goodness to me, forward them by the very first opportunity. There is a bottle containing a specimen of the new animal and three fishes, of one of which I formerly sent him the characters. There is one which he has never had before — the West India Parrot Fish.

There is a small bundle containing some insects and a skin of the new animal. There is, besides these, the large packet of dried plants; under the cover of which are your specimens of the Loblolly Bay. It will give me great pleasure to see these inserted into the new *Systema Naturæ*, and therefore I need say no more to you about dispatch.

You write me that you intend to have a drawing made of the large dried specimen of the new animal, which I sent you, and presented to the Royal Society. If this method is agreeable to you, it will be perfectly so to me, as every thing that you do is. But, unless to yourself and my friend Mr. Baker, I owe very little to that Society. Whatever respect I have met with has been from the honourable mention which you alone have been pleased to make of me, in the papers which you have presented; and for this I now and always shall consider myself as being obliged to you alone. But this by the bye, which I hope you will forgive. Allow me now to say that you may make any use of the specimen you please, and extract what you think proper of my letter to Linnæus, that regards its history; but I think you should publish my description along with it. I sent it open to you last year; and if I can get time to have it transcribed, I will send a copy to you now. Let me advise you to give it as lurid a look as possible, for it has one of the most lurid, torvous, threatening, surly, forbidding looks of any animal that ever I saw.

I should have been very happy, some years ago, to have heard what you say of Provincial Gardens. I always thought well of them, and could have wished, heartily wished, to have seen the measure take. But, alas! my friend, you and I have lived to see America aim at independency, and the die is fairly thrown. How far your popular idol has served Great Britain, and her colonies, a little, nay a very little, time will show. Even so little, and so short, that you will not believe it. We in America see it. We have seen our strength, and we know our power, and your famous Stamp Act showed our inclination. When your infatuation is over, you will view us in our proper light. To be sure, we have at present a most respectable idea of you and your gracious goodness, and the mild condescension of the present angelical, lamb-like Ministry.

We have ordered a statue of Mr. Pitt, our great hero and deliverer, and we may think of the thing bye and bye, but as yet we have not. Much might be said; but "*qui vult decipi, decipiatur.*" It is neither your business nor mine.

Plague on politics! they have driven every thing else out of my head; so that I must just subscribe myself yours for ever, with the greatest affection and esteem, ALEXANDER GARDEN.

I should have been glad to have had a letter from you by our new Governor, or from any of your friends, providing he has any turn for Natural History, as it may be much in his way to countenance and promote any plan, especially about provincial gardens. Indeed it always was my sentiment that it should be recommended from or by the Governor to the Assembly; then they would readily and at once do the thing, but they will never hearken to private persons on that head.

The Mercury stood at 91° to day (August 6, 1766). It now thunders and lightens as if it were to usher in the last trump.

Dear Sir, Feb. 2, 1767.

Your two last letters now lie before me replete with many curious observations and pieces of intelligence, and with many enquiries. Your friend, Dr. Turnbull, is now here, on his way to London. He has been in East Florida, is pleased with the country and lands, and has located 20,000 acres for himself, and as much for Sir William Duncan. He has a good opinion of the lands, and I believe with good reason, as they are well spoken of in general by our best judges, though no doubt there must be some bad lands amongst them. They certainly will yield some vegetables which ours will not bear, and I think they are not so sickly, at least on the sea side, as we are. They lie within the trade winds, which they have regularly during the day, and they are never so warm at nights; add to this, that they have not such sudden nor so great changes in the autumn and winter as we have. Now it is the suddenness and quantity of these changes that give us our autumnal intermittents, and winter inflammatory diseases. We are literally in the trawley rawley latitudes, which are the most changeable, and consequently most unhealthy. And on the contrary, their having a greater steadiness and equability of climate, will undoubtedly ensure a more steady state of health; and indeed experience has testified the truth of this for two or three years past, so that you may consider the above reasoning as *à posteriori*.

I greatly rejoice that America gains such an inhabitant as Dr. Turnbull, but I confess I envy Florida that pleasure, and wish he had settled with us, though I believe he has chosen the better part.

The doctor carries home some packages of East Florida plants, which you will see. I shall be very glad to know what you make of John Bartram's Tallow tree, and what you call that herb whose leaves look like the Fern Osmund Royal, while its seeds are large red berries in a cone, somewhat resembling the *Magnolia* in appearance. I shall be glad to know what you call these two.

I have begged Dr. Turnbull to make my apology about the male Cochineal; he will inform you how I have been disappointed.

You say Dr. Solander thinks that Catesby's *Phillyrea* is an *Olea*, so does Linnæus, but they must be mistaken, as it has male flowers only on one plant, and hermaphrodite on another. This I am certain of, though I did not observe it at first.

I do not know what return to make you for your care of my letters, &c. to Linnæus. I should be absolutely ruined if it were not for your assistance. I am on this, as well as on many other occasions, greatly indebted to your politeness and kindness.

What have you called the Loblolly Bay? Pray give it a name, and place it in your Transactions, that no other person may be before-hand with you. It is certainly new, at least it has never yet had a name.

I have sent you a small collection of seeds, but which I hope will be very acceptable as they are mostly rare and curious, and I shall be very happy to know that they please you. They will be delivered by Dr. Turnbull, who is so good as to take charge of them; several of these are seeds of new genera, whose characters you have, within the two last years, transmitted to Linnæus.

I was just now struck with the resemblance of one seed-pod to the Bohea plant; I mean the pod of it, sometimes found among the tea leaves.

The *Hopea* is beautiful.

I have given the doctor two large pods of a plant which I had from New York, called there the Horn plant, from the shape of the pods. It comes from Detroit. I have never seen it grow, and do not know what it is, if you know, pray inform me?

Yours, ALEXANDER GARDEN.

Dear Sir, April 18, 1767.

The business of this is to be conveyed to you by John Cree, a gardener, who was formerly bred by Mr. Aiton, and is a sober, careful, industrious fellow, with some small smattering of botany. He is modest and attentive, and has carried several things along with him, which I have advised him to show to you before he shows them to any body else. I hope he will mind what I have said to him. He is to return here again, and may be of use to you, as he is attentive and careful. It is upon this account only that I prevailed on myself to burthen you with him. I have seen Mr. Cook, who is to be employed here in making a survey of our province, to be paid by the Assembly here.

As I shall write to you soon, I shall add no more at present, but again to assure you of my friendship and best wishes to the best of men, by his devoted humble servant, Alexander Garden.

My dear Sir, June 2, 1767.

It is more with an intention to convince you I do not deserve the imputation of forgetfulness, which you lay to my charge, that I write you at present, than that I have any thing new to communicate. How could you say you should think I had forgotten you? My dear friend, it is impossible; never suppose me capable of it.

Before this time you have seen and conversed with

Dr. Turnbull, who must have given you great satisfaction, and you will have received my letters by him. I sowed the Horn plant, mentioned in my letter to you by him, which is now in flower, but I have not examined it. It seems to approach to the *Chelone* or *Martynia.*

What have you called the Loblolly Bay? please to name it. What is the Tallow tree? I am told it is of the *Petandria Monogynia.*

What is the Fern-like plant with a cone full of red seeds?

I have received Linnæus's letter, for which I thank you. He had not received the insects and *Siren* with the other animals; and I think by his letter he had not received mine. He asks some questions about the *Siren,* which I had particularly answered in my last, and seems to be ignorant of what I had there wrote of it, as far as I knew. I hope both the letter and insects, &c. arrived soon after.

I lately received a large specimen of the *Siren,* caught by a hook baited with a small fish. I was absent when it was sent me, and my servant killed and skinned it before I came home, otherwise you should have had it in spirits, but that loss shall soon be made up, as I shall procure some and send you, agreeable to your desire. I have sent you the dried skin of this; it measured 34 inches when just taken off, and is now 2 feet 7 inches.

What shall I say of a male Cochineal? The truth is, I have hunted all yesterday and to-day, and cannot find one. If I can possibly obtain one you may depend on having it.

I have sent you a few seeds, of the genera which I sent to Linnæus. They are only five, but curious enough.

No. 1. Is the Purple-berried bay of Catesby. Linnæus says it is a new and genuine species of the *Olea.* It has male and hermaphrodite flowers on different trees, which I think is not so in the *Olea.*

2. *Hopea,* a very pretty new tree.

3. The *Chrysophyllum* or *Sideroxylum,* according to Linnæus.

4. Linnæus calls a *novum et pulcherrimum* genus. It is indeed a beautiful shrub.

5. Is the plant which Roupel formerly drew. Linnæus says it is a species of *Styrax.* This surprised me not a little.

You have greatly mistaken me if you thought I wanted any particular compliment from your Society. My acquaintance there was so little, that I had no right to, nor the least expectation of, any such thing; but in consequence of what you wrote me about your intention of publishing an account of the *Siren* there, I told you I thought this would be needless, as I had sent it to Linnæus, with the characters, for the Upsal Society, to which I was indebted, and as I had no acquaintance in your Society, I still thought it better that it should be published by Linnæus, in the *Acta Upsaliensia.* But, my dear friend, if you choose to insert it in your Transactions, I shall most readily concur, and would further observe to you, that it seems to be an animal of prey. The one which I now send you, was caught with a roach for bait, and when it was thrown on the ground, it bit the hook into two pieces. These animals do not live long out of water, and are very brittle like a glass snake.

You will greatly oblige me if you would order Wilson to send me the new edition of the *Systema Naturæ*.

I have likewise often written to him for the 14, 15, and 16th volumes of Buffon, but the careless fellow does not mind my orders.

The things which I have sent you will be delivered by a young brother-in-law of mine. I have desired him to call at the Carolina coffee-house for you. He is to be with one Mr. Gibbons, a dissenting clergyman, in the city. I am, with the greatest regard, dear Sir, yours, &c. ALEXANDER GARDEN.

MY DEAR FRIEND, July 18, 1767.

It is not long since I wrote to you, but the pleasure which I always have in this intercourse is so great, that I take every opportunity of renewing it. Several of your obliging letters now lie before me, all filled with great marks of your esteem and regard for me, for which I am indeed greatly indebted to you. You give me a thousand incitements to the pursuit of Natural History, the dæmon of which constantly agitates me of himself; but, my dear friend, could you only conceive the troublesome, hurrying, and bustling life that I am obliged to live, you would heartily pity me. While on the one hand I am pressed with the greatest inclination to pursue Natural History, and fulfil my engagements to my friends, on the other I am distressed and oppressed by a fatiguing, worrying, uninterrupted constancy of other business.

It is with great reason that you reproach me about the Cochineal insect. I am ashamed of my indolence, or rather am sorry for my bad luck, for I have again and again searched for, but could not find, a male. For two or three years past I have not been able to spare half a day to go seriously in quest of such an animal.

You have likewise written to me about sea animals, and about my neglect of the *Toxicodendron* and the *Rhus*. It is true I have not sent them, but indeed I have not an hour to sit down and collect them, nor to dispose or arrange, either them or the observations which I made on them.

I have a brother at present in Edinburgh, where he has studied medicine for some years. He will be in London in 1768, summer time, when I will take the liberty of introducing him to you during his stay in London, which will be but two or three months, till he can get a passage for Carolina. Now I hope his arrival will set me free, for I intend to introduce him to my business; and as this will keep me a little longer in Carolina, I am determined to devote that time chiefly to Natural History. I shall have my confinement daily growing less as he becomes acquainted with the climate, people, and diseases.

I must beg your pardon for this account of my private plans, but as they are in perspective combined with my prospect of being left free to prosecute our favourite study, I therefore hope you will forgive me. Indeed the approach of this period, which I have been looking to for some years, gives me so much joy, that I cannot help teazing my friends with it.

I am very unlucky in never yet having seen Linnæus's last performance, though I have again and again written to David Wilson to send it me. I wish you would speak to him. I am extremely obliged to you for conveying Linnæus's last letter to me, and I am much pleased with his observations on the new plants. I hope when he receives the fishes and insects, it will draw another letter from him. I have sent you all the seeds of the new genera that I had by me. I suppose they are long before now delivered to you by James Peronneau, who had them in charge. I have several other matters to write to Linnæus about, and to send him, but at present I cannot find time to collect them.

The *Siren* is most certainly a new animal, I think beyond all doubt; but in order more fully to convince you, and to comply with your request of having a large one in spirits, I have sent you one by Capt. Gordon, of the ship Mary, in a pot of spirits. This was 32 inches in length when alive, as I then measured it accurately. It was very fierce, and threatened every thing around it. The negroes are much afraid of them, and say that they bite very severely. You will now have an opportunity of examining it carefully, both as to its external characters and the anatomy, in which I hope you will be accurate, as the specimen is large. I do not know whether this is male or female, but you may be able to discover by the dissection, which I suppose will follow the external examination. I have at different times sent you all the anecdotes which I knew of this uncommon animal, and lately I informed you that the one of which I sent you the skin, was caught with a hook and line, on which there was a bait of a piece of roach fish. Indeed it appears clearly that thay are animals of prey. I think the last plate which you have had executed seems to be very good, only when they are alive, the lateral and longitudinal membrane, which runs along each side of the tail, does not look so thin, for in that state it just appears like the body gradually growing thin on each side, but has no appearance of a fin, which I think yours has. I think the size of this specimen will convince you that it is no *larva*. We have no lizard, either land or water, that ever I saw or heard of, that is one third of the size, nay not a fourth or fifth of this size. Please to inform Linnæus of what you learn from the dissection of this large specimen.

I must have expressed myself very ill, and undesignedly, when I have given you occasion to say that you find I take amiss something in the Royal Society. I really never meant to say so, and you may be assured it was wholly owing to inadvertency of expression. I had neither right nor inclination to say any thing against so learned and respectable a body. I recollect well having said that as I had not any particular acquaintance or connection with that Society, it might be as well not to publish my account of the *Siren* there; but, my dear Sir, I meant nothing like an offence, and so declare; in the mean time, I am truly yours,

ALEXANDER GARDEN.

I have sent you, as a curiosity, a kind of Sponge which grows on our coast, in shape like a grenadier's cap. It always grows in this form, and I am told is met with frequently, though I never saw but this one.

Capt. Gordon who delivers the pot with the *Siren* and the Cap Sponge, belongs to Messrs. Ogilvie and Michie's house, and is engaged in this trade.

Mr. Ellis wrote to Dr. Garden, Feb. 9, 1768, enclosing a letter from Linnæus, and mentioning that Linnæus supposed the seeds of *Fungi* to be animated, but that he (Ellis) found the contrary by experiment; also desiring him (Dr. Garden) to correct Linnæus with regard to the *Rhus Vernix* of America being different from that of Kæmpfer, and the Copal. The true Copal, so valuable in varnishing coaches and snuff-boxes, grows in plenty on the Mosquito shore, as (being produced by) the *Hymenœa* of Linnæus, or *Courbaril* of Plumier. This tree does not produce the Gum Anime. Mr. Ellis also noticed Pallas's book on *Zoophyta*, and spoke of his own paper on this author's chapter of *Corallina;* as also of his answer and refutation of Dr. Baster's assertion, Phil. Trans. v. 52. 111, that *Corallinæ* are *Confervæ*. The paper upon Lord Hillsborough's new *Actiniæ* will be in this year's Transactions.

DR. GARDEN TO MR. ELLIS.

MY DEAR FRIEND, April 20, 1768.

I do not at present remember when I had the pleasure of a letter from you, or when I wrote one to you, so that I can say nothing on any of the former subjects.

The design of this is to beg leave to introduce to you the bearer, a brother of mine, who has a design of coming out to Carolina. I need not tell you that any countenance which you show him will be a singular favour done to me. I wish him to be four or five weeks at London, and I should be sorry that he were idle during that time. Permit me, therefore, my dear friend, to beg you would procure him a sight of the British Musæum, and of some of your botanical gardens and collections. I am much afraid he is very unacquainted with such things, but I think his visiting these places with you would inspire him with a true and real love of such studies; and, as he is destined for America, it might afterwards be of great service to him. I entrust him to you during his residence in London, and I beg your countenance towards him. Your advice, I know, will be of the greatest service, and may awaken his attention to what may be of the greatest benefit during his future life. I will say no more, but am, with a full heart, my dear friend, yours, ALEXANDER GARDEN.

If you could introduce my brother to Mr. Hunter, it would be a great favour done him, and it might give rise to an acquaintance between them.

MR. ELLIS TO DR. GARDEN.

MY DEAR FRIEND, London, May 10, 1768.

I wrote to you the 9th of February, and am in hopes of hearing from you every day. I have nothing new, but the experiment of the oak acorns preserved in wax, for 10 months or more, has succeeded extremely well. They were put into wax in February 1767, and given into the care of the Royal Society. In December following they were sent to Mr. Aiton, botanic gardener to the Princess Dowager of Wales, at Kew. He returned two pots of them in which he had sown 34. There came up 16 in one, and 9 in the other. By the beginning of March they were about six inches high, when he presented them to the Royal Society, and the manner in which I had preserved them was read. The great matter in preserving seeds for long voyages depends on their being full ripe, and properly sweated, as well as being wiped perfectly dry and clean, before they are enclosed in wax. Then, to avoid their being scalded by the too great heat of the wax, each seed should be rolled up in a coat of soft bees-wax, blood warm, which is very pliable. Afterwards they should be set for this to harden in the air. Then melt some bees-wax, and pour it into oval chip boxes, six or eight inches long, four deep, and three or four broad. The wax may be two thirds the depth of the box, and when it is so cool that you can bear your finger in it without pain, then place your seeds, already wrapped in wax, upon it in rows, till your box is full, the spaces being filled up with melted wax. Afterwards set the whole to cool where the air blows freely. If any cracks appear, they must be filled up with melted wax. When that is cool, put on the cover, keep the box as cool as you can, and your seeds will keep sound for a twelvemonth.

Please to send more Cypress, Pine, and Fir cones, and forward any parcels that may come to you from Pensacola, by the first ships.

DR. GARDEN TO MR. ELLIS.

My dear Friend, Received, July 6, 1768.

I have your letter of the 9th of February last now before me, full of friendship, full of intelligence, for which I am, and ever shall be, much indebted to you.

Linnæus's letter was extremely agreeable, and I rejoice to know that he found any thing useful which I sent him. Nothing can be more agreeable than the pleasure of adding one's mite towards the promotion of our knowledge of, and acquaintance with, the admirable works of nature; and to have Linnæus's approbation is indeed very gratifying.

I am sorry that I have not seen his new or 12th edition. I wrote to Mr. Wilson about it long ago, but he has forgot me; I wish you would remind him if you should perchance meet him any where.

I have, by Mr. Brewton, sent you a pot, containing three pretty large Sirens in spirits. One is for yourself; one for Linnæus, which I beg you would send him, with my respectful compliments, the first time that you write to him; and the other is for the British Musæum. I thought of sending you some alive, and if I can get any I will still try for it, as Mr. Brewton has been so obliging as to promise me to take care of them for you, and to deliver them safe to you. I have written into the country for them, and if they come in time, you shall have them. Mr. Brewton is a merchant of the first character and eminence in this place, and if you can show him any civility, it will be doing me a singular service.

I have received Dr. Pallas's two volumes some time ago, and wrote you before what occurred to me in reading him. I am but a very novice in all these marine productions, so that I can say little on this head, though I must observe that, as far as I could judge, his latinity is the best part of his book.

As to the quarto, I really think it is so glaring and gross a catchpenny, that I am amazed how he could have the effrontery to publish it, and attack Monsieur de Buffon, whose labours in that way must do him eternal honour, and confer infinite obligations on all the lovers of Natural History.

In your letter to me of the 9th of February, 1768, you mention some dissertations *de mundo invisibili*, which you sent me. I do not remember ever to have received any such, but only some occasional communications in your letters, about that matter, with which I have been highly pleased and entertained.

Some time ago I think I mentioned to you, that I have not lately examined the *Rhus* of any species, and that what I had formerly remarked of them had very much slipped my memory. At present I am so much embarrassed in medical practice, that it is absolutely impossible for me to attend to any thing, but as I expect my brother here this autumn, after having finished his studies, I hope he may be of service to me, and give me more command of my time. I will then, with renewed zeal and ardour, again attempt something in Botany and Natural History.

I wrote to you on the 20th of April, which letter will be delivered to you by my brother Francis, whom I have begged leave to introduce to you, requesting your advice and assistance to him in getting a sight of the British Musæum, which I hope will be an incitement to him to endeavour to cultivate that sort of study.

Mr. Gregg is again gone from hence. A difference which he had here with his former partners in trade, obliged him to leave this place, sooner I believe than he intended, so that I had not the pleasure of mentioning many things to him before he went, which I had intended to do, and to settle our future correspondence. I sent him your book on Corallines while he stayed here, and we had many conferences on Pallas's scheme. He expressed much pleasure in thinking that you would have satisfaction in some things which he was then sending to Lord Hillsborough.

Pray give my most respectful compliments to Dr. Solander, whom I consider so totally engaged in the Musæum, as not to admit of interruptions, otherwise I would have written to him.

You cannot easily fancy how much I am obliged to you for assisting Gosset with a medal of Linnæus. I should have been quite at a loss without him. I am ever yours. ALEXANDER GARDEN.

Of late I have dissected several specimens of the *Siren*, and have discovered several matters relating to it, which I did not know before, and which will be the subject of a letter to Linnæus very soon.

MR. ELLIS TO DR. GARDEN.

MY DEAR FRIEND, London, July 14, 1768.

I received your favour by the Nancy, Captain

Jordan, and thank you for the specimens of the Sirens. When they come, they shall be disposed of according to your wishes. I sent you 10 or 12 dissertations of Linnæus; among the rest, one on the *Siren*, and one *de mundo invisibili*. I forget to whom I delivered the parcel, but it was either to Mr. Ogilvie, or Mr. Karr at the Carolina (coffee house). This parcel was particularly directed to you, by Linnæus's desire. Another set of the same was ordered for Peter Collinson, and one for myself. I shall be proud of doing your brother Francis, or Mr. Brewton, any services in my power, on your account. As to Dr. Pallas, I am in no pain about vindicating myself; and have begun with him, as you will see, in this last volume (the 57th), of the Phil. Trans. As to the Corallines being vegetables, he now cries "*peccavi*."

Dr. Solander is going with Mr. Banks, a gentleman of £.5000 per annum estate, to the new-discovered land in the South Seas, with the astronomers, and no people ever went out better equipped for Natural History. After they have observed the transit of Venus, they go in quest of new discoveries, and come round by the Cape of Good Hope. Solander has leave to keep his place, and put in a deputy, till he returns.

I am sorry you are so busy that you drop Natural History. Pray do not forget the Deciduous Cypress cones. They are for Lord Hillsborough. I do not suppose it is a very difficult matter to procure a peck of them. I have received from Lord Hillsborough some specimens of the Nutmeg plants, that Mr. Gregg has sent from Tobago, the newly ceded island. They have no smell, but seem to be of that genus; for the same appearance of mace surrounds the nut, and the inside is marbled, like the East India Nutmegs; but they are very small.

Mr. Aiton of Kew says, the *Hopea* will not rise from seed. It should be sown in boxes with you first, as should the *Stuartia.*

I have just now received a letter from Mr. J. Gregg from Grenada, of the 15th of April, 1768. He has met with some specimens of the *Lophius Histrio,* or *Guaperva* of Marcgrave, p. 150, which he is in extasies with, as thinking it not already described.

I am much obliged to him for several specimens of the *Gorgonia ceratophyta,* with the polype-like suckers remarkably fine, and several others which he has taken great pains to preserve in spirits. He is the best collector I have, now you are so engaged. Lord Hillsborough is going to make a cabinet, or a little musæum, at his seat in Ireland.

DR. FRANCIS GARDEN TO MR. ELLIS.

SIR, Charlestown, South Carolina, Jan. 1769.

The great obligations I am under to you make it impossible for me to lose the first opportunity, after my arrival in America, to return my most grateful thanks for the many and repeated favours you were pleased to confer on me when in London;

but in particular, for the great trouble you so obligingly took upon you, in procuring me a sight of the British Musæum; and above all, for the honour you did me in showing me your own valuable and elegant collection of curiosities. The different *Animalcula* you showed me, by means of microscopical glasses, have ever since filled me with the greatest astonishment at the wonderful works of Nature, and admiration at the immense sagacity and penetration of those men who search into Nature's dark arcana; in which no man has made a greater progress, or can lay a juster claim to the high honour due to so great merit, than yourself.

I delivered your commission to my brother, together with the plates you sent him, for which he desires me to make offer of his most respectful compliments and thanks. At present, I shall beg leave to conclude, with my most respectful compliments, and those of the season, and am, with the highest esteem, Sir, your much obliged humble servant,

Francis Garden.

MR. ELLIS TO DR. GARDEN.

My Dear Friend, London, Jan. 14, 1770.

I should have written to you long ago, but expected every day to hear you had received a box of seeds from Mr. John Blowmart, by the packet from Pensacola, which I requested him to dispatch to your care, as it consists of some rare seeds for our nobility. He has enclosed me a copy of the letter he wrote to you, and told me he had sent you a specimen of the curious tree, discovered by me among a parcel of specimens, collected by accident, by the Chief Justice, Mr. Clifton's, black servant. It was found five years ago, and they never could get me any seeds till now.

I had a letter from Linnæus, with his respectful compliments to you. He longs much to hear from you. You have seen, no doubt, the Fly-trap, or *Dionœa muscipula,* which Mr. Young, the Queen's botanist, brought over. It grows in North Carolina, and is much esteemed here. I have sent Linnæus the characters of it, which gave him infinite pleasure, as the manner of its siezing and killing insects, exceeds any thing we know in the vegetable kingdom.

We have now got near 100 Tea plants, in different parts of the kingdom. Some of the seeds were brought in canisters, and sown at St. Helena. Many brought over in wax to England have succeeded. I hope you will soon have it in Carolina, where I believe it will thrive as well as in China. Tell me how your brother the Doctor does, and do not neglect to write to your old friend, who loves and esteems you, and is yours most truly.

J. Ellis.

Linnæus wishes to be informed of the cause of the luminous appearance of salt water.

I have a very good account of your relation Mr. Peronneau, whom I recommended to Dr. Hope.

DR. GARDEN TO MR. ELLIS.

January 15th, 1770.

My dear, my first, my chief Botanical Friend,

It is absolutely with shame and confusion of face, that I take up the pen to write to you. My long silence, my neglect in answering your affectionate letters, leave me not even the shadow of an excuse, and I can make no proper apology for either. Will you, can you, forgive me? If you do, I shall impute your excuse to your own goodness, of which I have already had a thousand proofs. Shamefully negligent indeed have I been. I have left and forsaken that study which gave the purest delight to my mind that ever it received, but I believe I only left it for a season, to return to it again in a short time, I hope, with renewed vigour and attention. For these three years past I have done nothing, neither read nor studied any branch of natural history. Indeed I have been sunk and lost in application to the practice of medicine alone. Closely confined to town, and having no intercourse with any person in that way here, the spark was almost extinguished. The perusal of some of your letters helped to rekindle it a little at times, and lately a fresh perusal of yours of May 1768, and June 1769, accompanying Mr. Pennant's, have roused me from my lethargy. It is not yet too late to return again to the ways of well doing. I have only to beg your kind fostering assistance to stimulate me to a fresh exertion of the opportunities with which Providence has kindly blessed me, in placing me in a land of wonders. Do not, my friend, forsake me. To you and Linnæus I owe my all in that way, and you must continue, by a continuance of your correspondence, to impell me to do you any services in my power, by making collections for you here.

An increased family naturally led me to seek an increase of provision for them; and in the hurry of that pursuit, I dropped the other object, but I have not lost sight of it, and though somewhat rusted in the study, yet I am willing to return to it. But in answer to yours of May 1768, which, I believe I never yet answered, I shall carefully attend to what you say on the method of preserving seeds. I had, by your direction, procured and preserved several of our best peach stones for Captain Hodgson, but I never heard of him nor from him, so they have perished. I shall be extremely happy to hear of the return of Mr. Banks and Dr. Solander, and to have some account of their important discoveries and observations. I hope you will inform me of them.

I shall remember what you say of the Pines.

I never saw the *Halesia diptera.* You shall certainly have the *Cypress, foliis deciduis,* next autumn.

I lately saw a plant from West Florida, which I fancy, from your description, must be the *Illicium.* I have sown some of the seeds, and if any come up you shall have them. What is known about it, or of it, as I am a stranger to it?

I come next to your letter, accompanying Mr.

Pennant's letter to me. I am much at a loss what reply to make to your obliging politeness, in thus introducing me to gentlemen of such characters as Mr. Pennant. I am sensible of what I owe you, though I have it not in my power to make any adequate return. It is with the utmost pleasure that I receive the honour of Mr. Pennant's acquaintance; but, alas! I can make him but a slender return. The idea which he has formed of my abilities to assist him, is by far too high, and many of his enquiries are on subjects that I am a stranger to. I will, however, take care to give what assistance I can in such things as I have a little knowledge of, particularly in Ichthyology. He may depend on the best collection of our fishes that I can procure for him, and I will endeavour to pick up what birds I possibly can, though I am but little conversant in Ornithology, and have no authors on the subject unless Linnæus's *Systema Naturæ*, and even of that I cannot get David Wilson to send me the latest edition.

I have inclosed a letter to Mr. Pennant, which I beg you will be so good as to forward to him, with my most respectful compliments, my cheerful acceptance of the honour of his correspondence, and an offer of the best service that I can render him in furnishing him with such hints as my situation can enable me to procure for him. From this day forward I begin to collect for him.

About two months ago I received a small box directed for you, from a gentleman in West Florida:

probably it may contain some of the seeds of *Illicium,* or some other curious seeds. I have given it in charge to Captain William White, whom you will see at the Carolina coffee-house every day. You will please to ask him for it, in case that he should forget it.

Last summer I sent my son to England, and placed him at Chiswick, under the care of my old master Mr. Rose. He is yet too young to be introduced to you, otherwise you may be sure that I would have taken the liberty of giving him a letter to you. When he has been three or four years more in England, I shall beg leave to introduce him to you.

At present I have nothing further to trouble you with, but only to beg that you would present my best respects to Sir Charles von Linné, when you write to him, and to my old friend Mr. Baker, when you see him; and believe me to be, with unalterable esteem, my dear friend, yours most affectionately,

ALEXANDER GARDEN.

MY DEAR FRIEND, Charlestown, May 12, 1770.

Though I have but little to send you of any consequence, yet I have trouble to give you. Inclosed you see a letter to Sir Charles von Linné, which I must beg the favour of you to forward to him by a safe opportunity. I was ashamed of having neglected writing for so long a space to that illustrious ornament of our favourite science, and could not help reproaching myself with ingratitude to him, especially as he had always treated me so honourably and with the greatest friendship. The same may be said of my behaviour to you, my dear friend, but I am persuaded if you knew the life I lead, confined to town, sickly as to health, embarrassed with a multiplicity of other avocations, you would pity me. I often resolve to quit this noisy nonsense, and give myself up to that favourite study, but four children put me in mind of my duty to them. The eldest of them I took the liberty to mention to you in my last.

Thinking of these matters the other day, I plucked up courage to write to Linnæus again, and I scraped together a collection of snakes, insects, lizards, fish, spiders, &c. and put them all into spirits, in a bottle, which I have directed for him to your care, and which I beg you will forward, with my letter and most respectful compliments, by the first good and sure vessel for Stockholm, or to whatever port you send his things.

I am now collecting for Mr. Pennant, from whom I had a letter the other day, in answer to one I inclosed in yours. He told me that you were well.

I have sent you, by Ball, a bundle containing some sponges, &c. And there are two very curious fish, but of what class they are I know not. You will know them at first. They were both caught among the Bahama islands, the one with the many claws, in very deep water, was taken with a baited hook. It appears to be nearly related to the

Asterias. It is at present all shrivelled up, but it looked beautiful, and had a thousand ramifications of its legs and arms.

The other was called to me a sea spider, and is certainly a curious animal. The gentleman who caught it took it as it was running on the clear bottom. It was in two or three foot water, and it ran, as he said, on its long legs like a spider, very swiftly. I shall be glad to know what they are.

I have, at last, got two of the *pericarpiums* of the *Beureria,* both full of seed, which I have sent to you. They came into my hands about a month ago, and I had not an opportunity of sending them sooner.

What did you with the Loblolly Bay, which you once intended to make a *Gordonia?* I think you should send it to Linnæus, or if you choose the characters to be re-examined and drawn, I will endeavour to do them in the best manner I can, and send them to you.

Nobody can be more unlucky than I am, in David Wilson for a bookseller. I have written to him three or four times within these two years, to send me the twelfth edition of Linnæus's *Systema Naturæ,* but I cannot get it. I shall be greatly obliged to you if you will be so good as to desire any bookseller of your acquaintance to send a copy of it to my friend Charles Ogilvie for me; he will pay them. I must quit Wilson, for, indeed, he has quitted me; and yet I think he never had any cause to complain. I am sorry to trouble you, but I want these books greatly, and cannot get them otherwise, so I hope you will excuse the trouble which I give you.

O my dear friend, how shall I rejoice when the time comes that I can calmly and quietly deliver myself up to Natural History, and enjoy more of your correspondence! but whether that time comes or not, you may believe me to be, with unalterable esteem, my dear Sir, your much obliged and very humble servant, ALEXANDER GARDEN.

MY DEAR FRIEND, June 20, 1770.

At the request of your former acquaintance, Mr. Henry Peronneau, I write this, to introduce to you his brother Mr. James Peronneau, the young gentleman who will deliver it into your hands. The abovesaid Mr. Henry Peronneau, who is his guardian, sent him to England some years ago, in order to breed him up for the bar. After finishing his grammatical education, he intended to have bound him apprentice to some attorney of character, and had him enter at the Temple. But the young man falling into bad hands, and being of a lively turn, is disgusted a little with study, and declines applying any longer. Mr. Henry Peronneau having his education much at heart, wants him either to apply as above mentioned to the law, or to go to either of the Universities, and study philosophy and the sciences there, for three or four years. He has given James his choice to do either, but from what he has learned of his wild turn of late, he is afraid that he will neglect both, and mispend his time in England, or come out to Carolina unqualified for any business.

Now, my dear friend, he desires that I would intercede with you in behalf of his brother, begging that you would take the trouble to converse with him, and be kind enough to advise him either to apply to the law, or to enter in either Oxford or Cambridge, and there to apply for some years, to get a liberal education, on which foundation he might afterwards build what superstructure he pleased. This last I think is the most eligible plan of the two, and I have written my sentiments on this head to him.

If you will be so good as to take this trouble, you will confer a very great favour on Mr. Peronneau, who begs me to present his respectful compliments to you; and you will at the same time add to the many obligations which I already lie under to you. I am, my dear friend, yours,

ALEXANDER GARDEN.

I wrote to you fully by Captain Ball, and inclosed my letter to Linnæus.

MY DEAR FRIEND, Charlestown, Dec. 24, 1770.

I received your obliging favour by Capt. White, and with your letter there came from our illustrious friend, a short line to me, which gave me great pleasure indeed. What a blessing it is that the life of this valuable man is preserved. May it still be continued to him for the good of mankind and ornament of science, until he has fully finished his great works. How much better are we acquainted with the productions of nature by this great man's means, than we were 50 years ago. How many additional arguments has he furnished for acquiescence in, and admiration of, the power, the wisdom, the goodness and providence of the Almighty Author of all!

I have one or two things which I think will please him; he shall have them by one of our spring ships; one of them is a species of Turtle, as yet nondescript. It is amazing how Catesby omitted this. It is found in abundance in the Savannah river, in Altamaha, and East Florida. It is a fresh water animal, grows to a great size, and is as delicate as the Green Turtle, having a large leathery cover over its back, and a head very like a Mole. I intend to send a copy of my account of this animal to Mr. Pennant, for his American Zoology, and if I can get a drawing of it copied, I will send him that. If I can obtain another Turtle, I shall send you one stuffed. It has a relation to the first species of Linnæus's last edition of the *Systema Naturæ*.

You may depend on my taking all possible care of any consignment for you from West Florida, or any where else. I know you have cause to suspect my carefulness, since my neglect for some time of the small box of seeds. By the bye, you never mentioned whether you got them. I should be glad to know; and will endeavour to be more careful.

What are the particular excellencies or beauties of the *Illicium anisatum*?

I have just received the flower of a plant from East Florida, about which I formerly wrote to you, and which I intend for Linnæus. I hope it will give him pleasure.

I have placed my son at a boarding-school at Chiswick, under the care of Mr. William Rose, my own school-master. He has been there about 18 months, and I had some thoughts of putting him to Westminster or to Eton after a while; but, my friend, I have had so many accounts of the little care that is taken of boys at either of these schools, and of the depravity of their morals and manners, that I am greatly at a loss what steps to take. I have the greatest confidence in Mr. Rose's friendship, and the highest opinion of his abilities and care. The question is, whether I shall leave him with Mr. Rose, to fit him for one of the universities, or send him to a public school? As you are on the spot, and must know the state of these schools well, may I take the liberty to ask your advice? I cannot urge you to give an opinion, but if you will be kind enough to let me have the favour of any assistance, to determine a matter of the utmost importance, to his and to my happiness, I shall be for ever obliged to you. I intended to introduce my boy to you some time hence, and till then I thought it improper to mention him to you, but my difficulties on this point oblige me to apply to you. He is now 13 years old, and has been in England about 18 or 20 months, under the care of my worthy friends Daniel Blake and Charles Ogilvie, in the City.

You are so good as to send your compliments to my brother Francis, but alas! he is no more. He died in September last, of a severe putrid fever, caught from bathing, in the heat of the day, after a long and hard ride. He was then about 70 miles from Charlestown.

Please to remember me to Mr. Pennant and to Linnæus, and I am, with the most affectionate regard, my dear friend, yours, &c.

ALEXANDER GARDEN.

MR. ELLIS TO DR. GARDEN.

London, Jan. 2, 1771.

The inclosed I received from our mutual friend, Dr. Linnæus, whom I find you have most highly obliged by your communications. He has desired that the Loblolly Bay may be called *Lasianthus*, but his letter came too late by a month, and I have called it *Gordonia Lasianthus* *. You must know that this elegant plant has flowered this year with us, and I have been much pressed to examine the new-blown flower, by a worthy friend of mine, Mr. Bewick, an eminent merchant, in whose garden it flowered. I have, with the help of the spe-

* It had already been called *Hypericum Lasianthus* by Linnæus, in *Sp. Pl. ed.* 1. 783. *ed.* 2. 1101; but is *Gordonia Lasianthus* of *Mant*, 2. 556, 570.

cimens you formerly sent me, and what have appeared here, determined it to belong to *Monadelphia Polyandria*, next to the *Stuartia*. This, with the *Illicium floridanum*, which I have described and figured, will appear in the next volume of our Transactions. The *Illicium* is a *Polyandria Polygynia*, and comes next to the *Magnolia*. You would do well to cultivate this plant. It exceeds most of our modern discoveries, as it seems to be hardy, and may be useful in physic, as the capsules have long been in esteem among the Germans, and are in all the foreign *Materiæ Medicæ*.

The absurdity of your news writers would have us think you have got the true China Tea. If this had been the case, no man would have known it so soon as you. I am persuaded you abound in quack botanists, as much as we do.

Philip Miller, the gardener of Chelsea, is turned out of his place, for his impertinence to the Apothecaries Company, his masters. They have got a much better, one Forsyth, late gardener to the Duke of Northumberland, who. has an excellent character, and will revive the credit of the garden, which was losing its reputation, and every thing curious was sent to Mr. Aiton, the Princess of Wales' gardener at Kew.

DR. GARDEN TO MR. ELLIS.

MY DEAR FRIEND, Jan. 26, 1771

Once more I set myself down to the agreeable task of writing to you, and though I have little new to communicate, yet I have a singular favour to ask of you, and however odd it may appear, yet I have that confidence in your obliging disposition, that I am in hopes you will not deny me this favour. The matter is just this. Two of my chief and much esteemed friends Daniel Blake and his lady have been in London for some time, and have once already visited the British Musæum. They are to return to Carolina next fall, but before that time I should be extremely glad that Mrs. Blake had an opportunity of seeing it again, attended by a person who could make such a visit an agreeable and rational entertainment to her. She is a lady of a most amiable and accomplished mind, and will highly relish the visiting of such a place, accompanied with your assistance. Now my friend, what I have to intreat of you is, that you would oblige me so far as to conduct Mr. Blake and her to the Musæum on any day when you can spare an hour. You will find Mr. Blake a very agreeable worthy man, and Mrs. Blake a lady of the sweetest and most amiable disposition in the world. They are my particular friends, for whom I have the most perfect esteem, and nothing could make me happier than to have it in my power to oblige them in any thing. I know nothing that would be more agreeable to Mrs. Blake than the visit which I propose, and I know no person but yourself of whom I could ask the favour which I now do of you, nor one who is so capable of making that visit so full of rational amusement to her as yourself. But besides this, if you will permit me to intrude further, I would likewise beg that you would accompany them to Kew Gardens, and show them what is curious or entertaining there. You will find her possessed of an excellent taste and relish for such entertainment. I have taken the liberty of inclosing a letter for you to my friend Mr. Blake; and as I shall beg of him to deliver it himself, I hope you will become acquainted. I need not say more. This will be a very great and singular favour done me, and I am persuaded you will be happy with the acquaintance of this worthy couple. I wish you could show her a good collection of shells. I will say no more on this head.

By your last obliging packet I had a short letter from Linnæus. I am indeed much indebted to him for the honourable mention he is pleased to make of the small matters which I have sometimes sent him, and at the same time I consider myself as greatly indebted to you for that share of his esteem and friendship with which he honours me. I rejoice that I had written to him before he wrote to me. I have now something else nearly ready for him, which I shall soon send home to your care.

I have a pretty tolerable collection of birds for our mutual friend Mr. Pennant, to whom I shall send them, by Liverpool. I think some of them will please him. Among the rest there is what I take to be a new species of Turtle*. It seems to me to have not

* See p. 336.

the smallest relation to any of the 15 species mentioned by Linnæus, excepting the first, and it only agrees with that in its leathery coat, but differs in every thing else. That sort lives in salt water, and this only in fresh rivers.

As I imagined that this would not be an unacceptable present to you, I have taken the liberty of sending a dried specimen, which has the head, feet, and body entire. You will observe that the neck is very long, but the animal, when alive, keeps it always drawn into the body, or at least within the covering, unless when it bites or wants to catch its prey or food, in which case it pushes the neck suddenly out to the length to which you see it extended. These Turtles are found in great plenty in Savannah and Altamaha rivers; but they live up in the freshes. They are excellent eating, being rather more delicate and luscious food than the Green Turtle. The projecting nose of this animal is a singularity, different from any other Turtle. Indeed the whole animal seems to me to be a singularity. What think you of it? If I can procure some small ones, I will send one to Linnæus for his new-built musæum on the hill. I wish much to have a ramble round the back parts of our province. I think there must be many new animals as well as plants. Bye and bye I shall certainly make such a tour.

I congratulate you on your appointment to the agency of Dominica, and I likewise congratulate Natural History, which will fare the better for this event.

The Turtle which I have sent you is not the one from which I took the inclosed characters, but nearly of the same size. You will see it unluckily had a cut across the nose, which disfigures it much.

The characters were a few notes taken at random, though I think they are very exact in regard to the weight, size, and number of the parts; but what I mean is, they are not well digested and arranged. I had them copied just as they were first written down.

The dried Turtle put up in brown paper, and directed for you, is given to the care of Capt. Ball, your old acquaintance, but as the old man may not be very careful in sending these things, I must beg you will order some person to call on board for them, lest the captain should omit or forget them. At the same time you will receive a square flat box, which was brought from West Florida by the packet, Capt. Clarke. He sent it to my house five days ago, and I have put it on board of Ball, being the first ship for London; but if the captain had given it to me when he first arrived here from Pensacola, I could have sent it to you a month ago.

It comes from governor Chester, and I imagine contains seeds. I hope they will get home safe, and in good time.

Whatever you want from West Florida, pray order it to come to my care, and you may depend on my punctuality in forwarding. What shall I say to this horrid war that is approaching? will it not interrupt and disturb every correspondence? I hope you will make some provision in case of french capture, as you did in the last war, when you may remember you desired your parcels to be directed to Mons. Du Hamel.

I forgot to tell you that I have a drawing of the fresh-water Turtle, and intend to get it copied for you. I shall send it by the first opportunity.

I am vastly pleased to hear of my namesake the *Gardenia* thriving so well with you, and should be very glad to get a plant of it, as I now have a piece of ground where I could cultivate it.

As to the Loblolly Bay, it is certainly a beautiful tree, and keeps in flower longer than any tree I know. It blossoms here from May till near the end of October, and is certainly a new genus. When it blossoms I shall send you a specimen of it, and one of the *Hopea* as you desire.

My brother Francis, to whom you so kindly desired to be remembered, died here about five months ago of a violent fever. This was a severe stroke to me; but I will not entertain you with so disagreeable and melancholy a subject.

I did myself the pleasure of presenting your compliments to Mr. Peronneau, who is much obliged to you for the trouble you took in speaking to his brother James. He desires me to offer his thanks and respectful compliments to you.

Under the same cover with the Turtle, you will find six nuts or beans, the seed of a tree in Surinam. They were brought here from Surinam by a captain, who told me the natives brought them to town and sold them, and people eat them as a fine nut. I have eaten some of them, and they seem to be very good, but I know not their name. I have sent you half a dozen, and have planted two of them. I shall be glad to learn what they are.

Now, my dear Sir, I think I have given you a pretty tedious scrawl, and if you have patience to get to the end of it, I shall be satisfied. I am, with the greatest respect and warmest regard, my dear friend, yours, &c. ALEXANDER GARDEN.

Mr. Daniel Blake is very often at the Carolina coffee-house, but lives, I believe, in Leicester-fields.

DEAR SIR, July, 13, 1771.

It is so short a time since I wrote to you fully by Capts. Ball and Gunn, that I have nothing now to say further on any literary subject. The only intention of this is to introduce to your acquaintance my very particular friend Henry Laurens, Esq. who will deliver it into your hands. You will find him so intelligent on every thing relating to America, I do not mean on politics, for that I know is not your concern; but on every subject where the interest of America, either regarding commerce or colonization, is concerned, that I think you will take pleasure in his conversation. He is minutely acquainted with the state of this and the neighbouring provinces, and as he has been in East Florida, I think he can give you some material information about that place. I am persuaded I need not say more to entitle him to whatever civilities you can show him; but as the chief intention of his visiting England at present is, to place his son, whom he carries with him, at some proper seminary of learning, so when that is once done, he will have many leisure hours, and then I doubt not but his acquaintance and yours may grow from a better knowledge of one another.

His son is a favourite of mine. I think he has genius, and I know he has application sufficient, under good masters, to enable him to become an accomplished man, and to make him a joy and delight to his father, as well as an ornament to his country. I must here acquaint you that he is the young gentleman who copied the drawing of the soft-shelled Turtle which I sent you by Ball, and he will be able to give you several anecdotes about it. He appears to me to have some inclination to the study of Natural History, and I well know you would rather encourage than damp any *penchant* to that delightful study. The accounts we have here of the state cf your universities, in point of government and morals, is not altogether favourable, but I hope things are not so bad as they are represented, and that my friend will be induced to place his son at one of them, which is what I really wish on the young gentleman's account.

I have now only to add that I am with the greatest esteem, dear Sir, your most obedient and very humble servant, ALEXANDER GARDEN.

MY DEAR FRIEND, December 10, 1772.

Some time ago I received your very obliging favour of the 6th of May last, inclosing a letter from Linnæus. Nothing gave me more pleasure than to hear of your welfare, and the continuance of your health, for the continuation of which I most sincerely offer my best wishes. I find you still pursue your favourite study of the Zoophytes, and I rejoice that you have so fair a prospect of living to put the finishing hand to this matter. It must give you great pleasure, and all the lovers of our science great joy, that you have been able to settle all matters relating to that dubious tribe of beings. I heartily wish that it had been in my power to have furnished you with any specimens from our coast, but not one has come into my hands ever since you first mentioned your design to me, otherwise, no person could have been happier than I should have been, in communicating them or sending them to you.

I am much obliged to you for your kind intentions in endeavouring to send me a Tea plant. It would indeed be most acceptable, and I doubt not but I should be able to make somewhat of it here with the help of your directions. When they become more plentiful, I will then again take the liberty to put you in mind of it.

If I should luckily catch another of those very extraordinary animals which Linnæus calls *Sireni simile*, I shall certainly send it to you for your Society. But I have never seen more than one of them myself, and though I have endeavoured to get some more, I have not yet been able. I own I think it is a new genus, and though of the same order as the *Siren*, yet totally different in many respects, as you would see by the characters sent along with it. Had you ever seen our soft-shelled Turtle before? What think you of young Laurens's draught of it?

What is become of Banks and Solander? What is become of all their immense collections? I think they must greatly encrease the animal and vegetable kingdoms in new genera; surely these new countries must have many yet unknown productions.

Permit me likewise to inquire after my friend Mr. Pennant. What is he doing? I hope he is well. Did he receive a letter, and bird with wax-like appendages to the secondary *remiges alarum*, which I sent him in March or April last year? I inclosed my letter for him in one to you of the 27th of March, and sent the bird to your care. I really had nothing else worth his notice to send him at that time, neither have I been able to procure any thing since; but I value and esteem his friendship so highly, that I will let slip no opportunity of cultivating it, when I have any thing worth notice to communicate.

When I consider that you, Mr. Banks, Dr. Solander, and Mr. Pennant, meet together sometimes in London, I often wish for Fortunatus's hat for a few minutes, to transport myself into a corner of the room. What an agreeable communication of ideas should I be instructed by, and what information should I receive! I long to enjoy such company, and of late have often wished to be of your Society, though you know that formerly I thought otherwise. You will oblige me if you will inform me of the manner of applying to be admitted a member, and I shall then beg the favour of you to assist me in promoting that application, and whatever the expences are, I shall desire my friend Mr. Ogilvie to pay them. If ever my good fortune should carry me to London, I should be much pleased to reap all the assistance and advantage I could, from the opportunities of attending the meetings. But you can best inform me whether I can be elected into the Society.

You would see by a paragraph of Linnæus's letter to me, that he thinks my East Florida Fern-like plant is his *Zamia*. I think he has been misled by Ehret and Trew's draughts. He begs to have one of the *pericarpiums*, and I have none at present, neither can I get one this year; but you may remember I sent you two of these by Capt. Curling last spring, and I then begged of you that you would be so good as to send one to Linnæus. I hope you thought of him, as I think it would give him great pleasure to be able to set that matter to rights. He begs to know whether there is one seed only, or more, in each *pericarpium*; he says Ehret paints two. Now you will easily remember that there are from 20 to 50, or thereabouts. The number is in general indeterminate, some more, some less, according to the size of the cone. Pray send him a cone and some seeds.

Some time this last summer I received a very polite letter from Dr. Laurence Theodore Gronovius at Leyden; I have now written him an answer, and sent him a box of fishes and insects. He did not inform me how to direct them, and therefore I have sent them to Mr. Ogilvie in London, to be forwarded by a safe and sure conveyance to Leyden. Let me intreat of you to give Mr. Ogilvie a proper address, or direction, to him, so that they may not miscarry, for I took much pains to collect them. I have desired Mr. Ogilvie to apply to you. May Heaven long preserve your life, and bless you with health! I remain, with great esteem, dear Sir, yours most truly, ALEXANDER GARDEN.

Charlestown,
May 15, 1773.

MY DEAR AND WORTHY FRIEND,

I am now at last about to answer your very acceptable and agreeable letter of the 4th December 1772. It was fortunate for me that I had not heard of your illness before your letter came to hand, which gave me the agreeable news of your recovery. This saved me much pain and anxiety on your account. I rejoice at this happy event, as well on my own account, as on that of all lovers of Natural History, that your life is still preserved. May gracious Heaven bless you with a continuance of good health!

On the 10th of December last, I did myself the

pleasure to write you a long letter, in which I mentioned several things on which I should be very happy to have your sentiments. Did Mr. Pennant get the bird with the wax-like tips to the secondary *remiges alarum?* I think this is a most curious and singular bird. I do not find it in Buffon's *Ornithologia.* Catesby has given a bad draught of it.

I particularly mentioned to you my inclination to become a member of the Royal Society, if you would be so obliging as to propose me, and the members should do me the honour to elect me. You know my sentiments on this matter formerly were different; but from what you have at different times said and written to me on this head, I have altered my opinion. I have directed Mr. George Ogilvie to answer your demand for the fees, or dues, on admission, in case the Society should judge me a proper person. I have further to beg of you to inform me what is generally expected from the newly-elected members, as I am quite a stranger to the practices and customs of the Society; at the same time I should be most willing and ready to comply with any thing in my power.

Allow me again to enquire after Messrs. Banks and Solander; are we to have any account of their voyage, or collections and observations?

In your last letter you are pleased to mention to me a proposal from West Florida, about a botanical garden, and to ask my opinion on that proposal. It is impossible for any person to be more willing to give you information than I am, if I were furnished with proper materials, but in this present case I really am not; however, what I know, I shall mention to you concerning this matter.

First, let me inform you that I am but little-acquainted with Mr. Romans. He appears to be a man little versed in Botany; he knows a few of the terms, but his knowledge, as far as I can judge, in the science, is both much limited and very superficial. He has been employed by the superintendant of Indian affairs as a surveyor, and I have seen some few things of his collection, but they were of no account or value. He is at present out of employment, and is here on his way to New York, where he intends to publish a chart of the Bahama islands and Cape Florida. He appears to me to be a man that would take pains, but I am afraid he wants knowledge for your purpose.

I am unacquainted with Dr. Lorimer, and only know from information that he has the character of an ingenious man, but whether he has a turn for Natural History I know not.

I am well satisfied that their Jalap is the *Convolvulus* you mention, or at least a species of the *Convolvulus*. We have many species of it that are cathartic, but none of them are the true Jalap. They are easily distinguished by the eye, from the one appearing always on the transverse section of the root to have concentric circles; and the appearance of the other is always radiated from the centre, without the concentric circles. Besides, I have tried the effects of almost every species of the *Convolvulus* on the human body, and they answer very well to the account of the *Materia Medica* writers of the virtues of the *Mechoacan*, but by no means to the Jalap. They are not only weaker than the Jalap, but have not the same specific virtues and effect.

1. As to what he says of the *Pistachia*, I cannot venture to pronounce any thing; I never saw it or any thing like it in this province.

2. The *Lentiscus* is not found with us.

3. The *Cycas circinalis* I have never seen here, or heard of it, in either of the Floridas.

4. His *Styrax officinale* must be as you conjecture.

Upon the whole, I am very suspicious of the justness of Mr. Romans's observations and remarks, and am afraid he speaks much on conjecture, regarding those plants mentioned.

I cannot advise you to trust much to his knowledge, though he might be willing to improve it.

As to provincial gardens in general, I think they would be extremely useful, if applied chiefly to the purposes of agriculture and commerce; kept on a large scale, and on a public bottom; but as to a botanic garden, I am afraid it will not as yet answer in any part of America, much less in these new and almost uninhabited provinces, where the expence would be great and the advantages small. Mr. Romans shewed me his estimate of the expence, which is not one fourth, fifth, or sixth part of what it would cost to carry it into execution.

When the present heats, heartburnings, and uneasinesses, in America, are over, and people give up the study of politics for the more sensible study of the cultivation of their lands, and natural police, then I am persuaded the idea of provincial gardens will be revived and carried into execution; when your catalogue of plants will become signally useful and serviceable to them, and their purposes of improvement.

I am now to thank you for your obliging present of the upland rice. I have distributed it to many, and have planted it three or four times myself, but hitherto none is come up, excepting one grain, which I will carefully nurse. I have heard that some others have succeeded better, from their seed being better preserved. If we can bring it forward, it will be a great addition to us, and our grateful thanks will be due to you. I have read and admired M. Poivre's book again and again. We shall have much reason to bless you and him for centuries to come; you will be one of our fathers and benefactors. I shall not fail to acquaint you with our success.

You write me that you think it possible to pack our rice so as that it shall not be subject to the weevil. Let me intreat that you will communicate your idea how this can be done; it is a knowledge that we at present have not, but stand much in need of.

I have taken the liberty of enclosing a letter for Linnæus under this cover, and I beg you will forward it to him as soon as possible. It relates chiefly to an East Florida plant, of which I formerly sent you a fruit with the seeds therein. I have now carefully examined it, and you will see the characters, which are somewhat like the characters of the *Zamia*, but yet I think different. I have sent specimens of this plant, and of all parts of it, in a bundle directed to you, and given to Capt. William White's particular care. You will greatly oblige me if you will desire him to send these things to your house, and then forward them to Linnæus.

You would see by his last letter that I came off conqueror in our dispute about the new genus *Anamelis* *, on which I plume myself not a little, but his candour charms me.

I have not been as yet able to procure another of the *Amphiuma means*, which he calls the *Sireni simile*. This appeared to me to be a still more singular animal than the *Siren*, as you might observe by my remarks and character of it in my last letter to Sir Charles von Linnè †.

It gives me the most sensible pleasure that I have had an opportunity to furnish this great and worthy man with any thing that has been acceptable to him; and I hope still to be able to get now and then somewhat further, that may not prove unworthy of his notice.

Let me intreat that in your next you will inform me very particularly of your health, good accounts of which will always give the sincerest pleasure and joy to, my dear friend, your most sincere friend and humble servant, ALEXANDER GARDEN.

* *Fothergilla.* See p. 340. † See p. 333 of this volume.

Some time ago I received Buffon's Ornithology, which I read with the greatest pleasure. At the same time I got the three volumes of coloured plates; they are truly superb.

MY DEAR FRIEND, March 21, 1774,

On the 4th of February last, I did myself the pleasure of writing to you, and since that time I had the honour of receiving your letter dated Hampstead, Nov. 3, 1773, in which you are so obliging as to communicate to me, in the most flattering terms, the account of my being chosen a member of your Society. As I had notice of this by the Secretary's receipt for the admission money, transmitted by my friend Mr. Ogilvie, so I took the liberty, in my letter above mentioned, to offer my best thanks to you for the exertion of your friendship with your members, in my favour. It is to you alone that I consider myself indebted for the opinion they have formed of me, and for the honour they have done me in obligingly making me one of their number. I do not know the usual mode or form of thanking them, but I beg you will permit me to entreat of you to present my respectful thanks to them, in such manner as you think proper, or that you would please to inform me of the common method followed on such occasions. This, at least, you may assure them of; my earnest endeavours to serve them in any thing I can, and to promote their plan in every way in my power.

You will oblige me much if you will be so good as to inform me whether I am to receive any certificate of my election, which I think you mention in your letter, if I understand it right. I should likewise be obliged to you to inform me of the manner in which I am to apply for the annual publications of the Society, whether and where I shall desire my friend to send for them and to pay for them? I am very unwilling to give you all this plague, but my total unacquaintance with your customs obliges me to be troublesome, which I hope, on this occasion, you will be so good as to forgive.

I have planted some of the Tea seed, which I got from Mr. Blake, but it has been in the ground so few days, that I do not yet know whether it will vegetate. The Chinese White Hemp, the *Tien Loam*, the *Tong-yaa-o*, and on looking just now, I think I see the *Isao Qui*, just peeping out of the ground. They were all planted on the 15th of March, and this is only the 6th day since that time. I think the seeds of the *Tong-yaa-o* seem to be a large species of the *Ricinus Palma Christi*, and it springs in the same way. By the look of the seeds of the *Sat shu*, or Lacker tree, I should really have taken it for our *Toxicodendron foliis pinnatis;* but I shall be able to speak more certainly when I see it growing, and then I shall be more full to you.

If Gordon has any particular art or method in raising the Tea plant, I wish we knew it.

The *Ou Cow* or Tallow-tree* will certainly succeed

* *Croton sebiferum, Linn. Sp. Pl.* 1425, now referred to *Stillingia.*

well here. It has stood the winter in the open gardens, and the plants that were out all the winter have thriven, and now look better than those that were housed.

The Cochin China rice seems in every respect the same as ours, only the grain is smaller, and I think it stools more than ours.

The account which you gave me of the tender state of your eyes affected me much, and I do most sincerely condole with you thereon; but I am in hopes that you will gradually get the better of this grievous affliction, and no person I think will more joyfully rejoice with you on so happy an event.

I have now, my dear Sir, to beg leave to introduce to you the bearer of this, my only son. He has been in England four or five years, but I could not think of troubling you with his visits while he was too young to profit from your conversation. He will now soon leave London to repair to a University; but as he will now and then visit London during the vacation times, I could wish that he had at such times your leave to pay his respects to you, to enquire after your health, and be permitted to see and converse with his father's most esteemed and valued friend.

I am yet entirely unacquainted with any particular *penchant* which he may have, or to what profession his genius may incline him. But I think I could wish him to have pleasure in looking at, considering, and admiring the works of his Creator, in the various forms in which they appear to us. This would never interfere with any profession, and it would be a source of benefit to himself. For this reason I should be happy to have him acquainted with, and introduced to, the curious gardens about London. I do not mean to give you any trouble with him, but if it should happen at any time to fall in your way to take him with you, in any of your walks to Mr. Gordon's or any other curious gardens, I should esteem it a great favour done me.

I have now only to add my best wishes for your perfect recovery, and the long enjoyment of your health; and I am, my dear friend, yours affectionately,

ALEXANDER GARDEN.

I have not heard from Linnæus since I wrote to him last, though I daily wish to hear his thoughts on the animal quadruped, and on my characters of the plant which he thinks is his *Zamia*.

The *Illicium* blossoms and thrives with us to a great degree.

MY DEAR FRIEND, March 12, 1775.

It is a long time now since I had the pleasure of hearing from you or of you, except what my son mentions in his letters, whom I have desired to call sometimes and enquire after your health, and inform me by every opportunity how you do. I have been greatly concerned for the distress which you have suffered from that complaint in your eyes, which I am in hopes may be happily removed before this time, on which desirable event I am certain none of your many anxious friends would take more real pleasure and joy than I should.

I wrote you a very long letter, and, I am ashamed to say it, a very confused one, dated 14 August, 1774, on the subject of electrical fish, which were brought from Surinam on their way to Great Britain. When I wrote that letter I was scarcely recovered from a severe fever, and was little able either to examine these fish accurately, or to digest what few trials I could make of their properties. However, as their appearance, structure, and properties seemed to me uncommon, I thought it proper to give you some account of them, with a desire that you might lay it before the Society, if you should judge it to be worthy of their acceptance or notice.

The person who owned these fish carried them from hence to England, but whether they arrived alive, I really know not. I desired him, if they arrived safe, immediately to go to you, and I gave him your address; but lest they should die by the way, I desired him to put them into a small kegg of rum, which I advised him to prepare, for the purpose of preserving them till he got home. I wish to hear both of the fate of my letter and of these fish, whether they arrived and reached you; and, if they did, what satisfaction they gave. I doubt not that when you are at leisure, you will give me the pleasure I desire of hearing from you.

Since I wrote you the above-mentioned letter, I

have had so very disordered a state of health, that I have not been able to do any thing in the way of procuring materials for fresh observations in Natural History. I have never been without the limits of Charlestown, seldom out of my house. If I again get strength, I shall employ part of my time in looking about me.

I have desired my son, who is about to leave London for a little time, to wait on you with this, and to present my compliments, and if you have any commands for this part of the world about August, he will take pleasure in conveying them for you.

I have lately received two volumes of your Transactions, sent me by I know not whom, nor at whose desire. I find many curious papers in them, particularly one on the electrical power of the Torpedo.

I wish that I had seen it before I wrote my letter to you, at least before I had seen those electrical fish. I find also a very curious paper on the reproduction of the Sea Anemonies, by the Abbé Dicquemare at Havre de Grace.

I will not fatigue you more at present with my scrawls, and only have to tell you that I remain with the most perfect esteem, my dear friend, your most obedient and very humble servant,

ALEXANDER GARDEN.

END OF VOL. I.

Printed by J. Nichols and Son,
25, Parliament Street, Westminster.

Printed in the United States
By Bookmasters